D0876284

BIOMASS ENERGY

From Harvesting to Storage

Proceedings of a workshop held at Marino (Rome), 19–21 November 1986, and organised by

—The Commission of the European Communities, Directorate-General for Energy
—ITABIA (Italian Biomass Association)
—AIGR (Associazione Italiana di Genio Rurale)

in collaboration with

ENEA–ENI—Regione Campania
RENAGRI—Società agricola forestale

BIOMASS ENERGY

From Harvesting to Storage

Edited by

G. L. FERRERO, G. GRASSI

Commission of the European Communities, Brussels, Belgium

H. E. WILLIAMS

Ecotec Research and Consulting Limited, Birmingham, UK

ELSEVIER APPLIED SCIENCE

LONDON and NEW YORK

ELSEVIER APPLIED SCIENCE PUBLISHERS LTD
Crown House, Linton Road, Barking, Essex IG11 8JU, England

Sole Distributor in the USA and Canada
ELSEVIER SCIENCE PUBLISHING CO., INC.
52 Vanderbilt Avenue, New York, NY 10017, USA

WITH 78 TABLES AND 107 ILLUSTRATIONS

British Library Cataloguing in Publication Data
Biomass energy from harvesting to storage.
1. Biomass energy
I. Ferrero, G. L. II. Grassi, G.
III. Williams, H. E.
662′.6 TP360

ISBN 1-85166-175-1

Library of Congress CIP data applied for

Publication arrangements by Commission of the European Communities, Directorate-General Telecommunications, Information Industries and Innovation, Luxembourg.

EUR 11045

Printed in Great Britain by Galliard (Printers) Ltd, Great Yarmouth

PREFACE

Collection, like transport and storage, is one of the basic technical and economic components of the processes of converting Biomass into energy. In recent years much progress has been made in R & D and industrial applications for the collection of all types of Biomass resources/feedstock, which ranges from orchard prunings to tree-felling waste, from straw to seaweed, from specific energy crops to arable-farming residues of any kind.

Though interesting results have certainly been achieved, many problems still remain, and their solution will largely influence the final use of these Biomass sources and residues. There are the problems of bringing out timber waste from inaccessible oak forests, and making the collection of algae and aquatic plants financially worthwhile ; should machines be specially designed for this work ; and what about the economic and social costs for a proper forest management policy such as forest fires, soil degradation and so on.

The need to assess what positives advances have already been made in the Community's member countries, so that Community action in this sector can be directed most efficiently, prompted the Directorate-General for Energy of the Commission of the European Communities to organize this workshop in collaboration with the other departments active in this field.

Apart from the Directorate-General for Energy's own demonstration programme on Biomass and energy from waste, other important activities are now in hand under the agricultural research programme of the Directorate-General for Agriculture and the R & D programmes on energy from Biomass and wood as a renewable raw material of the Directorate-General for Science, Research and Development.

This workshop was therefore intended to be a meeting-place for those active in the sector under EEC and national programmes, in order to assess the state of art of the technology and methodology of harvesting, to give a critical evaluation of the results obtained, to discuss the problems still outstanding, the aims to be achieved and what remains to be done in order to make better and more economic use of Biomass and residues which are not exploited at present. The reports presented, the discussions and contacts during the two days meeting produced answers to some questions. However, a lot remains to be done as the conclusions and recommendations testify.

Collection processes must be further improved if we wish to make economic use of Biomass itself and see wider application of energy conversion technologies.

V. ALLIATA G.L. FERRERO G. GRASSI

CONTENTS

SESSION IV - CROP BY-PRODUCTS, PRUNING

SESSION V - ECONOMIC EVALUATIONS, PUBLIC INCENTIVES

OPENING SESSION

Chairman: C. BALDELLI, Itabia

Welcome address

Opening address

The community energy demonstration programme : the context for biomass

Role and strategies of the ENEA in the biomass energy sector

European R & D programme of wood as renewable raw material

WELCOME ADDRESS

G. PELLIZZI
President, AIGR

In the name of the Italian Rural Engineering Society I am happy to welcome you all and thank you for participating in this workshop organized in close collaboration with ITABIA and sponsored by the EEC Commission.

For many years we have been discussing biomass as a potential energy resource, with specific reference to such by-products as straw and prunings, to forestal produce and energy crops.

Undoubtedly, significant technical progress has been made in several fields ; by now, we know what's available in every country, region by region ; we have set forth common procedures to evaluate biomass energy content, cost and profitability ; we have developed efficient conversion plants ; we are carrying out strict analyses and research on the actual potential of energy crops ; we are aware of the advantages and limitations of biomass conversion into liquid fuels.

In the field of thermochemical conversion, boilers of various sizes are by now commercially available and their use is spreading, while gasification is the subject of significant research and development. In many cases, many more biomass-fired boilers than currently in operation could be used and proved to be competitive from an energy and an economic angle in the rural world.

In spite of all this progress however, we must admit that widespread use of the available technologies is hampered by a number of barriers.

One of the toughest barriers (at times, a true bottleneck) is the lack of suitable mechanization chains for the operations starting with biomass collection and ending with its delivery at conversion plant inlet, and especially for harvesting, handling, drying, chopping, compacting, etc. This hampers the use of the technologies downstream.

Actually, the situation varies depending on the raw materials and country ; there is no doubt however that this problem should be tackled, analyzed and solved from both a technico-operative and an economic angle.

This workshop was organized to discuss the more relevant problems, collate the results, determine the weaker links of the chain that require special research and development, set forth common strategies and research projects to solve this problem once and for all, with reference to the social, orographic, pedoclimatic and structural situations, so that these barriers can be removed.

The stakes are high, and will be of benefit to those who develop technologies and manufacture equipment as well as, and especially, to the rural world, increasing its income and improving its standard of living.

I am certain that the discussions and contributions made by the participants will lead to significant operative conclusions and outline the precise road to be taken.

With this certainty in mind, I should once again like to thank the EEC Commission, ITABIA and all the Italian sponsors of this workshop, and wish you fruitful work and a pleasant stay at Marino.

OPENING ADDRESS

Carlo BALDELLI - ITABIA
Via Archimede 161 - 00197 ROMA

I am very honoured to open this workshop and I hope that all of you will find interesting topics in the next two work days.

I thank very much all the friends of the organising Committee and the sponsor firms that made this meeting possible.

At the present moment, when many politicians in Italy and in Europe are finally considering energy biomass and, generally, all renewable energies as important development factors, I hope that this meeting will contribute to the resolution of some of the problems linked to the many steps of energy biomass processing. I am particularly interested in this issue because I am faced with these same problems with the Robinia and Ginestra plantations in the farm "Civitella d'Arno", near Perugia. These plantations, shown in Fig. 1 and Fig. 2, are included in the EEC DG XII research project "PERUGIA BIOMASS", aimed at studying marginal land utilisation for energy crops.

According to the first trials the yield of these two species seems to be fairly good, since it is about 10-15 tons of dry matter per hectare per year. However in order to develop a market in this field it is very important to optimise all the steps of biomass processing between harvesting and storage.

Generally speaking I would like to point out two important problems :

- first the necessity that public incentives should encourage energy biomass production and not only conversion ;
- second it would be advisable to organise a task force to study whether it is possible to standardise the raw products of biomass harvesting. The aim would be to provide biomass conversion industries with detailed and standardised characteristics of the product to be processed.

ITABIA will be happy to cooperate with all organisations concerned with the above mentioned problems.

Fig. 1 - Ginestra plantation - 4 years old

Fig. 2 - Young Robinia plantation - 1 year old

THE COMMUNITY ENERGY DEMONSTRATION PROGRAMME: THE CONTEXT FOR BIOMASS

U. Zito and G.L. Ferrero
Directorate-General for Energy (XVII)
Commission of the European Communities

H. E. Williams
ECOTEC Research & Consulting Ltd

Summary

The aims of this paper are to outline in broad terms the scope of the Community Energy Demonstration Programme and to indicate some of the issues that it would be useful for the workshop to discuss.
The demonstration programme is the logical extension of the Community "Research and Development" to activities on the threshold of commercialisation. In the framework of this programme the "Biomass and Energy from Waste" sector is particularly important both because of the funds which have been allocated to it and because of the number and variety of technologies covered.
In this context problems related to biomass harvesting, storage and transport are of notable relevance to the technical and economic evaluation of all wood, wood waste and agricultural waste transformation processes.

1.0 THE COMMUNITY ENERGY DEMONSTRATION PROGRAMME

In June 1978 the Council of European Communities took the first of a number of decisions to grant financial support for projects developing alternative energy sources and energy savings. Since that time the energy demonstration programme has been modified (1) and has grown considerably with a steady increase, year by year, in the numbers of projects accepted for support.

The latest invitation to submit projects invites proposals in four main fields as shown in Table 1. As can be seen from that table, Biomass and Energy from Waste is only one part of the whole programme (although certain applications of biomass are to be found in the energy savings and solar energy groups). Nevertheless, as of November 1986 the Commission had decided to support 153 projects dealing with Biomass and Energy from Waste. This amounted a contribution of 70,606,105 ECU from the EEC or about 27% of the total cost of 260,917,496 ECU of these 153 projects. At the present time only 80 of the 153 accepted projects are on-going or completed. Thirty nine have been withdrawn for technical or financial reasons and 34 contracts are yet to be signed.

One may ask why biomass and energy from waste. The Community produces urban, agricultural and industrial wastes with an energy value of 100-120 Mtoe a year. The portion of this energy which could be exploited economically in the long-term is put around 50-60 Mtoe a year. Of this, 15-20 million toe could be on stream by the end of the century. In the context of alternative energy sources this is a significant resource compared to for example geothermal sources (3-7.5 Mtoe), solar energy (10-20 Mtoe) and wind energy (5-10 m toe). Moreover, from the above figures and Table 2 (2) it can be seen that biomass and energy from waste is a major untapped source of energy. In addition, there are a number of advantages of utilizing wastes for energy production which include in particular:

- the production of energy from materials which must be disposed of for environmental reasons, but whose disposal itself consumes a large quantity of energy
- reduction of the adverse effects of these wastes (e.g. animal slurries) on the environment
- a yield of by-products suitable for use as fertilizers or animal feed
- leads to technical advances in biotechnology
- improvement of the Community's energy balance
- has implications in the wider context of the Community's agricultural policy discussed below
- has wider environmental benefits associated with better forest management.

The Biomass and Energy from Waste sector consists of 11 sub-sectors as shown in Table 3. Each sector includes applications in one or more areas. Table 4 shows the numbers of accepted projects in each sub-sector which are applicable to the different areas. It can be seen that by far the largest number have been projects on the Production and Utilisation of Biogas and that the majority of these have been concerned with the use of urban, industrial or stock-rearing waste. In agriculture and forestry, where biomass harvesting and energy crops are of interest, accepted projects for energy production have primarily concentrated upon: direct combustion and thermo-chemical treatment for agricultural crops and crop by-products; and gasification and pyrolysis for forestry biomass. This does not, of course, mean that biomass harvesting and pre-treatment must always prepare biomass for such methods of energy production (much forestry biomass is used for direct combustion). However, it does serve to remind us that harvesting and pre-treatment methods and energy production processes need to be complementary.

2.0 SOME ISSUES IN HARVESTING AND STORAGE

As implied above, it is often stated that we require an integrated approach to the production, harvesting, storage, transport and utilisation of biomass. This means that the specifications which harvesting and storage methods are required to meet are derived from a number of sources:

- from what is technically and economically feasible
- from the energy conversion process used
- from the feed stock type and production methods
- from the scale and nature of the operation
- from the final use of the products
- from the environmental constraints

Clearly, the nature of the feedstock, method of utilisation and scale of operation can all vary widely. There are currently, within the European Community, few very large scale forestry biomass or coppicing operations. Nevertheless, associated with such approaches are technical problems such as continuous swathe harvesting and the handling of cross-falling trees, the storage and transport of large quantities of timber, problems of mould or spontaneous combustion in the storage of wood chips and so on. In fact, most forestry holdings in the European Community are small - indeed the average size of private forest holdings in the Community is probably about 8 hectares(3). In France, the average is even smaller, where 86 % of private owners holdings are on average only 0.84 ha(4). Should we be attempting to develop harvesting machines that are economical in such circumstances or is this infeasible and must we rely upon institutional solutions and such as encouraging co-operative management and harvesting between land owners ? Similar issues arise with respect to the harvesting and utilisation of many crop by-products - agricultural holdings vary in size, but many are quite small.

The harvesting of crop by-products or biomass from especially cultivated forestry (whether whole trees, coppices or only thinnings and trimmings are used) raises one range of technical issues and specifications. However, in many parts of the Community, there are opportunities for exploiting the biomass which arises from general management of forests - whether such management is primarily for reasons for fire prevention, soil erosion prevention or environmental improvement. The Mediterranean forests, scrub and maquis in Southern Europe in particular, are increasingly having to be managed for the purposes outlined above. In these circumstances the nature of the terrain and of the biomass that harvesting machines have to accommodate raise their own technical problems. The report referred to above, on the prevention of major natural hazards in France, notes that the reduction in clearance of brushwood is in part due to reduced time spent on forest management (compared with fire watch and fire-fighting) and that more effective clearance is hampered by the lack on the French Market, of available and suitable machines for the collection of brushwood etc.

It is not just the scale of the operation in terms of the production of biomass that is relevant but also the scale and nature of utilisation. If there is to be continuous consumption of large volumes, in centralised power generation or by large manufacturing operations (in the timber trade) then transport and the location of chipping/cutting/sawing equipment are issues to be addressed. Storage may also be a problem, both in relation to some of the technical issues mentioned earlier and in terms of the large amounts of space required (wood storage requires ten times as much space as the equivalent amount of coal or oil). However, much biomass is well dispersed of relatively low energy to bulk value and hence lends itself more to utilisation by many, small users such as farmers, rural householders and small rural communities. Here, too, storage problems arise, but more importantly there is the need to recognise that harvesting methods may have to serve very many different users who are unlikely to have common specifications for the feed stock they can use.

These comments serve to raise two general questions.

i) Is it possible to identify a limited number of types of integrated biomass production and utilisation systems that represent the variety likely to be found in the Community? Is it possible to assess which of these are or will be the most common in differing parts of the Community ?

ii) Is it possible to develop (or adapt existing) harvesting and pre-treatment methods which are flexible enough to be used widely within any one of the types of production/utilisation systems identified or even across a number of such systems ?

The answers to these questions, put alongside a discussion of the current experience of the technical and operational difficulties of harvesting and storage, will help to identify the range of applications for differing harvesting and storage methods.

3.0 SOME ISSUES IN EVALUATION

The main objective of biomass systems is, of course, the production of energy from new sources, hopefully at economic rates. However, a number of energy from biomass systems have the potential to provide wider environmental benefits in addition to energy production. This is particularly the case for biomass harvesting and utilisation associated with forest management. Management to remove dead growth and scrub, etc. can add considerably to the recreational and tourist value of forest areas. Forest management is also increasingly being seen as important in the prevention of forest fires (5). Indeed, M Tazieff's report on the prevention of major natural risks calls for a major compaign of fire prevention through clearing brushwood from forests(4). The Mediterranean forest is most at risk. In 1985 over 300,000 hectares(3) of forest were destroyed by fire - Spain and Portugal suffering losses equal to those in France, Italy and Greece together. Tragically fifty lives were lost in addition to a loss of capital equipment, subsequent soil erosion damage in areas left with no tree cover, visual damage etc. It has been estimated that forest fires in 1984 cost 25,228,310.00 Lire in Italy(6), however this is solely the cost of lost wood, and does not include all the wider costs alluded to above. Table 5 gives details of the number of fires and areas affected by them in Italy and France in the 1970's and early 1980's.

The French experience illustrates well the vulnerability of the Mediterranean forest in particular. On average 6,000 of its 10,000,000 hectares of non-mediterranean woods and forests are lost to fire each year whereas 34,000 hectares are destroyed by fire each year from the 4,700,000 hectares of Mediterranean forest (4).

If biomass harvesting can be the means to forest management for fire prevention then the wider benefits from such prevention should be taken into account when evaluating biomass harvesting and utilisation systems. How can such benefits be evaluated and what proportion justifiably be allocated to forest management through biomass harvesting ? Some answers to these questions would enable an economic evaluation of biomass harvesting and utilisation methods to take into account these wider environmental and social benefits, as is appropriate when considering public sector support.

Whilst there is already a role for biomass harvesting in forest management related to timber production, recreation, fire prevention, soil erosion and flood prevention (7) it is possible that, in the longer term biomass harvesting will have to be considered within a much wider policy context. The European Community is moving towards a forestry policy. In addition recent changes in the Agricultural Structures Directive (Regulation 797/85) allow payment to "farmers for managing certain "environmentally sensitive areas" rather than maximising production. More widely, the Community Discussion paper (8) on the future of agricultural policy recognises the need to maintain and expand economic activities and employment in rural areas and also examines the potential for biomass from forestry and energy crops. Changes in these directions may affect the price support structures for forestry and energy crops and/or the structures of income support to agricultural and forestry holdings thus altering the economic context of biomass systems. They also have wider, and not always positive, environmental implications.

Changes such as these are, however, a long way off. At the present time we can focus more fruitfully on the assessment of the wider benefits mentioned earlier: tourism, fire prevention, prevention of soil erosion, flood control etc. that can result from forest management using certain harvesting techniques. If these wider benefits make biomass harvesting and utilisation a good investment from the community view then the question arises as to how their use can be encouraged: for those who benefit may not be those who have to pay for the equipment and operation of such systems. Different approaches may be more or less appropriate in differing circumstances. For example:

i) At the level of individual communes co-operative, or partnership arrangements may be encouraged between a number of land holders and the commune fire services so that they combine to purchase and operate a system.

ii) Alternativly, centralised subsidy of development and marketing of machines might be considered or means found to support their use through national forestry and/or agricultural support structures.

iii) Attempts might be made to encourage/stimulate the market for biomass products - perhaps by encouraging communes to take up centralised common biomass utilisation plant for district heating, heating of localised industry or communal facilities. Another approach might be to encourage individual households to make use of similar types of heating systems so that a market is created for a given type of woody biomass.

These examples are put forward merely to stimulate discussions on the different ways of encouraging the use of biomass - there are no doubt many others and an assessment of the alternatives would be most useful.

It is implied by the above discussion that biomass systems may be, in narrow terms, uneconomic for the individual private investor and hence require a range of incentives to "internalise" the wider benefits. However, the rapid and widespread adoption of new technologies and approaches is not merely a matter of ensuring that they are commercially attractive. There are a number of other barriers to the diffusion such as a lack of information on the part of potential users and, even if they know of the possibilities, the reluctance of many users to change to new methods.

There are a number of ways such barriers can be overcome. The diffusion of information through easily understood material, case-studies, demonstration projects, workshops and seminars is one approach. The exchange of experience between groups with similar problems where those who are more advanced or quicker in adopting new methods can relate the experience to others is one effective approach: an alternative is to use existing advisory agencies to "spread the word" - for example most member states have agricultural advisory services who could perform this role. In addition, training agencies need to include new methods in their programmes. It is, of course, the case that people are most likely to adopt new approaches if they can see them being demonstrated (hence the Community Demonstration Programmes); try them out themselves; and if the proposed methods/ technologies are easy to operate and fit in well with the existing production and management practices of those who we expected to take them up. This latter point is extremely important; it is one that can only be answered by those who are designing, developing and demonstrating new approaches to biomass harvesting, storage and utilisation.

---

(1) In all, 4 Regulations have been adopted: Regulation (EEC) Nos: 1032/78; 13033/78; 1972/83; and 2126/84
(2) Energy from Biomass in Europe (1980) Editor Palz and Chartier p 215
(3) COM(85) 792 final
(4) Tazieff, H. "Rapport au President de la Republique presente par: commissaire a l'etude et a la prevention des risques naturels majeurs". Journal Officiel de la Republique Francaise 1983
(5) Seminar on Fire Prevention in the Mediterranean Forest. La Roque A'Antheron 1985
(6) G. Calasri, Head of Forest Fire Services, Ministry of Agriculture and Forests, Rome. "The Organisation of Forest Fire Prevention and Control in Italy".
(7) For example the planting of forests in France and Italy for soil erosion prevention and floor control under Regulation EEC No 269/79.
(8) COM (85) 333 Final

TABLE 1: MAIN SECTORS OF THE COMMUNITY PROGRAMME

ENERGY SAVINGS

- INDUSTRY
- BUILDINGS
- TRANSPORT
- ENERGY INDUSTRY

ALTERNATIVE ENERGY SOURCES

- SOLAR
- BIOMASS AND ENERGY FROM WASTE
- GEOTHERMAL
- HYDRO-ELECTRIC POWER
- WIND

SUBSTITUTION OF HYDROCARBONS

- USE OF ELECTRICAL ENERGY AND HEAT
- USE OF SOLID FUELS
  (COAL, LIGNITE, PEAT)

LIQUEFACTION AND GASIFICATION OF SOLID FUELS

- (COAL, LIGNITE, PEAT)

TABLE 2: ENERGY AVAILABLE FROM WASTES AND SURPLUSES -Mtoe (2)

	GROSS ENERGY CONTENT FUEL	ENERGY CONTENT SOLID	ENERGY AVAILABLE IN BIOGAS
PRINCIPAL AGRICULTURAL WASTES			
ACCESSIBLE ANIMAL WASTES			
- PIGS	2.87	-	1.5
- CATTLE	34.6	-	11.9
- POULTRY	1.98	-	1.13
TOTAL	39.45	-	14.13
CROP RESIDUES			
- GRAIN CROPS	33.47	23.48	4.09*
- GREEN PLANT MATTER	7.31	-	4.38*
- WOODY RESIDUES	2.7	2.7	-
TOTAL	43.48	26.18	8.47
TOTAL AGRICULTURAL RESIDUES	82.93	26.18	22.6
FOREST & WOOD WASTE	16.3	16.3	-
TOTAL	99.23	42.48	22.6

* From amount used for animal bedding - Figures take account of amounts consumed by livestock as well as the biogas conversion loss factor.

TABLE 3: THE BIOMASS AND ENERGY FROM WASTE SECTOR

	SUB-SECTOR	EEC PARTICIPATION (IN ECU)	NUMBER OF ACCEPTED PROPOSALS
1.	BIOMASS HARVESTING	150 889	2
2.	ENERGY CROPS	4 542 875	2
3.	TREATMENT OF WASTE	6 476 263	18
4.	PRODUCTION AND UTILISATION OF BIOMASS	18 717 870	56
5.	PRODUCTION AND UTILISATION OF REFUSE DERIVED FUEL (RDF)	9 255 384	14
6.	PRODUCTION OF HEAT AND ENERGY BY DIRECT COMBUSTION	17 043 290	31
7.	PRODUCTION OF GAS AND CHARCOAL BY GASIFICATION AND PYROLYSIS	8 926 690	18
8.	PRODUCTION OF COMPOST AND FERTILIZERS	1 148 117	3
9.	PRODUCTION OF ALCOHOL FUELS AND CHEMICALS BY BIOLOGICAL TREATMENTS	286 443	1
10.	PRODUCTION OF ALCOHOL FUELS AND CHEMICALS BY THERMO-CHEMICAL TREATMENTS	3 777 284	7
11.	PRODUCTION OF PROTEINS	280 000	1
	TOTAL	70 606 105	153

TABLE 4: APPLICATION OF SUB-SECTORS

	AREA OF APPLICATION
BIOMASS HARVESTING	
ENERGY CROPS	
WASTE TREATMENT	
BIOGAS PRODUCTION AND USE	
RDF PRODUCTION AND USE	
HEAT AND ENERGY BY COMBUSTION	
GASIFICATION AND PYROLYSIS	
COMPOST AND FERTILIZER	
ALCOHOL FUELS BY BIOLOGICAL TREATMENT	
ALCOHOL FUELS BY THERMO-CHEMICAL TREATMENT	
PROTEINS	
AGRICULTURE	
FORESTRY	
URBAN WASTE SEWAGE SLUDGE	
INDUSTRIAL WASTE	
STOCK-REARING WASTE	
RUBBER WASTE	
AQUACULTURE	

TABLE 5: FOREST FIRES IN ITALY 1970-1984 (5)

YEAR	NUMBER OF FIRES	TOTAL AREA AFFECTED (H.A.)	DAMAGE (LIRE x 1,000)
1970	6.579	91.176	5.144.088
1972	2.358	27.303	1.554.322
1974	5.055	102.944	12.815.580
1976	4.457	50.791	4.871.287
1978	11.052	127.577	17.955.412
1980	11.936	143.919	30.343.202
1982	9.557	130.456	52.982.949
1984	8.982	75.272	25.228.310

FIRES IN FRANCE 1973-1982

YEAR	NUMBER OF FIRES	TOTAL AREA* AFFECTED (H.A.)
1973	2.377	38.322
1974	1.915	33.313
1976	2.897	42.644
1978	4.362	39.210
1980	3.558	15.117
1982	3.962	52.735

* N.B. Total area including forests (approx. 30% of total); maquis and heathland

ROLE AND STRATEGIES OF THE ENEA IN THE BIOMASS ENERGY SECTOR

G. BIANCHI
Department for Alternative Sources and Energy Saving (FARE),
ENEA, Casaccia, Rome

The National Energy Plan (PEN) approved by Parliament in 1981, the general lines of which were reconfirmed in 1985, set a goal for the country of reducing its external energy dependence, which is one of the most longstanding among the industrialized countries. The use of national energy sources, together with the adoption of energy-saving techniques (which can be regarded as a virtual source of energy), has been the main approach advocated to reduce external dependence. I would like to stress not only the economic importance of this approach but also its impact on employment. To stop, or at least cut back on, purchases of fuels from abroad means not only reducing our political vulnerability but also replacing this expenditure with internal investment and the use of national labour and technologies.

An integral part of this approach, combining all its aspects and positive elements, is the exploitation of all the renewable forms of energy, particularly biomass, available in Italy. It is a commonplace often repeated in the case of renewable energy sources - and an argument that in fact is also reiterated in the case of nuclear energy - that such sources are not very important because they account for a low, almost negligible, percentage of our energy requirements. This view is totally erroneous : in the energy sector the criterion that large numbers can also be obtained as the sum of many small numbers is applicable. It is therefore right for attention to be devoted to all possible ways of exploiting energy without being concerned whether their contribution is small in percentage terms.

Another point which favours, and therefore ought to stimulate, the use of renewable forms of energy is the effect on the environment. In particular, the massive use of fossil fuels - oil, coal, gas - is a cause for concern on account of the substantial emissions of pollutants into the atmosphere : I need only mention the adverse effects that emissions of sulphur and nitrogen oxides and above all of carbon dioxide through the "greenhouse" effect can have on the atmosphere, and therefore the life itself, of our planet.

In contrast, the use of renewable energy sources is fully in keeping with protection of the environment and the combustion of biomass, if carried out with suitable techniques and technologies, does not alter the environment since it forms part of the natural cycle, thus helping to avoid the harmful effects that the uncontrolled disposal of waste can have.

These then are the reasons behind the commitment to use agricultural and forestry waste for energy purposes ; looking at some figures, it is readily apparent that the use of wood biomass, and more generally of cellulose waste, is no longer as insignificant as thought. In industrialized countries this source accounts for 3 % of energy requirements ; like all averages, this figure masks considerable differences from one country to another, such as 19 % in Finland and 11 % in Sweden, not to mention the developing countries, where the use of biomass meets very high percentages of energy requirements.

In Italy too, the figures are not negligible : wood and cellulose wastes constitute a theoretical supply of around 13 million toe a year, although only 15 % of this is used to produce energy. There is thus a vast potential waiting to be tapped. The size of the source and the comparatively little use made of it reflect, on the one hand, the underlying problems, namely the scattered nature of the source (in common with all renewable sources), the high costs of collection and use associated with undeveloped technologies, and on the other hand they indicate the paths that need to be explored in order to boost its use, namely the rational study of production areas associated with potential uses and the development of machines and technologies to exploit them.

The ENEA, which has been working for some time in this sector, has directed the research carried out to date along these two lines, namely study of the availability of technologies suited to the use of biomass, with analysis of the equipment to be used, ranging from automatic collection systems, storage equipment and equipment for preprocessing of agricultural and forestry waste to machines for final use, and study of demand, i.e. the uses that may be associated with this particular type of source.

In my opinion, the study of demand and its characteristics is the bottleneck that has to be unblocked if the applications are to become widespread. In this case too there are no general rules but a series of different cases to be studied and resolved. Bearing in mind the cost of energy produced from biomass and that of rival sources, the major difficulty is in fact to combine "availability" with "utilization". This is the field of study, the sector where the technological solutions have to be found. As already mentioned, the widespread introduction of applications constitutes the major difficulty because, while it is relatively easy to design or build a machine that can use agricultural waste effectively, it is more difficult to then find a whole range of applications and above all a large number of purchasers so as to make the particular energy production system economically viable.

We thus find ourselves in a situation where the study of ways of aggregating demand and the study of integrated demand/supply systems determine the application of the technology.

Following this line of approach in the recent reorganization of the Department for Alternative Sources and Energy Saving, the ENEA set up a specific project unit responsible for all activities relating to the use of biomass for energy purposes, including a project involving systematic studies of production areas and potential uses for the agricultural and forestry biomass source, and another project for the development or conversion of combustion technologies.

From the operational point of view the ENEA has tried to develop and stimulate the capacity to solve individual cases directly on the part of the users who are to apply them, resulting if possible in specific, localized successful projects and then to try to disseminate knowledge of these results so as to generate other successful projects, perhaps spontaneously.

The following major projects provide an illustration of the ENEA's activities in this field :

- A study on the quantification and utilization of the residual biomass from vine and olive trimmings was carried out in Apulia, with the development by local industries of suitable automatic machines for the collection and preprocessing of the biomass.
- A similar study is now getting under way to assess the potential of coppice forest wastes from the Bacino di Ormea in the province of Cuneo,

trying out the potential use of the biomass for a gasification plant feeding an 80 kW electricity generator.

- An experiment is under way, with help from the ENEA, for the use of rice husks in a cogeneration plant in a distillery at Saluzzo.
- Another study of the agricultural and forestry biomass inventory was carried out in the province of Enna in Sicily, with particular reference to the availability of straw and the possibility of alternative crops.

Recently Professor Colombo submitted to the European Parliament an idea for a project, called appropriately enough IDEA, which, based on consideration of the present surplus situation in European agriculture and the energy deficit, and of the environmental problems posed by the large-scale use of fossil fuels not only in Europe but throughout the world, raises the question of whether it would be a good idea to launch an international research programme on the use of agricultural products for energy purposes.

It is obviously a programme which involves in particular wide-ranging research on plant physiology and pathology, crop genetics, biotechnologies and technologies for converting agricultural products in the wide sense, agrotechnologies and all the other bioclimatic and engineering technologies that may have even wider repercussions with regard to the main aim of the project.

The project is still in a study and assessment phase ; the ENEA has recently set up a number of working parties to deal with the individual topics in greater depth, viz. :

- Biomass production
- Collection, preprocessing and conversion of biomass into energy
- Environmental and social impact of the production and utilization of energy biomass
- Overall economic assessments of the project
- A fifth working party has been set up to collect and store in a data bank the different experimental data and results and, in general, the information currently available and that which will be produced over the years.

The results of the working parties' activities are expected to yield information on the prospects for future studies and research.

EUROPEAN R & D PROGRAMME OF WOOD AS RENEWABLE RAW MATERIAL

F.C. Hummel
Consultant to Commission

Summary

The first phase of this programme ran from 1983 to 1986. There were about 130 projects of which eight were concerned with harvesting, which is the only research area of relevance in the present context. All the eight projects were concerned with the harvesting of small roundwood in thinning operations. The projects dealt mainly with the development of small skidders and of harvesting systems. Significant progress was made in the topics covered. The main general conclusion that has emerged is the need for more training facilities in order to ensure that recent developments are applied in practice. In the next phase of the programme which starts in 1987 there will be a switch of emphasis from individual projects to the promotion of the coordination of national research and development work in harvesting.

SCOPE

The first phase of this programme ran from 1983 to 1986 and consisted of six research areas, viz: Wood Production; Wood Harvesting, Storage and Transport; Wood as a Material; Wood Processing Without Modification of its Basic Structure; Processing of Wood and of Related Organic Materials into Fibre Products; and Wood as a Source of Chemicals. In all, there were about 130 projects of which eight were concerned with harvesting which is the only research area of relevance in the context of this conference.

All the eight projects dealt with the harvesting of small roundwood in thinning operations. Harvesting of mature timber was excluded, because this presents fewer problems and the harvesting of short rotation biomass plantations was also excluded as this is considered in the Community's energy from biomass programme.

The harvesting of small roundwood from thinnings presents special problems of Community concern. The cost of harvesting a given volume of small trees is higher than if the volume is made up of a smaller number of large trees because there is more handling; and the price obtainable either for industrial uses such as pulping or for energy are lower. As a result, thinnings are often uneconomic and are therefore neglected. That explains why millions of cubic metres of wood which are available for harvesting are left in the forest. Harvesting these thinnings would promote the growth of the better trees that are left standing and would help to reduce the Community's import bill in the timber sector which ranges between 15 and 20 billion ECU per year.

The projects are concerned with two particular aspects of harvesting. The first relates to machines. The main need here was for relatively small machines suitable for difficult terrain and small operations. The second aspect relates to harvesting systems, i.e. the efficient organization of the work in order to ensure optimum use of manpower and equipment.

RESULTS

The main results may be summarized as follows:

1. A French project was responsible for developing a narrow but stable tractor with flexible tracks. For economic reasons, it has been designed a

multipurpose machine carrying numerous and various equipments. The main features of the bases tractor are:

- ° Multi-purpose (2 engines 70 or 100 Kw, 3 coupling points, 2 power transmissions, 2 types of tracks)
- ° Stability (100 p.c. longitudinally, 60 p.c. laterally)
- ° Small size in spite of its power
- ° Low ground pressure
- ° Comfort and safety
- ° Easy maintenance.

It is possible to fit equipments for brush-cutting, skidding, felling, delimbing-cross cutting, harvesting and silviculture. Trials have confirmed the expected results. Following the prototype, ARMEF is studying the industrialization of the machine.

2. The prototype of another small very versatile tracked skidder was developed in an Italian project. It is designed to facilitate the extraction of small wood in hilly country.

3. Two projects, including the one mentioned above, were concerned with thinning beech coppice as a preliminary to conversion to high forest. The objects were first to organize the harvesting operations so as to reduce costs and secondly to obtain reliable figures on costs and returns as a basis for planning large scale operations. There are hundreds of thousands of ha of neglected coppice in mountainous country in Italy that could be converted into moderately productive high forest, while at the same time ensuring that the forest fulfils its important environmental functions. The Regional policy implications are also obvious.

4. Another two projects dealt with the organization of thinnings in coniferous plantations, one in Germany, the other in Italy. They both demonstrated the reductions in cost that can be achieved by making best use of machines and manpower through careful planning and organization and through the adaptation of silvicultlure to mechanization (use of systematic thinnings etc.).

5. A Danish project dealt with the separation of wood into three components: industrial wood for pulping, energy wood, the nutrient fraction. Pilot trials of flails, chains and steel brushes led to the fractionation of the industrial and energy fractions, and chains for the defractionation of the nutrient fraction.

6. Finally there was a study to determine the quantity and quality of woody material in branches and crowns which is not normally utilized, the object being to assess the potential for developing markets for this material.

CONCLUSIONS

The results obtained have made significant contributions towards rendering the harvesting of thinnings more economic. The projects have also led to more general conclusions, especially that the best available harvesting systems are not yet widely used in practice; there is therefore a great need for additional training and demonstration facilities.

THE FUTURE

In the second phase of the programme which is about to start in 1987 the emphasis in harvesting will move from individual projects towards promoting the coordination of national research and development work in harvesting. The Commission believes that, in the present circumstances, this will make the most cost-effective use of the limited funds that have been authorized.

SESSION I - FORESTRY

Chairman: P. MORANDINI, Istituto Sperimentale per la Selvicoltura
Rapporteur: G. SCARAMUZZI, SAF

Forest biomass for energy in EEC countries from harvesting to storage

Biomass production from coppice forest harvesting and first thinning of fast-growing plantations

Mechanised forestry chains in France

Evaluation of the work and objectives to be followed in France for the harvesting, storage and processing of forest biomass

Harvesting and storage of chipped wood for energy production in hill areas

Assessment of the energy balances for beech coppices in the Northern Appennines

Cooperative R & D in harvesting forest biomass in the IEA bioenergy agreement

Wood harvesting systems for energy purposes in developing countries

Session report

FOREST BIOMASS FOR ENERGY IN EEC COUNTRIES FROM HARVESTING TO STORAGE

Sanzio BALDINI
Istituto per la Ricerca sul Legno
Consiglio Nazionale delle Ricerche - Florence

Summary

The author provides a series of tables on wooded areas and timber production for energy and then describes the principal methods used for harvesting forest biomass from coppices or thinnings of coniferous forests in some Community countries. He stresses that in coppices it is preferable to use power chain saws for felling operations rather than hydraulic shears, since this helps to preserve the capacity of the stump to produce shoots and also to prevent damage to the remaining trees. Because of the considerable apparent volume of the material, skidding operations are carried out using agricultural tractors fitted with logging winches, or medium-sized or small forwarders, but on steep gradients skylines carriages with mobile power units are used. The biomass is transported to yards and, depending on the size of the stems and the harvesting method, it may be chipped or piled up in bolts.

Details are given of the various types of vehicle used to transport the material in the forest and to the timber industry. Harvesting and transport costs are given for some operations in Italy, France and the Federal Republic of Germany.

The last section of the paper deals with the drying and storage of biomass in its natural state in forests and in the form of chips in yards.

In the conclusions the author states that at the moment sophisticated automation procedures are not cost-effective, given the present structure of forestry holdings, particularly private holdings, the average size of which is 10.6 hectares in the Community countries. It is therefore more cost-effective to use smaller machines with a more limited operating capacity or - for skidding operations - converted agricultural tractors fitted with appropriate equipment such as winches, choppers and skylines, rather than highly specialized, excessively powerful machinery, which would merely result in very high operating and energy costs. As the cost of transporting biomass in its natural state and in the form of chips accounts for 25 - 35 % of the total cost of harvesting biomass, this aspect will have to be given careful attention.

1. WOODED AREAS AND OUTPUT

Wooded areas form approximately one third of that part of the earth's surface above sea level, i.e. 4 000 - 4 500 million hectares.

Broad-leaved forests cover a larger area (58 %) than coniferous forests, the largest of which are to be found in the Soviet Union, North America and Europe.

The total wooded area of the Community (Table I) is 55 745 000 hectares, approximately 13 % of the world total.

France, with approximately 15 million hectares, accounts for 27 % of the total wooded area in the Community countries.

An analysis of the area covered by coppices (Table II), the main source of renewable supplies of energy, shows that, within the EEC, France accounts for 12.55 %, followed by Italy with 6.48 % and Greece with 2.16 %.

Sources of energy, other than coppices, are the slash from final cuttings and the material obtained from thinnings.

Table III gives the tree density coefficient of each of the Member States.

Broad-leaved forests cover approximately 60 % of the total wooded area in the EEC countries, the most common types of tree being beech and oak.

The most common conifers are the Norway spruce, silver fir, black pine and cluster pine.

According to FAO statistics, world production of rough timber is approximately 2 700 million m^3, comprising 48 % fuel wood and 52 % timber for industrial purposes (woodpulp, saw wood, pit props, etc.).

As can be seen from Table IV, the volume of timber used for energy varies in relation to the degree of industrial development of the individual geographical areas.

In the industrialized countries, forest biomass covers approximately 3 % of energy requirements, although the percentages for Sweden and Finland (11 % and 19 % respectively) are significantly higher.

By contrast, in the developing countries wood (and particularly slash) is usually the main source of energy, sometimes even covering over 90 % of requirements.

In the case of the EEC countries, Table V shows that the highest percentages are to be found in France and Italy, both countries with substantial expanses of coppice, and in the Federal Republic of Germany (which does not have extensive coppices, however) (Table II).

TABLE I - Total wooded area (in '000 hectares) in the EEC countries (FAO 1982)

COUNTRY	TOTAL WOODED	HIGH FORESTS		COPPICES	TOTAL UTILIZABLE WOODED AREA
		CONIFEROUS	BROAD-LEAVED		
Belgium	680	283	138	179	600
Denmark	484	341	115	10	466
Germany (FR)	7,207	4,735	1,660	594	6,989
Greece	5,754	1,071	234	1,207	2,512
France	15,075	4,997	1,883	6,995	13,875
Ireland	380	298	49	-	347
Italy	8,063	1,935	813	3,615	6,363
Luxembourg	82	28	40	14	82
Netherlands	355	222	52	20	294
Portugal	2,976	1,383	1,026	218	2,627
United Kingdom	2,178	1,409	579	39	2,027
Spain	12,511	5,980	105	821	6,906
EEC TOTAL	55,745	22,682	6,694	13,721	43,086

TABLE II - Percentage of the total wooded area comprising coppices in the individual countries (*) and in the Community as a whole (**) (FAO 1982)

COUNTRY	TOTAL COPPICES ('000 HECTARES)	% (*)	% (**)
Belgium	179	26	0.32
Denmark	10	2	0.01
Germany (FR)	594	8	1.06
Greece	1,207	21	2.16
France	6,995	46	12.55
Ireland	-	-	-
Italy	3,615	45	6.48
Luxembourg	14	17	0.03
Netherlands	20	6	0.03
Portugal	218	7	0.39
United Kingdom	39	2	0.07
Spain	821	6	1.47
EEC TOTAL	13,721		

TABLE III - Tree density coefficient and wooded area per inhabitant in the EEC countries

COUNTRY	WOODED AREA			WOODED AREA PER INHABITANT (m^2)
	TOTAL ('000 HECTARES)	% OF AREA OF TERRITORY	% OF AREA USED FOR AGRICULTURE AND FORESTRY	
Belgium	615	20.0	21.4	628
Denmark	484	11.2	14.1	956
Germany (FR)	7,162	28.8	35.0	1,158
Greece	5,754	43.6	-	5,993
France	14,610	26.7	31.2	2,780
Ireland	333	4.7	6.6	5,402
Italy	6,325	21.0	26.5	1,132
Luxembourg	85	32.0	33.2	2,380
Netherlands	308	8.0	12.9	3,187
Portugal	2,976	33.5	-	225
United Kingdom	2,020	8.0	9.8	360
Spain	12,511	24.7	-	3,330
EEC TOTAL	55,745	23.9	-	1,602

TABLE IV - World production of fuel wood and respective percentages (FAO 1981)

GEOGRAPHICAL AREAS	FUEL WOOD CHARCOAL ('000 m^3)	%
Europe	51,491	3.7
Africa	258,928	18.9
America (Cent. & N.)	104,445	7.6
America (South)	242,669	17.7
Asia	663,209	46.2
Australasia	1,362	0.1
USSR	78,900	5.8
WORLD TOTAL	1,371,004	100.0

TABLE V - Production of fuel wood in the EEC countries and respective percentages (FAO 1981)

COUNTRIES	FUEL WOOD CHARCOAL ('000 m^3)	%
Belgium - Luxembourg	332	1.4
Denmark	268	1.1
Germany (FR)	4,000	17.0
Greece	1,790	7.7
France	10,406	44.2
Ireland	39	0.2
Italy	4,439	18.8
Netherlands	97	0.4
Portugal	500	2.1
United Kingdom	160	0.7
Spain	1,516	6.4
EEC TOTAL	23,547	100.0

I - SOME METHCDS OF HARVESTING FOREST BIOMASS (S. BALDINI)
1) First-stage extraction of whole trees and chipping in the forest ; 2) Skidding ; 3) Centralized chipping ; 4) Drying ; 5) Industries and conurbations

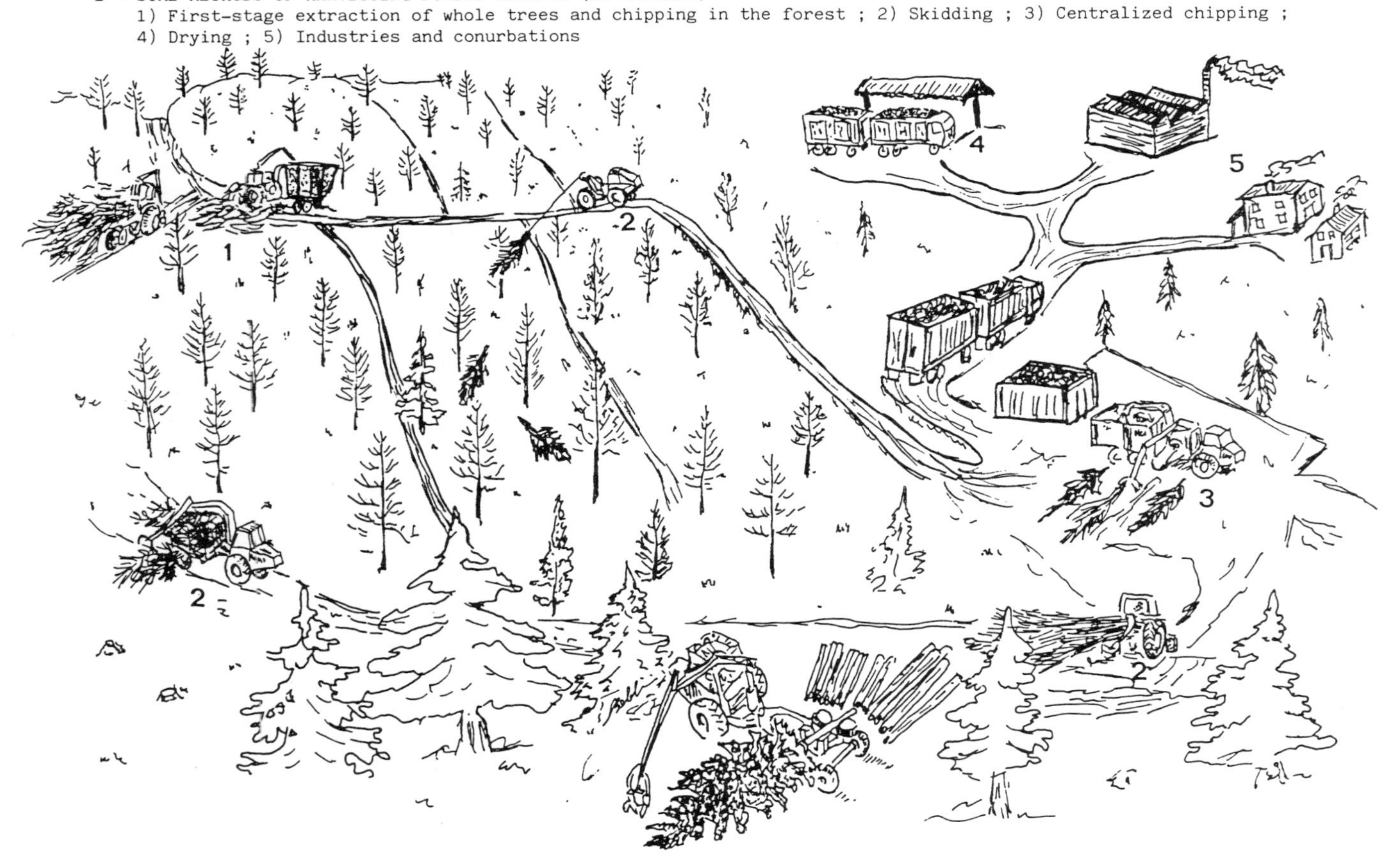

2. HARVESTING TECHNIQUES

Unfortunately I was unable to summarize the techniques employed for harvesting forest biomass in all the Community countries, because the requisite information was not forthcoming.

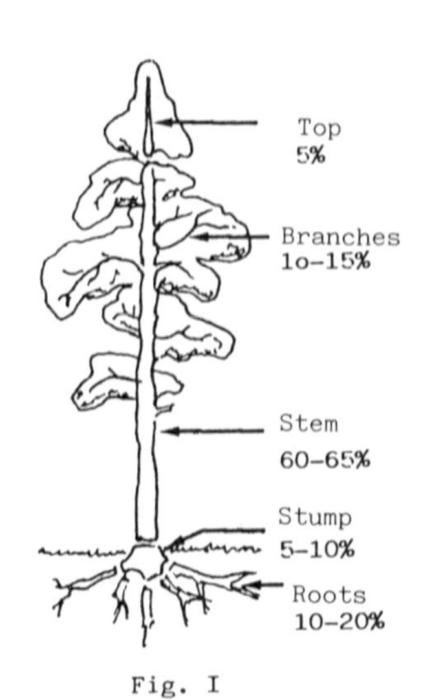

Fig. I

Harvesting all (or virtually all) of forest biomass entails choosing amongst different combinations of raw materials and deciding which harvesting technique to use.
The raw materials may be obtained from the harvesting of coppices, the first thinnings, the branches and crowns of trees harvested in subsequent thinnings or from the principal fellings (approximately 20 % of the total volume of trees) and from the harvesting of stumps and roots (approximately 15-20 % of the total volume of trees). Stumps and roots can be harvested only where no hazard is posed by surface water and when there is to be a new automated planting.
The real question at issue is the harvesting, by applying appropriate techniques, of that part of biomass which has to date been treated as slash. We shall see that these techniques are for the most part concerned with the whole tree, complete with branches and crown, the aim being to harvest the additional 20 % which has in the past been left in the forest.

2.1. Coppices

Given the characteristic configuration of the shoots close to the stump, and the life which the stump will have to sustain in the future, power saws (in capable hands) are still considered to be the best method of cutting the shoots, although in France consideration has been given to automated felling methods using hydraulic shears. The individual shoots are piled up in bundles parallel to the hauling lines.

Skidding is carried out using forwarders with articulated frame steering, fitted with winches (winch skidders), or logging winches mounted on the hydraulic lift linkage of an agricultural tractor, actually at the felling point if the gradient is less than 30 % and from the extraction routes in cases where the gradient is steeper. For steep gradients skylines/carriages with mobile power units are also used if the extraction routes are no further than 300-400 m away ; tractors can be used only when the routes are no more than 100-150 m distant.

On level ground and in individual cases when the diameter of a tree is more than 20-25 cm at the butt, the trimming operation is carried out (and the timber cut into sections if necessary) in the forest using power saws. Given the irregular shape of the stem of broad-leaved trees, it is only in rare cases that an automatic trimming device can be used. The lop and top are loaded onto forwarders with articulated frame steering by means of a grapple loader and transported to a yard which is accessible to lorries or road trains, and chipped. The chips produced are blown straight from the machines into containers, to await loading at a later stage, to avoid the transport vehicles being at a standstill ; the transport vehicles are fitted with multilift or ampliroll systems, so that the driver can perform the loading operation single-handed.

The bolts are later skidded by the same type of forwarder or an agricultural tractor fitted with a logging trailer.

The power of the machinery depends upon the volume and size of the timber, the density of the forest track network and the relief of the terrain.

The timber is chipped along the forest tracks by self-propelled hoggers fitted with a dump body, in cases where it cannot be chipped in the yards because they are too far away.

In addition to the aforementioned equipment, a machine called a "Scorpion" has recently been designed in France ; this makes full automation of all systems possible in coppices, since the timber is felled and chipped immediately at the felling point (Mr Morvan will give more details on this method).

The costs of producing chips with semi-automated systems range from FF 155 to 300/t = 23-45 ECU/t (25 % humidity) in France, depending on whether the logging is done by the owner of the forest or by a logging company, while the costs of production with fully-automated systems range from FF 250-300/t = 37-45 ECU/t (45 % humidity).

In Italy the costs vary from Lit 30 000-58 000/t = 20-39 ECU/t (approximately 45 % humidity) if the logging is done by the owner of the forest, to Lit 60 000-85 000/t = 40-57 ECU/t (approximately 45 % humidity) if the logging is done by Government bodies.

The - unfortunately - high cost of producing wood for energy is in some cases attributable to the high costs of constructing machinery which is not always used to capacity, the absence of a proper network of forest tracks, and in some instances to inadequate training in logging (owing to lack of vocational training centres for loggers).

It is rather difficult to use sophisticated types of automation for the first thinnings for the conversion of coppices into high forests, owing to the large number of trees per hectare which are left.

It is also virtually impossible to harvest the total crown biomass because of the danger of damaging the remaining trees. After the first thinning 800-1 300 trees/hectare are left.

For the most part the timber, trimmed and partially cut into sections, is skidded uphill by logging winches which are radio-controlled or fitted to agricultural tractors and downhill by polyethylene (PVC) timber chutes.

For the second thinning, if the gradient permits, automated felling methods are used. Given the substantial spaces between the trees left standing, the material can be skidded uphill from the track by winch or skyline.

2.2. Thinning of coniferous forests

More sophisticated forms of automation are used for thinnings in coniferous forests than for those in coppices, since each tree is isolated from the others, and the shape of the stem is more regular. Where gradients are under 30 %, the felling operation is almost always carried out by machines which also carry out the first-stage extraction operation over distances of under 50 m. In some cases, for trees with a diameter of less than 10-15 cm at the butt, a power saw, with or without a special frame, is used, and where gradients are over 30 %. Power saws are still the most efficient means of performing the felling operation.

The trees or faggots are laid in the direction of the skidding routes, to avoid damaging the trees which are left.

Yields will be increased and the cost of felling and skidding operations reduced only if traditional methods of silviculture are abandoned. Geometric thinnings will result in more productive felling and skidding operations than traditional selection thinnings, irrespective of gradient.

Unfortunately, traditions also considerably slow down the development of new technologies ; for instance in the Federal Republic of Germany it is only recently that the use of wood chipped in the forest has been considered as a possible source of energy.

Approximately 10 million m^3 of slash (top and lop, bark, stumps and roots) is available each year in the Federal Republic of Germany ; 40 % of this material could be harvested and, if it were combined with the wrack from sawmills, the timber industry could achieve a high degree of self-sufficiency in energy.

Whole-tree logging is still economically viable where trees have a diameter of 22-30 cm 1.30 m above the ground, even if only part of the plant is to be used to produce energy. Trimming is carried out in the yards using automated methods and the slash (lop and top and defective trees), is chipped at the same time. Trimming and topping are carried out at the felling point only when the diameter is greater than 22-30 cm and, where the gradient and relief of the terrain permit, the slash is loaded onto forwarders with articulated frame steering and transported to the milling yards to be chipped up (Table III).

In the Federal Republic of Germany it is estimated that the costs of harvesting material for energy range from DM 60-75/m^3 = 28-35 ECU/m^3 (in the case of freshly-harvested biomass).

Some data for Italy (provided by the Istituto per la Ricerca sul Legno - CNR) are given in Figures I and II.

Fig. II - Cost of harvesting in Lit/t, for the first thinning of Norway spruce forests (32-year old stand)

Felling + skidding	Felling + skidding Lit/t	Operations at the stacking area	Operat. at stack.area Lit/t	Total Lit/t
	34.105,8		20.015,5	54.121,3
	35.912,2		16.785,0	52.697,5
	42.387,5		7.762,5	50.150,0

FIG. III - Cost of harvesting, in Lit/t, for the main felling of cluster pines (52-year-old stand)

Felling + skidding	Felling + skidding Lit/t	Operations at the stacking area	Oper. at stack.area Lit/t	Total Lit/t
	15.857,2		11.017,1	26.874,3
	79.858,5	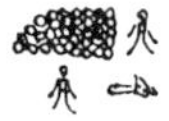	6.328,6	86.187,1
	21.008,6		6.850,0	27.858,6
	20.524,3		6.850,0	27.374,3

Cost of harvesting in Lit/t, for the last thinning of cluster pines (52-year-old stand)

Felling + skidding	Felling + skidding Lit/t	Operations at the stacking area	Oper. at stack.area Lit/t	Total Lit/t
	37.608,6		12.545,7	50.154,3
	32.504,3		12.890,0	45.394,3

II - Machinery and equipment which can be used for the harvesting of biomass in coppices

a) Felling with power chain saws fitted with chipper blades helps to increase productivity by 20-30 % on sloping ground (Baldini).

b) Where the gradient is not too steep and where distances are under 100-150 m, skidding operations are carried out using agricultural tractors converted for logging purposes, fitted with logging winches. (Photo Baldini).

c) Where gradients are over 30 % and distances under 350-400 m, skidding operations are carried out using skylines carriages with mobile power units. (Photo Baldini).

d) Chipping operations are carried out in forests using medium-sized or small choppers operated by agricultural tractors. (Photo Baldini).

e) Where tracks cannot be negotiated by lorries, the chipping operation is carried out in the forest using tractors fitted with choppers and containers (Photo Baldini).

f) In harvesting areas chipping is carried out using hoggers with autonomous motors, often of over 194 kw. (Photo Baldini).

III - Machinery and equipment which can be used for harvesting biomass from thinnings of coniferous forests

a) Where the gradient is less than 25-30 %, felling operations are carried out using hydraulic shears (Photo Baldini).

b) For distances of over 100-150 m, where gradients are less than 40 %, small tractors, often fitted with special caterpillar tracks, are used for the first-stage extraction operation. In the foreground can be seen the prototype tracked mini-vehicle designed by the Istituto per la Ricerca sul Legno - CNR. In the background there is a 52 kw medium-powered forwarder with articulated frame steering also designed by the Istituto (Photo Baldini).

c) Medium-power forwarders with articulated frame steering are used for skidding the biomass when the quantity of material is such as to make their use viable.

d) Logging versions of agricultural tractors, fitted with logging winches with one or two drums, as in the illustration, are used in other cases (Photo Baldini).

e) In the harvesting area the timber may be trimmed and cut into sections with machinery fitted to agricultural tractors (Photo Baldini).

f) Branches and whole trees with a diameter of less than 15-18 cm are chipped (Photo Baldini).

3. TRANSPORT

It has already been stated that it is not cost-effective to transport the total forest biomass(including crowns) over long distances because of its considerable apparent volume ; the distance between the felling point and the milling yard should therefore be as short as possible.

Research conducted in Italy shows that where distances are greater than 1 000-1 500 m to tractor logging, it is advisable to load the material onto trailers.

Material in the form of chips and bolts or stumps and roots is transported along the main skidding extraction routes and cart tracks by agricultural tractors fitted with logging trailers or by forwarders with articulated frame steering, and on truck hauling roads by lorries, road trains or articulated vehicles.

When the consumers' premises are not more than 10-15 km from the place of production, forwarders or agricultural tractors may also be used on truck hauling roads.

3.1. Agricultural tractors with logging trailers

50-80 kw four-wheel drive agricultural tractors with a standard cab are used.

A one or two-axle logging trailer is attached to the tractor, the double axle helping to prevent sharp jolts in transit.

The capacity of the trailers ranges from 5 to 10 t, according to type.

For the transport of bolts 1-2 m in length the trailer is usually fitted with accessories and a grapple loader, plus a dumping mechanism to facilitate the unloading of the timber.

For the transport of chips the trailer, if it is stationary during the loading operation, is fitted with a body that is covered at the top with metal mesh and has at least one panel that can be opened, so that the chips can be unloaded quickly with the aid of the dumping mechanism.

3.2. Forwarders

These are a form of tractor with articulated frame steering, in which the rear carriage has been replaced by a semi-trailer, usually with a double oscillating axle.

They are fitted with a hydraulic loader and/or a dumper body.

When it is not possible for lorries to approach the logging area, these vehicles, like agricultural tractors with trailers, collect the chips or other material from the forest and unload them into larger containers on surfaced roads or into rail wagons, or take them direct to the industry. The maximum radius which can be covered by vehicles of this kind is also approximately 5-10 km.

3.3. Lorries, road trains and articulated vehicles

For transport within a radius of 40-50 km, the usual practice is to use lorries which can do two or three runs per day (8 hours) carrying 7-10 t of material each time.

For distances of 80-120 km road trains and articulated vehicles are used ; these can carry 25-30 t of material, although they make only one journey per day.

These vehicles are fitted with a hydraulic loader, if they are used for transporting bolts, or with multilift or ampliroll loading mechanisms for the transport of containers.

Loading costs can be reduced if, instead of being piled up at the edge of the road, the bolts are put straight into special containers which

IV - Transport

a) Transport of bolts along forest tracks on logging trailers (Photo Baldini).

b) Transport of bolts by forwarders with articulated frame steering along negotiable roads (Photo Baldini).

c) Transport of branchwood in special forwarders with articulated frame steering (Photo Baldini).

d) Transport of chips in containers on trucks. When the container is full, it is loaded onto the truck by the driver singlehanded (Photo Baldini).

can be loaded on to the vehicles with the ampliroll or multilift device.

Some manufacturers of lorries and articulated vehicles fit a special type of conveyor belt, the same width as the vehicle platform, to the lower part of the body, so that the chips can be unloaded quickly without the use of the dumping mechanism.

Costs per tonne of transport by road train

Distances	FF	ECU	LIT	ECU
60	30	4.46	6 000	4.05
60-120	120	17.84	14 000	9.46

4. DRYING AND STORAGE

The drying of the biomass is important because it is a means of increasing its output in use. The process can be carried out near the forest or in the sorting yard, depending on the type of material to be produced.

4.1. Pressed into granules

Material reduced to small chips (of a few millimetres in length) is dried, sifted and fed into a press which reduces it to granules, to obtain this product. The press will process only timber which is in small particles, with less than 10 % humidity.

The resultant product has the appearance of cylinders 6-8 mm in diameter and 15-20 mm in length.

The granules must be stored in a dry place. In damp environments they disintegrate and, if they are handled excessively, they are liable to produce dust and possibly cause accidents.

They are used in small automatic units.

As a rule, the heat needed for the drying process is obtained by burning some of the granules or other slash.

4.2. Pressed into briquettes

The material used must be clean, have a low humidity content and be in fine particles. The chips almost always have to be dried and ground before being fed into the press.

The increase in pressure and temperature during the process gives the product a glazed appearance.

The briquettes measure 20-100 mm by 30-300 mm.

4.3. Billets and quartered timber

Experience has shown that :

- quartered timber has a greater surface exposed to the air than billets, and therefore dries more quickly ;
- short pieces dry more quickly than pieces one or more metres in length ;
- the humidity of timber which is piled up in the open air falls to 40-45 % after four months, 30-35 % after six months and 20-30 % after a year. The humidity of the timber on the outside of the pile is always lower than that of the timber inside the pile.

4.4. Chips of freshly-harvested biomass

Forest products and by-products (poles, billets, branchwood and shoots) are piled up at the side of the extraction routes in yards or transported to the industry, in order to provide consumers with a regular supply of chips.

It has been noted that products which have been chipped can easily be stored for a few weeks.

The material is reduced to chips in the forest or at the yard at a later stage, as and when it is required.

4.5. Chips of dried material

The material (poles, billets, branchwood and shoots, slabwood and edgings) is piled up in its natural state to dry.

It is piled up in the open air in the forest or at the yard. The drying time ranges from 4 to 12 months. In some cases the piles are put under a waterproof sheet or roof to protect them from bad weather. They are then chipped at a later stage.

The resultant product, being less damp than freshly-harvested biomass, is stored for a longer period (some months).

It is always preferable to be able to supply products whose size and humidity are such that they are suitable for immediate use.

Two different methods are used for storing chips, one for chips to be used in the short term and another for chips to be used in the long term.

Short-term use : In this instance users receive replenishment supplies of chips every 10-15 days, and the piles must be not greater than 20-30 m^3. The timber is dried out rapidly by the air. The chips are protected by a roof structure, open on two or four sides.

Long-term use : In this instance replenishment supplies are not provided continuously, and the piles may even exceed 100 m^3. In some cases the timber is packed into enclosed premises. The chips must be dried out before storage, to prevent fungus infections or spontaneous combustion due to fermentation. In this instance the most economical method is to dry out the timber naturally by putting it under a roofed structure or in a yard during the spring-summer period and stirring the chips every 8-10 days.

If the chips consist of freshly-harvested biomass, their humidity is 55-60 % at the time of arrival and if they are heaped in piles not more than 3 m high and covered by a roof, this will fall to 20-22 % within four months during the summer and within six months during the winter.

With the use of artificial drying methods, the drying process could be completed in a few days but would be more costly from the energy standpoint. Experiments of this kind have been carried out in Germany. The timber is chipped and spread in a layer 80-100 cm thick on a trailer with a capacity of 8-10 m^3, which has a mesh floor raised above the floor proper. Warm air produced by burning waste and some of the chips is blown into this space, and within approximately eight hours the original 60 % humidity is reduced to 16-18 %. The chips can then be stored even in closed premises without any danger of deterioration.

Research conducted at the CNR's Istituto per la Ricerca sul Legno by Orlandi and Gambetta in 1985 on two piles of mixed chips of 40 t and 85 t respectively showed that after a week of storage in the open air the maximum temperature recorded inside the piles was 65°C. The temperatures remained high for approximately two months and then fell slowly. The outer 50-80 cm layer, however, reflected variations in the ambient temperature.

The weight loss of the chips after 16 months averaged 20 % inside the pile, with only slight variations at different heights, whereas the weight loss near the surface ranged from 8 % (at the bottom of the pile) to 15 %.

To date no economically viable means of storing chips of freshly-harvested biomass in the open air has been found. To reduce the risk of

V - Drying and storage

a) Loose bolts left in the forest to dry, in preparation for chipping (Photo Baldini).

b) Bolts piled up in the forest to dry, before being sold in that form (Photo Baldini).

c) In the background a pile of chips beside an agricultural vehicle. After drying the chips will be burned in special furnaces (Photo Baldini).

spontaneous combustion, in warm countries, the height of piles is kept under 7 m, and the duration of storage is reduced.

In France the storage period is normally not more than three weeks. In the Scandinavian countries chips to be used for energy purposes are stored for only a very short period, approximately one week.

The influence of the size of the particles on the deterioration of the product has not yet been studied in sufficient depth. Experiments in France have shown that the degree of fermentation and the temperature recorded in piles consisting of chips of over 20 mm in length are lower than those recorded in piles consisting of smaller chips.

Experience to date shows that the condition of the chips after storage depends upon :

- the way in which they have been stored : in the open air with some makeshift covering or under roofs or in silos ;
- the place of storage : dry floor, relative air humidity, rainfall, ventilation, climate ;
- the type/species used ;
- the percentage of bark and leaves ;
- the size of the particles ;
- the height of the pile ;
- the initial humidity of the product.

It has therefore been possible to draw the following conclusions :

- Freshly-harvested chips produced from whole trees with crowns do not store satisfactorily and deteriorate very rapidly. Within 2-3 weeks the chips may start to deteriorate.
- Chips with bark, piled up in the open air, deteriorate three times more rapidly than chips of timber without bark.

The bark and crowns are therefore the cause of this rapid deterioration (Table V).

5. OBSERVATIONS AND RECOMMENDATIONS

The research or demonstration programmes currently being implemented in the Community countries aim to achieve a consumption of 50 Mtoe by the end of the century.

Of the European countries, discounting those with a traditional interest in forestry (Norway, Sweden and Finland), the country most anxious to develop national forestry resources for energy supplies is France, which aims to use 9 Mtoe of wood by 1990 and 15 Mtoe by the end of the century, and is also proposing to encourage cooperative ventures.

This means that all the countries will unquestionably have to step up research, which must not be confined to the individual stages in the process but must encompass the whole process including the final utilization of the product both on an industrial scale and on a small scale for the heating of logging companies and small and medium-sized communities.

The following conclusions have been drawn on the various aspects considered :

Harvesting techniques.

This subject is the most fully documented and from an analysis of the data it is apparent that the most cost-effective methods of harvesting forest biomass for energy from coniferous forests, thinnings and coppices entail the use of semi-automated techniques for skidding whole trees complete with branches and crown.

The timber is chipped along the forest extraction routes or in yards, provided these are within a radius of 1 000 m. If the yards are not within this radius, this method is not to be recommended since it would increase the cost of the skidding operation, as the weight per cubic metre of the branch-wood would be less than that of an equal volume of chips.

Sophisticated automation procedures do not appear cost-effective, given the present structure of forestry holdings, particularly private holdings which average 10.6 hectares in the Community countries. It is therefore less costly to use smaller machines with a more limited operating capacity or, for skidding, specially adapted agricultural tractors fitted with appropriate equipment such as winches, choppers and skylines, rather than highly specialized or excessively high-powered machinery which would merely result in very high operating and energy costs.

Transport

A study of literature on the subject shows that transport costs account for 25-35 % of the total cost of harvesting biomass. As this is a significant percentage, this aspect needs to be studied from the standpoint of both the transport of the material from the forest to the main storage area and the distribution of the dried chips to consumers. The cost of transporting chips direct from the forest to consumers should also be given consideration, and it will then be possible to determine the cost and energy consumption of the various types of vehicle and loading method.

Drying and storage

The figures supplied are acceptable for the few countries which provided them, although more attention needs to be paid to the size of the chips, since this has a significant impact on the deterioration of the product.

In this connection it would be well worthwhile to establish, on the basis of the various parameters (type of timber, harvesting period, climate of area, volume of material, size of chips), the thresholds within which the energy used for drying exceeds the energy obtainable as a result of the consequent reduction in the humidity of the product.

Another aspect which needs to be investigated further is the weight loss of the timber attributable to biodegradation, caused by high temperatures and humidity, which results in a reduction in the energy produced as heat. This is a drawback since it is now generally acknowledged in all the countries that good-quality chips, produced from forest biomass, are a valuable source of energy.

BIBLIOGRAPHY

(1) BALDINI, S., BAGNARESI, U., BERTI, S., MINOTTA, G. (1986). Ricerche sui bilanci energetici di utilizzazioni forestali. Seminario interno delle unità operative del P.F. E/2, Pisa 3-6 novembre 1986.

(2) BALDINI, S., CIVIDINI, R., CURRO' P. (1977). La riunione FAO-ECE-ILO in Finlandia sulla raccolta e utilizzazione della biomassa forestale. Cellulosa e Carta n. 12.

(3) BRIGANTI, G., ELIAS, G. (1986). Prolusione al seminario delle UU.00 del P.F.E. Tecnologie energetiche : stato dell'arte. Pisa 4-6 ottobre 1986.

(4) LAUFER. (1986). Bilan des travaux et objectifs à poursuivre en France pour la recolte, le stockage et le conditionnement de la biomasse forestière. Workshop "Energy-Biomass" Rome, November 20-21, 1986.

(5) MARINELLI, A. (1982). Aspetti dell'economia forestale e del legno nei Paesi della C.E.E. - I.N.E.M.O. Roma, Collana Studi e Statistiche n. 1.

(6) MORVAN, J. (1986) Mechanized forestry chains in France. Workshop "Energy-Biomass", November 20-21, 1986.

(7) PATZAK, W. (1982). Reduction of biomass losses in thinning operations in coniferous stands through whole-tree chipping. Technical and Economic aspects. CEE/FAO/ILO Seminar on reducing forest biomass losses in logging operations. Moscow (USSR) 4-11 December 1982.

(8) STREHLER, A. (1986). Handling and storage of straw and woodchips. Workshop "Energy-Biomass" Rome, 20-21 November 1986.

BIOMASS PRODUCTION FROM COPPICE FOREST HARVESTING AND FIRST THINNING OF FAST-GROWING PLANTATIONS

P. CURRO - S. VERANI
Centro di Sperimentazione Agricola e Forestale - Rome
Società Agricola e Forestale (Gruppo E.N.C.C.)

Summary

Enhanced utilization of coppice forests and management of the forest plantations established since the last war are the major problems in the Italian forestry sector, in terms of both production and fire control.

As an aid to solving these problems, a series of trials was conducted on different forest formations with varying soil morphology, aiming at the development of suitable mechanized harvesting systems capable of substantially reducing operational costs.

The report summarizes the results of trials of coppice harvesting and first thinning in conifer plantations with total or partial production of chipped material for industrial applications (particle or fibreboard, paper pulp) or energy uses. Two oak-coppice stands, two plantations of *Pinus radiata* and one of *Pseudotsuga menziesii* are concerned.

Coppice harvesting was organized as follows : felling (by power saw) and bunching, first-stage extraction (by skyline or winch skidder), skidding (by winch or grapple skidder), chipping. Thinning was by both systematic and selection methods, with the following sequence of operations: felling (by power saw or feller) and bunching, first-stage extraction (by winch skidder, skidding (by grapple or winch skidder), chipping.

In the coppice harvesting trials, productivity was 1.2-2.5 t/h for felling, 3.8 t/h extraction by skyline, 17.3 t/h for skidding, 4.9 t/h for combined first and second stage extraction by skidder, 7.3-8.0 t/h for chipping.

Productivity in systematic thinning was 10.9 t/h for felling by power saw and 8.6 t/h for felling-bunching by feller ; in selection thinning, productivity was 5.0 t/h for felling by power saw and 4.0-4.3 t/h for felling by feller, 2.3 t/h for combined first and second-stage extraction and 17.0 t/h for skidding only ; it ranged between 8.3 and 10.5 t/h for chipping.

The differences in productivity were affected by the equipment used, the organization of the forestry operations, the skidding distance and the average load transported.

1. INTRODUCTION

Enhanced utilization of coppice forests and thinning are the major problems of the Italian forestry sector.

Covering a surface area of over 3.6 million hectares, i.e. about 57 % of all Italian woodlands, coppices are an important part of the Italian forest stock. As a result of rural depopulation and the crisis in fuelwood, which is the traditional outlet for coppice products, the yield from Italian coppices fell from about 10 million m^3 in the post-war years to less than 3 million m^3 in the period 1973-77, subsequently rising to 5.6 million m^3 in recent years.

At the same time, the high cost of manual labour frequently led to incorrect management of the conifer plantations established since the war, one of the results being a much greater risk of forest fires.

As a contribution towards solving the above problems, a series of trials was carried out on different forest formations with varying soil morphology, with a view to developing suitable mechanical harvesting systems capable of substantially reducing operational costs.

This report summarizes the results of trials of coppice harvesting and first thinning in conifer plantations, with total or partial production of chipped material for industrial applications (particle or fibreboard, paper pulp) or energy uses. These trials were in part carried out under the energy and raw materials programmes of the Commission of the European Communities.

2. COPPICE HARVESTING

2.1. Features of the stands harvested

The trials described were carried out on two formations, one located on Monte Peglia, in the Province of Terni, and the other in the Cerva District, in the Province of Catanzaro.

The first was a shelterwood coppice about 25 years old, 90 % of which consisted of Turkey oak (*Quercus cerris* L.) and pubescent oak (*Quercus pubescens* Willd.), while the remaining 10 % was European hop hornbeam (*Ostrya carpinifolia* Scop.) and flowering ash (*Fraxinus ornus* L.). The stand density was 5 523 stems/ha and the weight per hectare 176.8 t. The site gradient varied from 10 to 30 %. The second stand was a coppice 30 to 35 years old consisting mainly of Turkey oak, with varying percentages of Italian oak (*Quercus farnetto* Ten.), Evergreen oak (*Quercus ilex* L.), European hop hornbeam (*O. carpinifolia*) and Italian maple (*Acer opalus* Miller). The stand density was 5 088 stems per hectare and the weight per hectare 239 t. The gradient ranged from 30 to 100 %.

2.2. Working methods

The working methods adopted in the trials, in which whole tree chipping was practised, may be described as follows :

a) felling by power saw and bunching of the trees ;
b) first-stage extraction of the felled product by means of skidder-mounted winches or skyline ;
c) final extraction to the processing area by means of a skidder ;
d) comminution of the trees by chipper.

Whether first-stage extraction was carried out by skidder or skyline depends on the gradient and general situation. Generally speaking, skyline logging was used on slopes in excess of 20 %.

In the harvesting method chosen, the working team consisted of 6 to 7 men : two workmen for felling and bunching of the trees, two men employed on chokering the bundles for extraction, one operator for the first stage of extraction, one operator for skidding (if first-stage extraction is effected by a winch skidder the same man performs both functions) and one chipper attendant. In the processing area a small tractor with a front-mounted stacking blade is required to position the trees in front of the chipper and to clean up the processing residues.

2.3. Working time and productivity

Felling and bunching

Felling was carried out by power saw. The productivities achieved in the two trials and the average time taken for the various phases of the operation are shown below.

	Monte Peglia (TR)	Cerva (CZ)
Mean shoot weight, kg	32	47
Productivity, t/h	1.18	2.5
Average times for the various working phases :		
- felling	0'39	0'22
- bunching	0'23	0'14
- moving	0'10	0'30
- clearing of stools and undergrowth	0'17	0'06
- refilling power saw and sharpening blades	0'29	0'17
- downtime	0'44	0'23
	1'62	1'12

The higher productivity in the trial at Cerva is explained by the greater unit weight of the coppice shoots. The difference between the two trials for the time spent on "moving" is accounted for by the fact that the site at Cerva sloped more steeply than on Monte Peglia, while the difference in "clearing of stools and undergrowth" is caused by the fact that the undergrowth on Monte Peglia was denser.

First-stage extraction

First-stage extraction was effected with a Timberjack 225 skidder fitted with winches and with Koller K 300 or Hinteregger skyline systems. Since the two skylines were equally efficient, only the figures for the Koller K 300 will be quoted in the interests of brevity.

The productivity and the average time taken for the various sub-phases of the first extraction stage are indicated in the following table.

No figures are provided for Monte Peglia since first and second stage extraction were here carried out as a single operation.

	Cerva (CZ)
Mean load, kg	247
Mean distance, m	44
Productivity, t/h	3.76
Mean times for the various operational phases :	
- travelling without load	0'27
- chokering	1'72
- hoisting	0'45
- travelling under load	0'53
- unloading	0'69
- downtime	0'36

	4'02

The longest time was that for "chokering" because of the steep, rocky nature of the ground, which impeded movement of the workers, and because of the difficulty of synchronizing this phase with the "hoisting" operation.

Skidding

The following figures show the productivity and mean time for the various operational phases in skidding, which in both areas was carried out with a Timberjack 225 skidder equipped with winches (Monte Peglia) or a grapple (Cerva).

	Monte Peglia (TR)	Cerva (CZ)
Mean load, kg	2000	1640
Mean distance, m	373	240
Productivity, t/h	4.9	17.3
Mean time for the various operational phases :		
- travelling without load	2'91	1'60
- manoeuvring	1'32	0'50
- chokering/picking-up	1'80	0'91
- first-stage extraction	5'00	--

- travelling under load	4'04	1'82
- unloading and manoeuvring	4'17	0'74
- downtime	4'99	0'63
	-----	----
	24'23	6'20

The substantial differences in travelling time and productivity are explained by the different gradient of the extraction racks (15 % at Peglia, negligible at Cerva), the skidding distance and the fact that the work was differently organized ; in Cerva the skidding operation consisted in picking up the materials concentrated by skyline at the side of the extraction route with the skidder's grapple, while on Monte Peglia the same skidder was used for the first and second extraction phases with the result that more time was required for the successive phases of "chokering", "first-stage extraction" (which was non-existent at Cerva) and "unloading and manoeuvring".

Chipping

This operation was carried out with a "Morbark 12", the average productivity being 8 t/h on Monte Peglia and 7.3 t/h at Cerva.

2.4. Discussion

The trials demonstrated that the method of work adopted gave good results. In particular, the two skyline systems (the "Koller" and "Hinteregger") and the Timberjack 225 proved very effective means of timber extraction.

An analysis of productivity in the various operations shows that the first-stage extraction figures are disappointing. In the skidding operation on Monte Peglia (which, as already mentioned, included the first extraction phase) the average loads transported seem rather low in relation to the capacity of the equipment. Where the lie of the land permits, productivity could be improved by forming larger bunches at the previous stage of "felling and bunching".

Observations of damage to the stools (destruction of bark) as a result of mechanization indicated that such damage was negligible when first-stage extraction was effected by skyline (1.7 % of stools damaged) and was greater (14 %) if skidder was used.

3. FIRST THINNING OF CONIFER PLANTATIONS

3.1. Features of the stands thinned

The trials described were carried out on two plantations of *Pinus radiata* D. Don and one of *Pseudotsuga menziesii* (Mirb.) Franco, located in Central Italy. One of the first two plantations is at Tuscania (VT) and the other near Grosseto ; the third is at Rincine in the Tuscan Appennines, in the Province of Florence.

The plantation of *P. radiata* at Tuscania was nine years old at the time of the trial. The trees were planted in staggered rows (hexagonal pattern) with a spacing of 2.5 m, 1 516 trees and 175.7 t per hectare, the mean stem diameter at a height of 1.30 m was 16 cm and the average tree height 11.5 m. The ground was level. The *P. radiata* plantation near Grosseto was 13 years old at the time of cutting and the stand characteristics were as follows : planting pattern hexagonal at 2.5 m spacing, 1 500 trees and 233.7 t per hectare, mean stem diameter at a height of 1.30 m 18 cm, average tree height 15.3 m. The terrain was flat. The plantation of P. menziesii was 14 years old at the time of the trial with a hexagonal planting pattern at 3 m spacing, 1 172 trees and 110.0 t per hectare, average stem diameter at a height of 1.30 m 15.7 cm, average tree height 10.2 m. The slope varied from 20 to 50 %.

3.2. Working methods

In all the plantations, combined (systematic and selection) thinning was carried out so that the entire operation could be mechanized. Between 40 and 50 % of the trees were cut. In the two P. radiata plantations one row in four was removed in the systematic thinning operation, and selection thinning was carried out within the three remaining rows.

In the plantation of P. menziesii one row in six was removed in the systematic thinning phase and selection thinning was carried out in the five rows retained. In the latter case the overall proportion of thinnings taken was lower (30 %) since the stand was on a mountain site and there was therefore a risk of snow damage.

The sequence of operations, which in the two P. radiata plantations involved the production not only of chippings but also of round timber for papermaking or pallets, was as follows :

a) felling with a feller mounted on a tractor, a mini-tractor equipped with a feller-buncher or a power saw ;
b) bunching of trees (on average 5 or 6 at a time) along the corridor formed by the systematic thinning ;
c) first-stage extraction with winches mounted on a skidder ;
d) skidding with a skidder equipped with winches or a grapple ;
e) chipping of the trees at the processing area.

For the production of round timber, other operations were carried out such as trimming, cross-cutting and barking.

3.3. Working time and productivity

Felling and bunching

In the plantations of P. radiata, systematic thinning was carried out with a feller mounted on a Fiat Allis 345 B and selection thinning by means of a Makeri 33 T. In the plantation of P. menziesii the trees were felled by power saw because of the steepness of the slope. Only small trees were bunched.

The figures for productivity and average working times are as follows.

Systematic thinning

	Tuscania (VT)	Grosseto	Rincine (FI)
Tree weight, kg	163	196	135
Productivity, t/h	7.7	8.6	10.9
Mean times for the various operational phases :			
- approach to the tree	0'25	0'26	0'32
- cutting	0'06	0'08	0'17

- manoeuvring	0'30	0'33	--
- bunching	0'32	0'40	--
- movement between rows	--	--	0'12
- refilling of power saw and blade sharpening	--	--	0'03
- downtime	0'20	0'30	0'10
	1'13	1'37	0'74

Selective thinning

	Tuscania (VT)	Grosseto	Tincine (RI)
Tree weight, kg	81	120	92
Productivity, t/h	4.0	4.3	5
Mean times for the various operational phases :			
- approach to the tree	0'40	0'53	0'47
- cutting	0'07	0'10	0'14
- manoeuvring	0'20	0'30	--
- bunching	0'41	0'50	0'33
- movement between rows	--	--	0'02
- refilling of power saw and blade sharpening	--	--	0'04
- downtime	0'12	0'25	0'10
	1'20	1'68	1'10

Productivity in felling and bunching by mechanical means was similar in the two types of thinning but lower than for felling with the power saw. This is explained by the fact that nearly half the total time required for the mechanized operation is taken up with "manoeuvring" and "bunching", whereas these phases do not occur in felling with a power saw or account for a smaller proportion of the total time.

Skidding

The following are the productivities and mean travelling times for the skidding operation, which was carried out at Tuscania and Grosseto with a Timberjack 380 and at Rincine with a Holder A 60.

	Tuscania (VT)	Grosseto	Rincine (FI)
Mean load, kg	1404	1734	698
Mean distance, m	342	155	198
Productivity, t/h	10.4	17.4	2.3
Mean times for the various operational phases :			
- travelling without load	2'28	1'49	2'29
- manoeuvring	0'75	0'35	0'97
- chokering/picking-up	0'79	0'79	3'94
- first-stage extraction	--	--	1'86
- travelling under load	2'54	1'95	3'35
- unloading and manoeuvring	0'89	0'60	3'29
- downtime	0'87	0'80	2'20
	8'12	5'98	17'90

The marked differences in the productivity of the Timberjack 380 and the Holder in the various areas arise from three main factors : differences in the equipment used (the Timberjack is much more powerful than the Holder), the quantity of material transported and differences in the way in which the work is carried out (with the Timberjack, skidding was carried out with a grapple, while winches were used on the Holder).

As a result, skidding with the Timberjack did not involve any preliminary extraction phase and the times for "chokering/picking-up" and "unloading and manoeuvring" were much lower than with the Holder.

Chipping

The productivities for the chipping operation, which was carried out in all three trials with a Morbark 12, were as follows: Tuscania 10.58 t/h, Grosseto 8.38 t/h, Rincine 8.70 t/h.

3.4. Discussion

The trials show that the thinning operation can be completely mechanized if a combined method (systematic and selection) is adopted.

Particular importance attaches to systematic thinning (complete removal of a row of trees). This makes it possible to establish "corridors" within the plantation, which can subsequently be used as extraction racks. Constraints on complete mechanization may be the tree spacing (if below 2.5 m), especially in respect of selection thinning, and steep gradients (if in excess of 25 %), which precludes or at least severely restricts the use of skidders.

An analysis of the data obtained shows that satisfactory rates of productivity were attained for the various operations. With regard to the data on the felling operation, it is to be noted that the highest productivity in systematic thinning was attained when cutting was carried out with a power saw (P. menziesii). At the same time, it must be emphasized that when this practice was adopted, no bunching was carried out. Bunching is, however, of great importance to the subsequent skidding phase, the total time for which is enormously increased, as shown by the experimental data, if the products have not been bunched and there is a corresponding drop in productivity.

PRINCIPAL FEATURES OF THE MACHINES USED

Koller K 300 skyline

Consists of a winch unit with two drums (one for the skyline rope and the other for the haulback and feed rope), a steel-section tower, a carriage and the power unit. The characteristics of the components are as follows:

- Power unit : Perkins diesel engine rated at 53 kW (72 CV) with hydrostatic transmission.
- Tower : steel-section, height from ground 7 m, retraction during transport by integral Tirfor.
- Skyline rope : length 400 m, diameter 16 mm, 216 WS construction, breaking strength 160.8 KN (16 400 kg), maximum requirement for tensioning 4.5 t.
- Haulback and feed rope : length 400 m, diameter 9.5 mm, 114 Seale construction, breaking strength 52.5 KN (5 360 kg).
- Carriage : Koller SKA 1 with two pulleys riding on the skyline ; automatic operation with hydromechanical control ; weight 150 kg, maximum capacity 1 000 kg.

The total weight of the unit is about 1 600 kg and movement is by towing.

Mini-Urus skyline

Unit with independent engine mounted on a towable chassis ; consists of a tower whose base houses the drums for the skyline and haulback and feed ropes ; retraction by hydraulic winch. The characteristics of the components are as follows :

- Power unit : Volkswagen petrol engine, 1 200 cc, rated at 22 kW (30 CV), 4-speed gear box with mechanical reversing gear.
- Tower : height from ground 4.7 m.
- Skyline rope : length 330 m, diameter 16 mm, 216 WS construction, breaking strength 160.8 KN (16 400 kg).
- Haulback and feed rope : length 350 m, diameter 8 mm, Seale construction, breaking strength 37.4 KN (3 850 kg).
- Carriage : similar to that mounted on the Koller K 300.

Total weight 2 200 kg.

Holder A 60 skidder

Articulated skidder with Holder 37 kW (50 CV) diesel engine, 12-speed gearbox and reversing gear. Length 3.45 m, width 1.50 m, height 2.1 m, minimum ground clearance 40 cm, weight 3 600 kg.

A small stacking blade is mounted at the front end and two winches at the rear with a capacity of 70 m of 14 mm cable.

Timberjack 380 skidder

Articulated skidder with GM 100 kW (136 CV) diesel engine ; length 6.74 m, width 2.64 m, height 3.53 m, minimum ground clearance 56 cm, 8-speed gear box with reversing gear. A stacking blade is mounted at the front end and at the rear a grapple with a capacity of 0.8 m^2 and an Eaton winch carrying 138 m of 14 mm cable. All-up weight 9 307 kg.

Timberjack 225 E skidder

Articulated skidder with Perkins 55 kW (75 CV) diesel engine ; 8-speed gear box with reversing gear ; low-pressure tyres ; stacking blade fitted at the front end and at the rear an Esco grapple with capacity approximately 0.8 m^2 and winch carrying 70 m of 14 mm cable. Weight 6 600 kg.

Makeri 33 T mini-tractor

Mini-tractor of Finnish construction designed for felling operations, especially selection cutting. Fitted with a Deutz 26 kW (35 CV) diesel engine, height 2.75 m, length 3.66 m, width 1.62 m, minimum ground clearance 41 cm. Equipped with tracks spanning 3 tyred wheels ; ground bearing pressure 0.40 kg/cm^2.

Feller-buncher unit mounted at the front end for felling and trimming of trees up to 25 cm in diameter at the butt.

Fiat Allis 345 B

Wheeled loader fitted with a Morbark feller to cut trees with a diameter of up to 30 cm at the butt. Characteristics : Perkins 59 kW (80 CV) diesel engine, height 3.02 m, length 4.60 m, width 2.23 m, minimum ground clearance 38 cm, weight 7 200 kg.

Morbark 12 chipper

Mounted on a chassis with two double wheels ; comprises operator's cab, boom loader, chipper opening, dirt discharge spout (for bark, leaves, etc.) and chip discharge spout. Fitted with a John Deere 140 kW (190 CV) diesel engine. Chipper unit comprises a large chipper disc fitted with two knives.

BIBLIOGRAPHY

Currò P., Eccher A., Ferrara A., & Verani S., 1981 - Prove di primo diradamento su Pinus radiata D Don presso Tuscania (Early thinning trials on Pinus radiata D. Don at Tuscania. Cellulosa e Carta 32 (9) : 3-53.

Currò P. & Verani S., 1984 - Prove di diradamento e sramatura totalmente e semimeccanizzata (Trials of fully and partly mechanized thinning and trimming). Quaderni di ricerca S.A.F. n. 1, Roma.

Currò P. & Verani S., 1984 - Tempi di lavoro e rendimento di esbosco con 'Timberjack 225' in un ceduo di cerro (Working time and productivity in skidding with a Timberjack 225 in a Turkey oak coppice). Quaderni di Ricerca S.A.F. n. 4, Roma.

Currò P., & Verani S., 1986 - Prove di concentramento del legname in ceduo di cerro con due tipi di gru a cavo (Trials of first-stage timber extraction in Turkey oak coppices with two types of skyline). Quaderni di Ricerca S.A.F. n. 10, Rome.

Currò P. & Verani S., 1986 - Development of mechanized thinning systems for fast-growing conifer plantations. Final report Contract ECE No Bos-133-I(S)

Scaramuzzi G., 1984 - Utilization of coppice forest biomass for fuel and other industrial uses. Final Report Contract ECE No ESE-R-035-I(S).

Harvesting of coppice forests :

1. View of part of a forest after felling and bunching of shoots

2. Koller K 300 skyline for first-stage extraction of felled product

3. Removal to the extraction rack by skyline

4. Extraction of the product by means of a Timberjack 380 skidder, fitted at the rear with a winch and grapple

First thinning of conifer plantations :

1. Detail of a <u>Pinus radiata</u> plantation after systematic thinning

2. Trimming with power saw after felling by means of a feller mounted on a Fiat Allis 345 B tyred vehicle

3. Makeri 33 T mini-tractor in the trimming operation

4. Timberjack 380 skidder in the processing area following extraction

MECHANIZED FORESTRY CHAINS IN FRANCE

J. MORVAN
Association pour la Rationalisation et la Mécanisation
de l'Exploitation Forestière

INTRODUCTION

France possesses about 15 million hectares of forest and some other wooded areas, which represent approximatively 1.7 million cubic meters.

This includes 6 million hectares of coppice and coppice under high forest, which is underexploited.

With this kind of renewable resource, it is normal to think of using wood energy to face the energy crisis, but forestry works only use the old logging method in one meter long processed billets, not adapted to automated heating boilers.

Chipping started but was used in small units for limited volume local requirements. Chipper feeding was by hand. This work is hard and expensive if it's done by payed manpower. It was necessary to mechanize wood fuel logging and different systems were perfected.

In this paper we describe existing systems and new machines and we specify if they are prototypes. There are of course some other ideas, which are still projects.

One system was operated at Landes forest sylviculture in South West France, where there are more than one million hectares of maritime pines.

LOGGING SYSTEMS

1. CHIP HARVESTING

1.1. Full integrated system : only one machine

"SCORPION" fells, chips and hauls chips to the road ; trees and brushes are cut by two discs fitted with knives, then driven by vertical and horizontal drums and after this, chipped by a drum-chipper.

Chips are thrown up in a high-tilting bin which can be dumped into trucks and trailers or in the bin of a special forwarder.

The machine has a hundred and sixty kW engine and an hydrostatic drive on six wheels and on all felling, feeding and chipping functions.

It can cut brushes, small softwood poles or coppice of only fifteen centimeters in all terrains.

It works by lane and its operating efficiency changes with the material from fifteen to fifty five m^3 of chips.

It is built in France, by the CIMAF, designer M. Van Landeghem.

1.2. Two machines system used in maritime pine

In this case felling is an independent operation of harvesting and transport.

It is used in early selective thinning, when poles are too small (about forty cubic decimeters each) to be sold to pulp mills.

Thinning is very important for sylviculture of maritime pine, but it is expensive for the owner who can't sell this wood. The object of this system is to pay for the thinning by selling the chips.

The felling machine was designed by ARMEF and sold by SIFER. It is a shear on a knuckle boom supported by a sixty kW farm tractor with four driving wheels.

The tractor is driven in reverse ; so the driver can see the work he has to do and select the trees.

With an experienced driver, it is not necessary to blaze the trees to fell.

Trees are planted or seeded on lines and there is always a three or four meters free track for the machine to pass between seeded lines.

This machine is swift because the tractor has supplementary hydraulic equipment and it fells more than three hundred poles per hour. The average yield per year of work was measured at two hundred and sixty poles per hour. The driver puts the poles in the free track to make the work easier for the second machine.

Chipping and transport can be done by two different machines :

- one combined machine SIFER/ ARMEF,
- one BRIMONT machine.

The combined SIFER/ARMEF machine is a farm tractor with a chipper on the power take-off a 3 point linkage and a dump trailer. The tractor is driven in reverse with the chipper in front. The dump trailer is coupled to the front hook of the tractor. On the roof of the tractor, there is a loader to feed the chipper. The driver feeds the chipper with poles when it is advancing along the tree line.

The tilting bin dumps the chips sideways in a container along the track.

The machine can be separated by :

- removing the dump trailer and the chipper, the tractor and the crane could be used for hauling with a trailer,
- removing the loader, the tractor could be used in sylviculture, for planting.

The combined machine is eleven meters long, 2.4 meters wide and weighs seven tons. The chipper capacity is twenty eight centimeters and the content of the trailer is ten cubic meters. The average yield in maritime pine is twelve to fifteen m^3 of chips per hour with a hauling distance of six hundred meters.

This yield is not high but ARMEF, who designed this combined machine, looked for a cheap means, which could be used for several months each year.

Last year technical costs, without overhead costs, wood price and profit, were about hundred to hundred and twenty F per ton or thirty five francs per m^3 loose volume.

The combined machine, built for Landes country, was used in Normandie, Massif Central and Pyrénées but the farm tractor can work only in flat grounds and when it's working in coppices, it can not overcome high stumps.

The BRIMONT machine

It is a hundred and twenty kW forwarder with a front chipper and a small loader and on the far side there is a high tilting bin of twelve cubic meters. It works like the combined SIFER machine but has a better cross country ability. Its yield is about five tons per hour but it is difficult to know its prime cost because we do not know its selling price. It is made in France by BRIMONT.

1.3. Heavy systems : five machines

This method was in common use in France between 1974 and 1980. The combined system of five machines and the repair truck were called "circus" and more than ten circuses worked in coppices, for pulp-mills.

Only one or two circuses are still working ; because at present the pulp-mills do not want whole tree chips with barks and also because big felling areas are uncommon in France.

Felling and piling is done by a Morbark head fitted on the Bobcat front loader wheeled tractor or on hydraulic shovels on full-tracks. This mounting provided the best results because of its reliability and its cross country ability.

Chipping is done on the road side or on a skidding ride used by sixty cubic meters self unloading semi-trailers. Chippers were almost all Morbark eighteen or twenty two inch models on semi-trailers, with a Cummins or Fiat two hundred and fifty or three hundred kW engine. The twelve inch Morbark chipper was not successful.

To feed the chipper bunches, hauling is done by grapple skidders, principally Timberjack type and by clambunk skidders. These machines, able to carry twice as much wood as grapple skidders, were able to collect wood from greater distances in the cutting area.

The five machines, two felling machines, one grapple skidder, one clambunk skidder and a mobile chipper, produced about three hundred to four hundred cubic meters of chips per day.

This method is advantageous for cutting areas bigger than ten hectares but in France the average cutting area is about three hectares, it is too heavy a method for the investment and for the costs.

2. SIMULTANEOUS HARVESTING OF PULPWOOD AND CHIPS

Pulp-mills were afraid of concurrence of fuel wood so ARMEF decided to satisfy the two customers. We tried to collect logging residues after a traditional coppice cutting but this method was not economically advantageous.

Then we tried to harvest pulpwood and fuelwood at the same time.

2.1. Simultaneous harvesting in coppice with sorting of the products when they are cut

Felling is done with a special hardwood shear felling head designed by ARMEF.

The driver feels and picks out poles : the small ones are piled in bunches, the biggest ones are piled beside.

Processing is done by a man who crosscuts poles into four or six meter long logs and cuts the tops off big branches.

Hauling is done by a forwarder which collects pulpwood. A chipper hauler comes after to chip tops and bunches of small poles. Sparses branches are not collected.

The system works with a felling machine, two men processing, a forwarder and the Brimont chipper hauler.

The felling machine yield is about forty to sixty tons per day, the logger yield is about thirty tons each per day and the chipper hauler about twenty tons per day.

This method is advantageous for a pulp-mill which gets an opportunity to utilize fuel chips, logs of four or six meters long, and works in cutting areas less than one hundred kilometers from the mill.

2.2. Simultaneous harvesting without sorting of felled products

We tried another method with a chain saw felling head (but a chain saw without bar) able to cut all the clump at one time.

Then a forwarder comes with a grapple saw which bucks the bunch without delimbing.

The crows remain complete. After hauling all the crowns stay piled for the chipper hauler.

With this method we harvest less pulpwood than with the other traditional harvesting method and it is necessary to have a good-sized coppice, but chip harvesting is easier. In any case, an energy wood opportunity is necessary and tests have stopped now because there are no opportunities. The feller is only a prototype.

2.3. Simultaneous harvesting in coniferous tree thinnings

We tried to collect only crowns after the work of the harvester designed by ARMEF.

In the case of a fifteen years maritime pine thinning operation, we harvest twenty tons of crowns per hectare and we have the same conditions of a non commercial thinning operation which we described above.

CONCLUSION

In France there are methods adjusted to the different cases of the French forest, but the actual energy costs (fuel, gas, electricity) are not propitious to their expansion.

However, we think that it is possible to use the energy if the conditions are advantageous :

- big and cheap quantity of standing wood,
- shorting distance of transport,
- cheap logging costs,
- harvesting combination of fuel chips and a campaign against fire forest, or pulpwood.

This requires a development of new methods and the adaptation of forest workers to these new methods.

1. CHIP HARVESTING

1.1. Full integrated system : SCORPION

Felling

Chipping
hauling

Transport

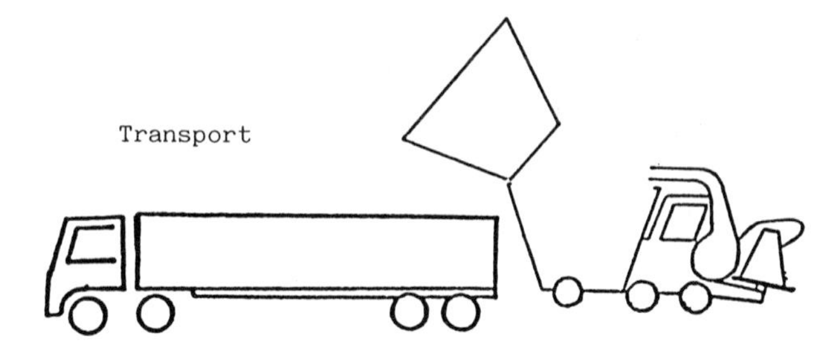

1.2. Two machines system

Felling

Chipping hauling

Transport

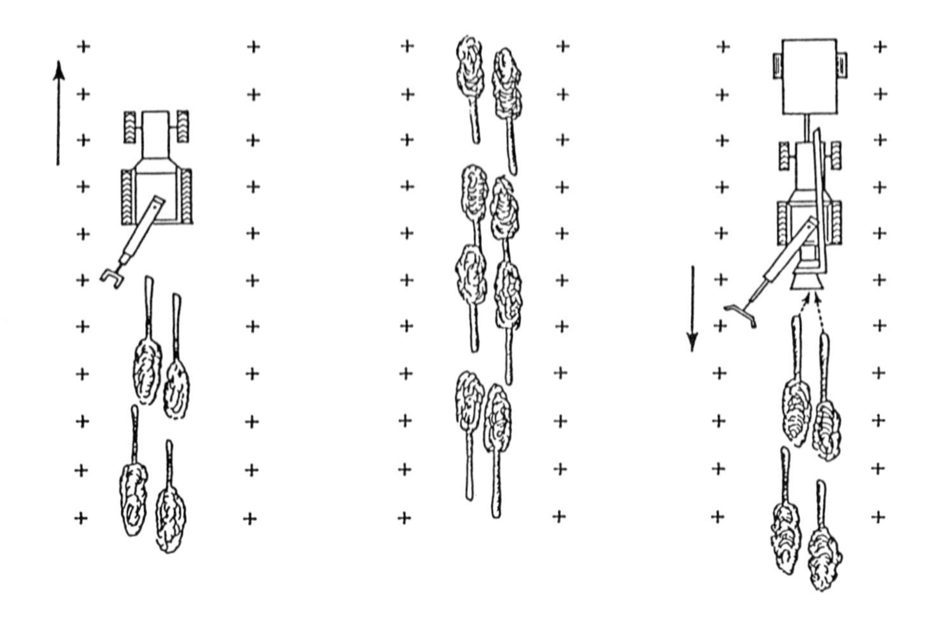

1.3. Heavy system : 5 machines

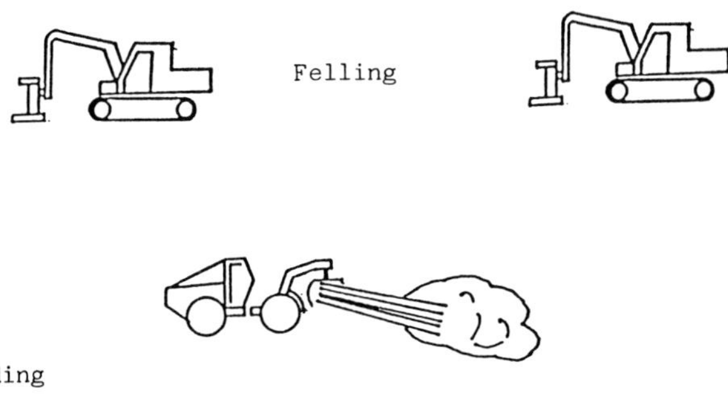

Skidding

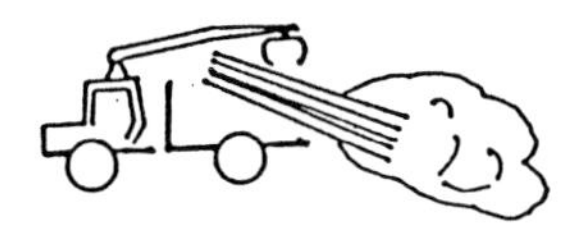

Chipping

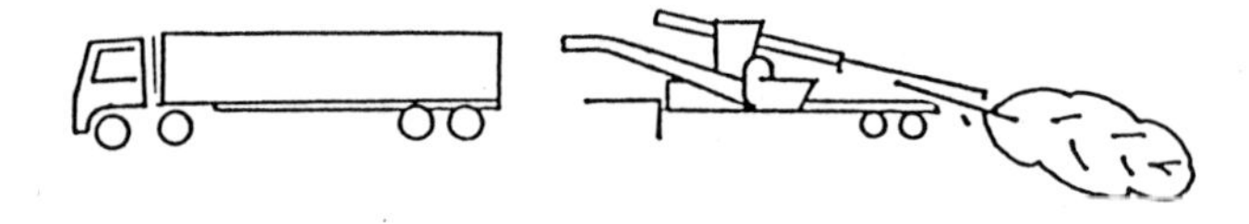

2. SIMULTANEOUS HARVESTING

2.1. With felling and sorting out

Felling and

sorting out

the products

Processing pulpwood

Pulpwood

energy wood

2.2. Without felling sorting out

2.3. Simultaneous harvesting in softwood thinnings

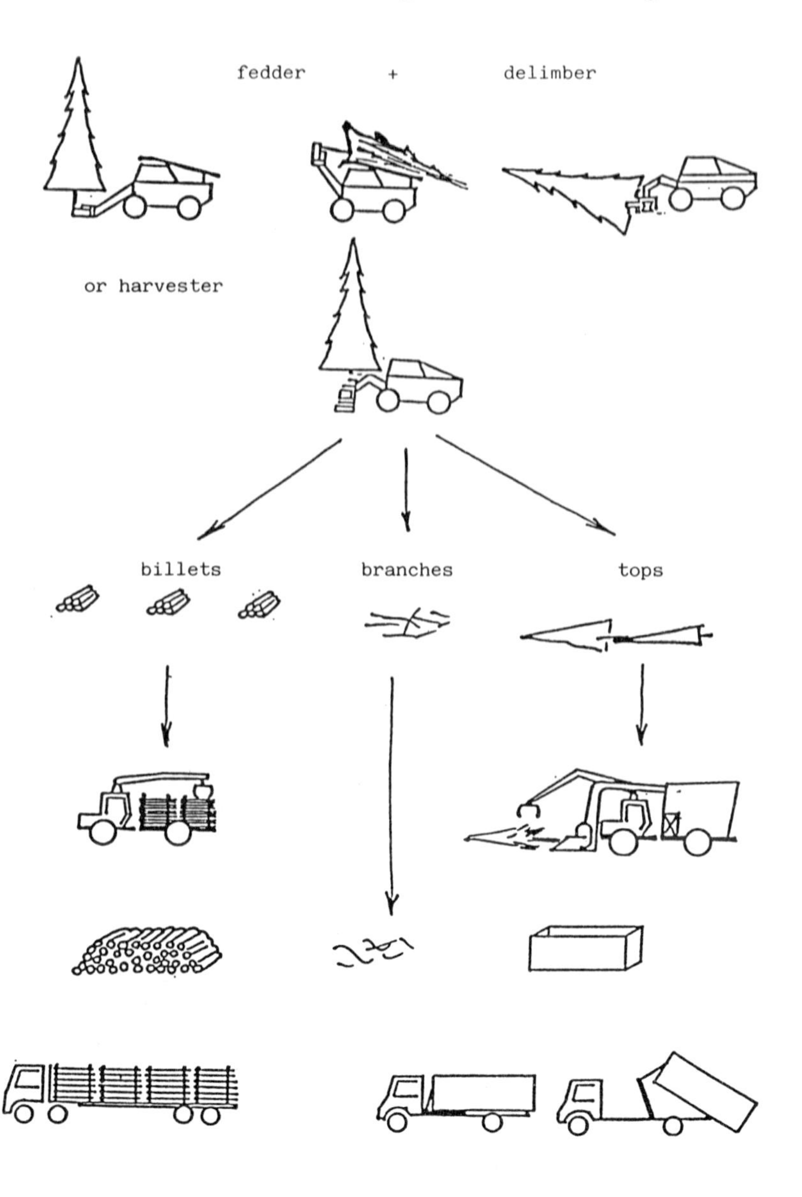

EVALUATION OF THE WORK AND OBJECTIVES TO BE FOLLOWED IN FRANCE FOR THE HARVESTING, STORAGE AND PROCESSING OF FOREST BIOMASS

Pascal LAUFER
A.F.M.E., Paris

1. INTRODUCTION

France, which has been working on the production of ligneous biomass from fast-growing coppice since the beginning of the 1970's, has, since 1980, applied itself to the development of techniques and materials for the collecting of forest biomass. Today, whether used by individuals or industries, efficient and innovating materials make it possible to collect small sized forest biomass.

However, the collecting of forest products is only carried out in existing forest, since even if fast-growing coppice now exists, it is not yet ready for collecting. Today, about 285 hectares of short-rotation coppice have been planted on the initiative of the French Energy Agency (AFME), by the INRA (45 hectares in basic research) and the AFOCEL (240 hectares in applied research). This surface area will be extended to 450 hectares by 1988, mainly due to a campaign of multi-local tests already begun (240 hectares) covering 400 hectares spread over the whole of French territory. This was consigned to the AFOCEL by the AFME. We are testing 12 species.

Thus, the techniques and objectives to be followed in France for the harvesting, storage and processing of forest biomass applied to plantations of ligneous biomass will stem from techniques used in existing forest.

With what we know today, we must remember that fast-growing coppice can be collected partially in the form of logs for which the exploitation techniques are well-known, and completely or partially in the form of chips.

2. COLLECTING CONDITIONS

Fast-growing plantations set up on soils of good potentiality in relation to the species used, will be on ground where the slope may be up to 30 %. Machinery such as agricultural tractors will be able to work on the flattest plots ; elsewhere it will be necessary to call on 6 or 8 wheel drive machines which are capable of manoeuvring at an inclination and on slopes.

A priori, fast-growing coppice will be collected in France every 7-12 years at the most, according to the species. The machines will pass relatively frequently, be it those used for plantation and maintenance or those used for harvesting.

In order to preserve a good soil texture, the tools used will have to be quite light especially since ground in France is fragile and unstable under weight for a large part of the year.

Since the rooting often remains superficial, it will be important to make sure that the cutting techniques used (chain saw, shears, knife disks or knife rollers) do not uproot the trees.

Moreover the cutting height will probably have to be a little higher at each rotation, which presupposes the availability of machinery equipped for such a purpose.

3. EXISTING COLLECTING METHODS AND MACHINERY

France has a solid experience in the collection of forest biomass. Nowadays, there are a number of firms which produce chips from varied plantations for use in power or industry (chipboards, metallurgy). On the other hand, numerous collecting tests with innovating techniques or machines are being or have been realized and some of these have led to industrial construction.

As far as the short circuit production channel is concerned (tractor + trailor + integral chopper) the cost of the final product (value of standing timber + material + labour + transport + storage) varies from 22 to 44 ECU per delivered ton (with 25 % moisture content on raw timber) :

- 22 ECU in autoproduction by a farmer during his idle time
- 36 to 44 ECU for a small firm collecting residue felling or standing coppice.

I point out the case of the communal recovery of wood fuel. For example, the commune of MARCHESIEUX in NORMANDY. The farmers cut each 15 years the hedges and keep the logs. The commune buy and harvest the branches.

For a firm of industrial size, the cost price of a delivered ton (with 45 % moisture content on raw timber) is about 33 ECU.

An experiment led by the ARMEF (Association pour la Rationalisation de la Mécanisation et de l'Exploitation Forestière) in thinning out maritime pines on flat ground in the Landes, with the use of 2 successive machines (1 agricultural felling tractor + 1 tractor equipped with a cant hook, a small chopper and a trailor) makes it possible to envisage the production of undried chips of which the cost price delivered for heating or industry would be 20 ECU per ton.

France is also working on the development of integrated, manoeuvrable, light machines, capable of working on sloping ground for the harvesting of standing biomass or residue of forest plantations, cut into chips and transported by the machine itself. The AFME helped some research centres and companies as ARMEF - CEMAGREF - BRIMONT and CIMAF.

This is the case of the SCORPION (240 HP. - 10 tons - 2.10 m x 6 m - hydraulic transmission) constructed by the firm CIMAF and fitted with a collecting head and a 6 m basket by means of which the delivered ton at 45 % raw moisture content has a cost price between 36 and 44 ECU. This commercialized machine carries out forest work (clear cutting of coppice plantation maintenance, systematic thinning) and clears undergrowth (fire fighting in the south of France).

On the other hand, the CEMAGREF (a machinery research centre) has constructed a prototype wheel drive carrier (6 tons - 1.6 m x 15 m) with hydrostatic transmission and seat correction, on which a collecting head for soil biomass (residue) or standing biomass and a 6 m storage basket can be adapted.

We tested another machine with the BRIMONT Company.

The importance of the work which has been carried out in research and development of material, as well as the cost prices of the mobilization of the chopped wood by firms which commercialize this product today, indicate that the cost of the final product of forest biomass is still too high if sufficient market openings for small forest wood are to be secured as for the products from the plantation of fast-growing coppice. This, compared to the low price of the oil today.

It must be ensured that research in forest plantation makes it possible to lower substantially the cost of biomass production by mechanization.

This should be carried out by means of an optimization of the existing machines and exploitation systems on the one hand, and also by the development of more efficient machines.

It can also be thought that the double exploitation of biomass in logs (for paper manufacturing or energy) and in chips (wood for industry or energy) directly from the standing tree or from bundles collected beforehand, should bring about a lowering of costs for the longest rotations (7-15 years). Mixed collecting techniques should therefore be developed, either with existing material or new systems.

The material used in the fast-growing ligneous plantations will have to permit, (in satisfactory economic conditions), the harvesting of young wood at different heights at each rotation without uprooting the stumps. There too, tests and development of machines will be necessary.

Finally, since French soils do not permit a very great ground pressure in winter (soaked soil) and wood harvesting must take place in winter, machines fitted with wide caterpillar bands and high resistance, low pressure tyres will have to be used. Tests and development of materials will also be necessary in this field.

4. DRYING - STORAGE - TRANSPORT

Outside the Mediterranean zone, the chips put into piles do not dry. On the contrary, they accumulate surface moisture and loss of material is high.

Storing the chips in piles outdoors cannot alone permit the establishing of a buffer stock, constantly renewed and necessary for a commercializing firm.

The harvested biomass will be mainly stored in the form of poles or bundles and processed afterwards.

However, for small production units of forest biomass (a few thousand tons a year) a simple wooden hangar with a concrete floor (about 50 ECU/m) constitutes quite a modest storage cost of about 2 ECU a ton. Drying the 3 m high piles takes 3 months in summer and 6 months in winter and lowers the moisture content of the raw chips from 45 % to 23 %.

It is certain that as far as transport is concerned, the costs should be lowered. Studies and tests should be carried out in this field, all the more so considering that it is one of the most important items making up the cost of the final product (4 ECU-17 ECU a ton of chips).

5. CONCLUSIONS

It seems that if a great deal of progress has been made in the harvesting of forest biomass over the last few years, a certain number of points remain to be improved in order to make a product, which still risks being too expensive to harvest, more competitive : the harvesting machines themselves, and the techniques of harvesting, storage and transport.

It is certain that for some future uses of forest biomass (sucro-chemistry, paper pulp) barking the wood will be necessary. This technique has not yet been fully developed for small wood and chips and so it will be necessary to add to existing research or to initiate new studies in this field.

However, we must not lose sight of the fact that the collecting techniques will be all the more efficient when the plantations produce more biomass. The effort of producers is therefore closely linked to that of the harvester.

HARVESTING AND STORAGE OF CHIPPED WOOD FOR ENERGY PRODUCTION IN HILL AREAS

L. LISA
Institute for Agricultural Mechanization of the Italian National Research Council (CNR) - Turin

Summary

With the use of winches and chippers at well organized timber sites, the harvesting of timber in hilly areas can once again become viable. Storage of the wood can, however, be difficult because of its very high moisture content on harvesting.

Felling areas have been studied with a view to optimizing the stages of extraction and chipping of timber. The storage of wood chips in wire-mesh cages of varying heights, according to the size of the chips, has been studied to determine the possibilities of natural drying and storage of the different species of wood with minimal wastage.

The less expensive storage of wood in small heaps protected from the rain also proved successful.

Natural drying and storage of chipped wood was improved by felling at the end of autumn and not skidding or chipping the wood until the end of winter to reduce the moisture content of the wood chips.

1. INTRODUCTION

In Italy, coppices are particularly common in hill and piedmont areas where orographic conditions prevent the land being used for any other purpose.

With growing production surpluses of the principal crops in the European Community and the steady increase in the discrepancy between production costs in hilly areas and plains, it makes increasing sense to use the steeper areas and small irregular plots, where the difficulty of mechanization makes traditional crops less and less competitive, for timber.

It is therefore extremely important to find methods of harvesting and storing timber for energy production so that a resource which has been more or less abandoned in recent years can once again become viable.

In the last few years, therefore, the Institute for Agricultural Mechanization has been carrying out extensive research on the use of winches and chippers and on methods of storing chipped wood for energy production in hill coppices.

2. COPPICE PRODUCTIVITY

In the hill area under consideration in Albugnano (Asti), the coppices are exploited at an average age of 18 to 20 for vineyard poles, fencing and firewood. The main species are oak (Quercus pubescens), ash (Fraxinus ornus), common acacia (Robinia pseudoacacia), chestnut (Castanea sativa) followed by hazel (Corylus avellana), field maple (Acer campestris), elder (Sambucus nigra), etc.

Note : The latin name of the trees, in brackets and underlined, is indicated only the first time the species are cited in the paper.

In the sample areas, the average yield on felling 19- to 20-year old trees not under cultivation was approximately 150 t/ha, subdivided as follows :

- poles 22.5 t/ha 15 %
- brushwood 20.7 t/ha 13.8 %
- not chippable 22.5 t/ha 15 %
- other 84.3 t/ha 56.2 %

Using conventional methods, brushwood with a diameter of under 4 cm at the base (13.8 %) is left in the forest, a fraction (15 %), especially of chestnut and common acacia, is used for poles and the remainder for firewood (106.8 t/ha, i.e. 71.2 %).

Using the chipping machine, even brushwood can be used ; the base sections of the largest shoots which cannot be chipped by medium-sized machines (economically viable in these areas) and cannot be used for poles because they are too short or crooked, are used for firewood. Apart from poles, 105 t/ha (70 %) is thus used as chipwood and 22.5 t/ha (15 %) as firewood. If the most prevalent species is unsuitable for poles, the percentages of chipwood and firewood increase.

The stools were composed of an average of 1.8 shoots in the case of common acacia, 3.7 for ash and 4.2 for chestnut. The base diameters of the shoots were on average 5 cm for ash and approximately 9 cm for chestnut, oak and common acacia, with an average height of 6.3 - 9.1 m.

With an annual fuel consumption for heating a one-family farm of approximately 18 t of dry wood (26 damp) and an average cycle of 18 years, which will supply 105 t/ha of damp chipwood (72 t/ha dry), a surface area of 0.25 ha per year will be required, from a total area of a least 4.5 ha of coppice available.

3. COPPICE HARVESTING METHODS

At Vezzolano farm, Albugnano (Asti), the use of winches and chippers for harvesting was monitored for several years at different sites, with the collaboration of the CNR'S Institute for Timber Research. The work was carried out at felling areas with average slopes of 35 %, to a maximum of 60 % well provided with forest roads.

The trees were felled by power saw, away from the direction of skidding and were skidded in bundles using a winch. With crews of two, capacities of 2.14 t/h were achieved over the total time worked.

For first-stage extraction alone, using a lightweight 42 kg portable winch, with a 4.5 kW motor and a pull of 1000 kg, operated by a two-man crew, average loads of 0.25 t were achieved with a productivity per man of 0.86 t/h and idle time of 20.5 % over an average distance of 21 m. For the same work performed using a winch mounted on a crawler tractor, weighing 290 kg and with a pull of 4500 kg over an average distance of 23.4 m for mean loads of 0.40 t, a productivity of 1.35 t/h per man (+ 36 %) was recorded with idle time reduced to 5.2 %. With this second type of winch, the skidding of bundles to the processing site (an average distance in this case of 143 m) could be carried out at the same time, with an overall productivity per man of 0.87 t/h.

Of course, the real working productivity on the site varies according to the skidding distance. Using a tractor mounted machine, for first-stage extraction alone, work capacities varied between 2.2 t/h for 50 m to 3.8 t/h for 10 m, without taking waiting time into account. For skidding alone, capacities varied between approximately 1.3 t/h for a distance of 500 m, 2.2 t/h for 200 m, and 3.6 t/h for 50 m. For first-stage extraction and skidding combined, however, capacities fell to approximately 1 t/h for a distance of 500 m, 1.54 t/h for 200 m and 2.1 t/h for 50 m. However, the two operations should preferably be carried out simultaneously to avoid longer idle times and the duplication of effort involved in coupling and uncoupling the bundles.

The shoots were skidded to a processing area where they were cross-cut and the pieces large enough to make posts selected. The remainder was chipped straight away. If this solution was not possible, the smaller pieces had to be piled to be chipped later. Work capacities in this case were 0.64 t/h per man for piling as opposed to 5.56 t/h for preparation alone followed by chipping.

4. CHIPPING

Several types of disc and drum type chippers were tested, both carried on the tractor and hand-fed.

In a felling area where preparation and chipping were carried out simultaneously by a four-man crew, actual working capacities of 2.56 t/h were achieved, rising to 3.10 t/h without taking into account waiting and idle times and to 3.69 t/h with the team of four working only on chipping timber already prepared and piled. In the latter case, productivity during skidding was lower than with the simultaneous method, because of the waiting time of those working on skidding with the winch while those preparing the wood piled the timber. In fact, the entire process of simultaneous skidding and chipping gave productivities of 0.43 t/h per man, as against 0.25 t/h when the operations were performed separately.

In sites using simultaneous methods, it is not always possible to gauge work efficiencies, which are to some extent dependent on the skidding distance, because waiting times can reduce overall productivity. In the various cases examined values were obtained of 15 to 20 and even 30 %. When skidding distances are particularly long, part of first-stage extraction should be carried out separately at the edges of the skidding tracks to avoid waiting times at the preparation and chipping sites.

The medium-sized chippers used absorbed power varying from 10 to 30 kW, with peaks of 60 kW. Fig. 1, for example, plots the power input and the number of revs for a disc chipper cutting 16 mm sized chips from two different sized shoots.

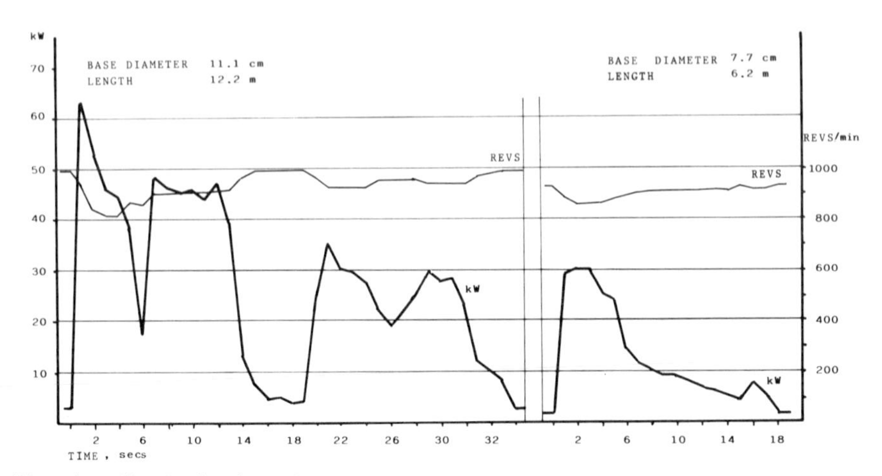

Fig. 1 - Chart showing the number of revs and power input into a disc chipper processing two shoots of different sizes.

With the larger shoots, rapid decreases in the power absorbed occured, caused by big branches slowing down the progress of the trunks between the feed rollers. It is therefore necessary to use more powerful tractors (60-75 kW) to make maximum use of the machines and facilitate transport even in difficult areas. The power input varies according to the length of the chips, the species, the moisture content of the wood (more power is needed for dry wood) and in particular the diameter of the shoots. There is in fact a correlation between the latter and the power input. With shorter chips (5-6 mm), the effect is less marked, but productivity is considerably reduced.

The costs of timber harvesting, chipping and transport to the farm have been analysed for the hill areas studied, on the basis of an annual working time for the winch of 250 h and for the chipper of 350 h (used by several farms) and for the other machines the working times of an average farm. The cost of materials was taken to be that applying in spring 1986, while labour costs were assumed to be 10.400 Lit/h.

Table 1 shows the average costs and working times for the entire operation from felling to transportation to the farm over an average distance of 500 m, using conventional methods of felling and cross-cutting by power saw, hand sliding down to the valley, manual loading and unloading of the trunks, as compared with a mechanized method using a tractor mounted winch and a medium-sized chipper.

The cost falls from 96,463 Lit/t, at 8.25 h/t using conventional methods, to 81,765 Lit/t and only 4.70 h/t using mechanized methods. If the chipper and winch were used only by a single medium-sized farm, the costs would rise to 139,180 Lit/t.

Table I - Average costs and working times for felling and harvesting operations in hill areas

	Conventional method			Mechanized method		
	Costs (Lit/t)	Working times(h/t) machines	Working times(h/t) labour	Costs (Lit/t)	Working times(h/t) machines	Working times(h/t) labour
Felling	19,385	0.55	0.63	20,318	0.59	1.22
Skidding	61,568	-	6.67	21,902	0.63	1.48
Chipping	-	-	-	28,758	0.70	1.45
Transport	15,510	0.20	0.95	10,787	0.18	0.55
Total	96,463	0.75	8.25	81,765	2.10	4.70

Figure 2 shows the rise in costs of felling and chipping in 1984 according to skidding distance at two sites, performing the first-stage of extraction separately from or simultaneously with skidding and chipping. Under these circumstances, it is more convenient to carry out the two operations simultaneously over a distance of 120 m. With the operations performed separately, beyond a skidding distance of approximately 210 m, costs increase (though still remaining below those at the other site), because of waiting time at the chipping stage rather than at the skidding stage. As labour costs increase, overall harvesting costs also increase more sharply with the conventional method than with the more mechanized method.

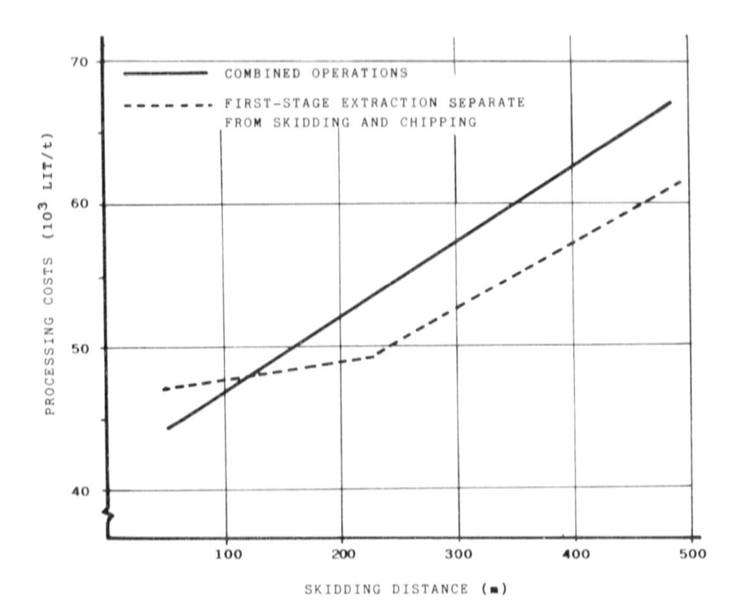

Fig. 2 - Increase in felling, skidding and chipping costs according to skidding distance at two sites.

5. DRYING AND STORAGE OF THE CHIPPED WOOD

The moisture content of the wood on felling varies according to the season, the tree species and the part of the plant. Studies carried out on coppice at Vezzolano showed average moisture contents on felling of approximately 34 % for flowering ash and common acacia, 39 % for oak and 43 % for chestnut. The moisture content of the trunk at the end of autumn is 2 to 3 % lower than at the end of spring, particularly in ash and common acacia ; in the branches, where moisture content is several percent higher (as much as 10 percent in flowering ash and common acacia), the difference is less obvious. The top of the trunk also has a higher moisture content.

Storage of wood chips with this level of moisture can lead to strong internal heating and loss of dry matter, with a risk of spontaneous combustion in large piles.

Extensive research is therefore under way to find drying methods, including natural ones, to avoid these problems.

5.1. Artificial drying of chips

Artificial drying of chips is possible. In tests carried out using a batch drier with ventilation from the base to about 1 m up, using air heated by a gas oil burner (temperature increase of 10-20° C), it took 25 hours to bring the moisture content down from 36 to about 15 %, with gas oil consumption of 57 kg/t of dry matter and an efficiency of approximately 7 MJ/kg of evaporated water. Using air heated by solar collectors, similar results were obtained, temperatures rising by 5 to 22° C, depending on the season (March to May) and the time, with the drying process complete within 20 to 25 hours. The use of simplified solar collectors (black tube, 1.5 m in diameter and 100 m long) using a 16 m^2 platform, the temperature was raised by 5 to 15° C and drying took 35 to 40 hours.

The use of a wood chip burner to heat the air in an installation with a 20 m^2 platform and mechanical-aided unloading, gave similar results with less waste.

The analysis of the costs of a plant of this type used mainly for drying forage gave values of approximately 25,000 Lit/t of dried chips (4).

5.2. Natural drying in wire-mesh cages

For natural drying the use of wire-mesh cages 1.5 m high and between 0.5 and 1.5 m wide was studied. As soon as the wood is chipped, the cages are filled using an elevator and covered to protect them from the rain. The chips can be stored until use. The front wall of the cages can be dismounted for removal of the chips.

Various lengths of chips were considered for different types of wood and different moisture contents over a period of four years of tests. The first comparison of storage in cages of varying width (50, 100 and 150 m) gave good results, with only 6 % loss of dry matter even for the widest cages (Table 2). In 1983-1984, the comparison between cage widths of 75 and 150 cm with chips at a slighter lower humidity (30 %) gave good results for the wider size with losses of only about 2 %. Reducing the length of chips from 16 to 6 mm, raised the maximum fermentation temperatures because the initial moisture content in this case was higher (35 %) ; losses were also greater. This was also for oak, which has a higher moisture content.

Table II - Initial moisture content, maximum temperatures and losses of dry matter during storage in wire-mesh cages. Average values

Species and method	Width of cages cm	Length of chip mm	Initial moisture %	Maximum temperature °C	Loss of dry matter %
1982-83					
Mixed	50	12	33.0	-	5.4
Mixed	100	12	36.5	-	4.7
Mixed	150	12	34.7	-	6.7
1983-84					
Oak	-	10	33.2	45.5	2.2
Ash	-	10	28.6	24.0	2.0
Common acacia	-	12	28.7	57.0	1.3
-	150	6	35.4	42.0	4.8
-	150	10	30.0	33.0	1.9
-	150	16	28.6	37.0	1.2
-	150	10-12	29.5	42.3	1.8
-	75	10-12	30.1	42.0	1.8
Winter felling	150	10	30.0	33.0	1.9
Early felling	150	10	25.0	19.5	4.1
1984-85					
Oak	150	-	32.2	38.7	3.7
Ash	150	-	29.2	23.7	2.6
Common acacia	150	-	32.5	44.0	3.4
-	150	8	30.7	38.0	3.6
-	150	12	31.6	32.0	3.5
-	150	16	30.7	26.5	2.2
1985-86					
Oak	150	10	37.7	43.7	7.5
Ash	150	10	32.5	21.7	5.9
Common acacia	150	10	33.7	33.3	6.3
Chestnut	150	10	42.0	36.3	12.0
Chestnut	150	16	36.1	33.0	8.9
Autumn felling	150	-	37.5	26.6	10.0
Spring felling	150	-	37.4	37.6	8.3
Early felling	150	-	33.3	36.6	6.1

In 1984-1985, with initial moisture contents varying between 28 and 33 %, good storage results were again obtained in cages 150 cm wide with little wastage (between 0.8 and 4.3 %) and the same phenomena as before. In this case, however, by increasing the chipping length from 8 to 10 or 16 mm, the fermentation temperature was steadily decreased without affecting moisture content.

In 1985 and 1986, however, with higher initial moisture contents (from 33 to 44 %), in the 150 cm wide cages the chips showed clear signs of browning and moulding as a result of considerable heating with greater dry matter losses (between 5 and 15 %), the higher percentages being for chestnut, which its 44 % initial moisture content. With early felling in autumn and chipping in spring, moisture contents were lower, losses were reduced to 6 % (although fermentation temperatures were rather high). This meant that the chips could be more successfully stored, except in the case of chestnut, which still had 39 % moisture content and 10 % loss of dry matter.

With the wider cages (150 cm) it is therefore necessary to keep moisture content below 32 % to avoid excessive fermentation and loss of dry matter.

5.3. Natural drying in covered heaps

It proved possible to dry and store the chips in heaps 1.5 to 2.5 m high protected from the rain.

A series of tests compared different heights of chip heaps with 30-36 % moisture content, including and excluding a turning treatment. Even repeating this operation three or four times the results were poor because, despite the fact that the rapid rise in temperature to 55-60° C and even 67° C at the top of the pile was interrupted, it recommenced after a few days, although at slightly lower temperatures, due to oxygenation. Storage was good, however, even though the turned chips were browned and showed more signs of moulding. Without turning, the top layer of approximately 15-20 cm was brown and sometimes mouldy, due to condensation during heating, while the central part had a moisture content of 20-22 % (13-18 % on the outside) and storage was successful. Only the part below approximately 30 cm from the ground, remained rather wet.

Another test with heaps 2 m high, without turning and at 30 % initial moisture content, gave good storage results, with a final moisture content at the centre of the heap of 13-17 %.

5.4. Early felling of the timber

To alleviate the problem of natural drying and storage of the wood chips, early felling at the end of autumn was studied for several years, in collaboration with the Forestry Institute at the University of Turin (2), with skidding and chipping delayed until the end of spring to decrease the initial moisture content on storage.

Of the three felling periods - autumn (immediately after leaf fall), beginning of winter and beginning of spring - felling in autumn or the beginning of winter was found to be preferable, because of the slightly lower moisture content on felling and the greater reduction in moisture content, although environmental conditions at the beginning of spring mean greater evaporation.

Tests carried out in 1982-1983 and 1983-1984 on three species (ash, common acacia and oak) showed average reductions of 6.2 and 9.2 % moisture content respectively (Table 3), depending on the weather in the season in question. The reduction in moisture content varies according to the species, being greater in oak and chestnut, which had higher moisture contents on felling ; in fact, the evaporated water as a percentage of the initial amount (mean values over 2 years) is 32.5 % in oak, 28.4 % in ash and 26.9 % in common acacia.

Table III - Initial and final average moisture contents in common acacia, ash and oak timber and evaporated water as a percentage of the initial amount, with felling at different times of year

Date of felling	Moisture initial %	Moisture final %	Difference	Evaporated water %
23.09.82	34.9	28.9	6.0	24.2
14.10.82	34.7	27.5	7.2	28.7
07.03.83	35.9	30.7	5.2	21.0
Average	35.2	29.0	6.2	24.6
12.10.83	35.7	26.4	9.3	35.4
11.11.83	36.8	26.5	10.3	38.2
10.04.84	37.5	29.6	7.8	29.7
Average	36.6	27.5	9.2	34.4

6. CONCLUSIONS

The research as a whole has enabled a set of methodologies to be defined for the harvesting and storage of chipped wood, including natural drying, which makes the use of this energy resource viable once again.

The advantage of using chipped wood in automatic heating installations for farm buildings and also driers has been confirmed and its use is now becoming widely established (5).

Medium-sized chippers need to be fairly intensively used and this is not always possible for small farms. These will therefore need to share or subcontract machines.

BIBLIOGRAPHY

(1) LISA L. - Preparazione e conservazione della legna sminuzzata. Macchine e Motori agricoli (1984) 42 (4), 35-40.

(2) LISA L., QUAGLINO A. - Taglio anticipato della legna da sminuzzare per utilizzazione energetica. Annali Fac. Scienze Agrarie Univ. di Torino (1984) 13, 201-247.

(3) THORNQUIST T. - Logging residues as a feedstock for energy production. Drying, storing, handling and grading. Swedish University of Agricultural Sciences. Depart. of Forest Products, Report n. 152 (1984), 1-104.

(4) GHIOTTI G., LISA L., PICCAROLO P. - Forage drying using a furnace fired with wood chips. 2° Conferenza Internazionale Energia e Agricoltura, Sirmione 13-16 ottobre 1986.

(5) LISA L., FINASSI A. - Harvesting, storing and utilization of wood biomass and rice husk. 1° Convegno FAO-CNRE : Calore ed energia elettrica della biomassa legnosa. Ormea 17-19.10.1986.

ASSESSMENT OF THE ENERGY BALANCES FOR BEECH COPPICES IN THE NORTHERN APPENNINES

U. BAGNARESI*, S. BALDINI, S. BERTI, G. MINOTTA*
Istituto per la Ricerca sul Legno - C.N.R. - Florence
*Istituto di Coltivazioni Arboree, University of Bologna

Summary

The authors describe the results of a study of the energy balances for beech coppice harvesting carried out on ten experimental plots in the northern Appennines. The following types of operation were considered : shelterwood coppice (clear felling with reserves), conversion to high forest, coppice selection. Various methods of carrying out the work were investigated for each type of operation to determine the energy efficiency of the various practices. The results obtained are set out in tabular form in the text.

1. INTRODUCTION

Beech coppices represent 6.5 % of Italian woodlands and cover 414 559 ha. They have traditionally been managed mainly as shelterwood coppices. Where this method, which is geared to the production of fuelwood, charcoal and chips, ceases to be attractive because of increasing labour costs and the difficulty of mechanization, these coppices are generally converted into high forests. Another form of practice, which is, however, less widespread than the above, is the coppice selection system. The present project (Progetto Finalizzato C.N.R. Energetica/2) uses the results of surveys carried out on ten experimental plots to calculate the energy outputs and inputs for the methods of harvesting such beech woods. Various operational criteria were obtained by adopting different extraction practices, but always on steep slopes, since Italian beech coppices are generally grown under such conditions.

The research was restricted to harvesting operations extending as far as the stacking bay and thus does not include data on transport.

2. STUDY METHODS

The surveys and the felling, topping and trimming and skidding operations were carried out during the winter.

The output was calculated by analysing the relationships between the stem diameters at a height of 1.30 m and the above-ground dry-matter ligneous biomass partitioned among the bark and wood (considered separately) of the stem to a diameter of 4 cm, the bark and wood of the thick branches (diameter over 4 cm) and the small branches (diameter under 4 cm), using the equation $\text{Log}_{10} Y = a + b \log_{10} D_{1.30}$

On the input side, direct and indirect energy consumption were calculated for the various types of machinery and equipment used as far as the truck hauling road.

It was decided that energy consumption should not be used as a measure of labour input because of wide discrepancies between the values quoted in the literature. This is in line with the practice followed by other authors.

Two parameters were, however, included to assess the human work input on the various types of site, viz. the number of hours of work per hectare and the relationship between the output and the number of hours of work required to achieve it.

3. RESULTS

The results are shown in the following tables. Table I shows the figures for shelterwood coppice systems and conversion to high forest, while Table II contains the figures for coppice selection systems.

4. DISCUSSION

The values of the energy output/input ratios seem high even though they relate to harvesting on very steep slopes.

The output/labour ratios vary from 2008 to 3118 MJ/h for coppice selection systems, from 3031 to 3154 MJ/h for conversion to high forest and from 4398 to 4948 MJ/h for shelterwood coppices.

In more general terms, the output/input ratio in respect of the quantity of biomass removed is much better for shelterwood coppices than for the coppice selection system.

The use of polythene dry logging flumes virtually doubles the output/input ratio in downhill extraction, while the energy efficiency of skidders with winches is greater if they are used for uphill extraction.

In the coppice selection system, the output/input ratio for the removal of whole trees was equal to that for topped and trimmed stems only if a small logging winch was used. However, this method gave the lowest output/labour ratio of all the techniques employed.

Table I - Camugnano District : Shelterwood coppice, predominantly beech with cherry and willow, on soils derived from an alternating arenaceous/marly formation, age 32 years, altitude 950 m above sea level, exposure north-west (compartments a, b, d, e) and north (compartments c, f), mean annual overground increment 5.4 m^3/ha (compartments a, b, d, e) and 7.3 m^3/ha (compartments c, f), shelterwood coppice (compartments a, b, c) and conversion to high forest (compartments d, e, f).

COMPARTMENT	PREDOMINANT GRADIENT %	ABOVE-GROUND BIOMASS BEFORE CUTTING (FRESH WEIGHTS) t/ha	ABOVE-GROUND BIOMASS AFTER CUTTING (FRESH WEIGHTS) t/ha	EQUIPMENT USED FOR FELLING AND EXTRACTION	DIRECTION OF SKIDDING	PRODUCT SKIDDED	OUTPUT MJ/ha X 1000	INPUT MJ/ha	OUTPUT / INPUT	LABOUR h/ha	OUTPUT / LABOUR MJ/h
a	60	177.2	15.1	Power saw Polythene flumes	Downhill	Topped stem and large branches	1,388	6,255	223.0	291	4,770
b	60	214.0	45.4	Power saw Skidder Logging winch	Uphill	Whole tree	1,860	13,265	140.2	376	4,948
c	60	229.1	6.6	Power saw Skidder Logging winch	Downhill	Whole tree	2,357	19,341	121.9	536	4,398
d	60	179.6	89.1	Power saw Polythene flumes	Downhill	Topped stem and large branches	745	4,710	158.2	239	3,118
e	60	173.0	75.9	Power saw Skidder Logging winch	Uphill	Topped and trimmed stem	763	7,195	106.1	242	3,154
f	30	192.8	90.1	Power saw Polythene flumes	Downhill	Topped stem and large branches	806	4,427	182.1	266	3,031

Table II - San Piero in Bagno District (FO) : pure beech coppice managed by the coppice selection system on soil derived from an alternating arenaceous/marly formation ; age of last age class 36 years, altitude 980 m above sea level, exposure north-east

COMPARTMENT	PREDOMINANT GRADIENT %	ABOVE-GROUND BIOMASS BEFORE CUTTING (FRESH WEIGHTS) t/ha	ABOVE-GROUND BIOMASS AFTER CUTTING (FRESH WEIGHTS) t/ha	EQUIPMENT USED FOR FELLING AND EXTRACTION	DIRECTION OF SKIDDING	PRODUCT SKIDDED	OUTPUT MJ/ha X 1000	INPUT MJ/ha	OUTPUT / INPUT	LABOUR h/ha	OUTPUT / LABOUR MJ/h
a	10	272.6	168.9	Power saw Small logging winch Skidder Logging winch	Uphill	Whole tree	1,078	10,910	98.8	537	2,008
b	20	229.7	100.7	Power saw Skidder Logging winch	Uphill	Whole tree	1,342	17,506	76.6	492	2,727
c	20	230.4	106.6	Power saw Tractor Agricultural trailer	Uphill	Topped and trimmed stem	970	10,213	95.0	311	3,118
d	20	214.4	102.4	Power saw Skidder Logging winch	Uphill	Whole tree	1,167	14,835	78.6	422	2,765

BIBLIOGRAPHY

(1) BAGNARESI, U., CANTIANI, M., MINOTTA, G. (1984). On energy balances of planted poplar stands in high-water bed areas. Paper presented at the First Technical Consultation on Forestry Biomass for Energy, Garpenberg, Sweden, June 25-27.
(2) BALDINI, E. et al. (1982). Analisi energetiche di alcune colture arboree da frutto (Energy analyses of fruit tree plantations). Riv. Ing. Agraria N° 2, pp. 201.
(3) BLANKENHORN, P.R., BOWERSOX, T.W., MURPHEY, W.K. (1978). Recoverable energy from the forests : an energy balance sheet. Tappi 61 (4), 57-60.
(4) BLANKENHORN, P.R., BOWERSOX, T.W., WEYERS, R.E. (1982). Energy relationship for selected cultural investments. Forest Sci., Vol. 28, n° 3, 459-469.
(5) CAVAZZA, L. (1983). Output and input analysis in energy balance of agricultural production. Riv. di Agr., 17.
(6) JARACH, M. (1985). Sui valori di equivalenza per l'analisi ed il bilancio energetico in agricoltura (Equivalent values for energy analysis and the energy balance in agriculture). Riv. Ing. Agraria N° 2.
(7) MATTSSON, M.L.J.E. (1978). Consommation d'énergie par machines utilisées en forêt et l'ensemble de la foresterie suédoise (Energy consumption by machines used in the forest and the entire forestry sector in Sweden). Séminaire sur les aspects énergétiques des industries forestières. Udine 13-17 November.
(8) MONTEITH, D.B. (1980). Energy production and consumption in Wood Harvest. In CRC Handbook of energy utilization in Agriculture, Ed. D. Pimentel, 449-464.
(9) PARDE', J. (1980). Forest Biomass. Forestry Abstracts, Vol. 41, n° 8, 343-361.
(10) PELLIZZI, G. (1985). La legna fonte di energia rinnovabile (Wood as a renewable energy source). Inf. Agr., N° 5, 27-30.
(11) RANGER, J. (1978). Recherches sur les biomasses comparées de deux plantations de Pin Laricio de Corse avec ou sans fertilisation. (Research into comparative biomass yields of two plantations of Pin Laricio de Corse with and without fertilization) Ann. Sci. Forest, 35 (2), 93-115.
(12) SPUGNOLI, P., ZOLI, M. (1983). Costi energetici della meccanizzazione agricola (Energy costs of agricultural mechanization) : la Conferenza Internazionale Energia ed Agricoltura, Milan 27-28-29 April, 4/1-4/25.
(13) SVANQVIST, N. (1985). Fuel consumption as cost indicator. The Swedish University of Agricultural Sciences. Department of operational efficiency, Garpenberg, Sweden.
(14) ZSUFFA, L., BARKLEY, B. (1984). The commercial and practical aspects of short-rotation forestry in temperature regions : a state-of-the-art review. In Bioenergy 84. Proceedings of Conference 15-21 June, Goteborg, Sweden, Vol. I, 39-57.

COOPERATIVE R&D IN HARVESTING FOREST BIOMASS IN THE IEA BIOENERGY AGREEMENT

C P MITCHELL
Department of Forestry
University of Aberdeen, AB9 2UU, UK

Summary

Cooperative R&D has been conducted within the frame of the IEA Forest Energy Agreement since 1978. Recently the terms of the Agreement were enlarged to include all sources of biomass for energy.
In this paper the main results of cooperative research in the area of harvesting, processing and transport for the period 1978-1985 are described.
The objective of the new IEA Bioenergy Agreement is to carry out a programme of cooperative R, D&D and exchange of information on biomass energy. There are three tasks; Task III is concerned with 'Development of improved methods for harvesting, processing and transport of forest biomass for energy from conventional forests'. It's objective is to develop improved methods and harvesting systems for producing wood for energy from various assortments (logging residues, stumps, small and cull trees) from conventional forests by cooperative R&D in the following research areas; (a) forest operations in conventional forests, (b) preparation, drying, storing and internal handling (c) standardization of measurements.
The Task will operate for three years from 1 May 1986. Participants contribute to the funding of the work both in-cash and in-kind. In-kind contributions consist of such funds and man-power necessary in order to execute the work.
There are currently eight participating countries; Canada, Denmark, Finland, Norway, New Zealand, Sweden, United Kingdom, United States of America.

1. INTRODUCTION

The International Energy Agencies Forest Energy Agreement was initiated to foster cooperative R&D in the field of Forest Energy. Forest energy was defined as 'the use of short rotation forestry biomass and forestry residues to produce clean fuels, petrochemical substitutes and other energy intensive products'.

The Implementing Agreement contained one Task with the objective to develop a programme of voluntary cooperation and coordination in the planning and execution of national R&D programmes.

The cooperation was initiated in late 1977 with four participants (Canada, Ireland, Sweden and USA). Subsequently the following countries also joined; Austria, Belgium, Denmark, Finland, Norway, New Zealand, Switzerland and the UK. The last two countries withdrew from the cooperation two years later.

Initially four Planning Groups were established to implement the objective of the Task;

PGA - Systems analysis
PGB - Growth and production
PGC - Harvesting, processing and transport
PGD - Conversion

The work was funded through a common fund which was contributed to by all participants. This money was used for information exchange, conducting state of art reviews and publishing reports and the Forest Energy Newsletter (1).

Actual research was conducted through cooperative projects. These involved several countries working together to solve a particular problem. The participants of each project provided the required funds on a shared basis.

From 1 January 1986 the scope of the Agreement was enlarged to embrace the whole field of biomass energy and its title changed accordingly to the IEA BioEnergy Agreement.

The objective of the Agreement remains as before but with a change in emphasis - to carry out a programme consisting of cooperative research, development and demonstration and exchange of information on biomass energy.

The overall programme is based on three functions;
(1) information exchange on national programmes
(2) voluntary coordination of national programmes
(3) cooperative research, development and demonstration projects.

There are three Tasks concerned with forest energy;
Task II - Improvement of biomass growth and production technology in short rotation forestry for energy
Task III - Development of improved methods for harvesting, processing and transport of forest biomass for energy from conventional forests
Task IV - Improvement of methods for converting biomass feedstocks into useable energy forms.

The following countries are signatories to the Agreement although not all participate in each Task; Austria, Belgium, Canada, Denmark, Finland, Ireland, Norway, New Zealand, Sweden, United Kingdom, United States of America. The participants of Task III are Canada, Denmark, Finland, Norway New Zealand, Sweden, UK and USA.

In this paper the results of work undertaken in the area of harvesting under the Forest Energy Agreement will be discussed and the programme of work in harvesting in the BioEnergy Agreement will be described.

2. FOREST ENERGY AGREEMENT

In the period 1978-1985 cooperative research was carried out with a number of countries participating in a joint project, Initially, work was concentrated on three subject areas: mechanization of short-rotation forestry; harvesting systems for small trees and logging residues; and optimization of wood preparation with regard to harvesting, transport and conversion. A list of projects undertaken is given in Table I.

2.1. Mechanization of short rotation forestry

Two projects were conducted in this area; one concerned with establishment and the other with harvesting.

Typical requirements of stand establishment for energy plantations are careful site selection, agricultural quality of soil preparation, pre-planting weed control, planting, post-planting weed control and cultivation. The mechanization requirements for all these operations were reviewed (2) and it was found that in general equipment already exists which can be used directly or be readily adapted for most of the plantation operations except planting. The main work of this project was in the design of an automatic cuttings planter (3). This design consists of a modular unit which is mounted on the rear of a farm tractor. Up to six modules can be attached allowing six rows to be planted, the spacing can be adjusted for any design of plantation.

Cuttings are held in a magazine until needed. Each cutting is handled separately and injected into the soil. The minimum planting rate is 2000 cuttings/hr/unit. Unfortunately the design was not built and tested.

2.2. Harvesting short rotation plantations

Each of the participating countries developed its own harvesting system in close collaboration with the other participants. A number of prototype machines have been developed and tested but none has yet reached the stage of commercial production. These have all been documented (4).

2.3. Harvesting fuelwood from conventional forests

Two main opportunities for the recovery of wood for energy from conventional forestry have been examined: (a) harvesting forest residues and residuals, (b) harvesting small trees.

2.3.1. Residues

Twenty nine different methods for harvesting forest residues and residuals, four types of processes which can be performed at road-side or landing and three categories of trucks that can be used to haul the forest biomass from the landing to the plant were identified (5).

Harvesting methods can be divided into integrated (single pass) and non-integrated (double pass) methods; the first involves the harvesting of logging residues concurrently with the harvest of conventional forest products; in the second method, forest residues are harvested independently from the traditional logging operation. Each method has been described and illustrated; costs, productivity and fuel consumption figures are given based on case studies published in technical reports.

Computer simulation was used to identify promising integrated harvesting systems in five of the participating countries (6).

Of the systems studied the tree section system and the pulpwood and residue chipping system were the only systems specifically designed for integrated residue recovery. Both of these systems performed best in their country of origin, indicating that the differences in operating conditions largely account for the different harvesting technologies chosen.

Five promising areas for development were identified as a result of the analyses; compaction of tree sections, full trees or residues; delimbing systems for tree sections; residue handling at the landing; attachments to delimbers and delimber-buckers to catch the residues; chip benefication systems.

2.3.3. Harvesting small trees

Small trees from thinnings represent a vast potential resource of biomass for energy providing economic methods for harvesting them can be found.

The whole range of operations involved in harvesting small trees has been studied systematically and a manual produced in which the various methods practised in each of the participating countries are described and illustrated (7). Harvested products examined included whole trees, tree sections, solid firewood, chips, chunks, bales or compacted bundles of trees or tree parts.

Descriptions are given of motor/manual and machine felling and hand, horse, machine and winch bunching of whole trees in selective and combined thinnings. Information is given which enables estimation of the expected productivity (in terms of trees/hr in relation to dbh) in a given situation and pattern of operation.

2.4. Preparation

A survey of existing woodfuel conversion equipment revealed a wide diversity of systems requiring different fuel forms for optimal operation; homogeity and low moisture content appear to be the main desired characteristics for a forest biomass fuel (8). Five kinds of processes are used to prepare forest biomass as a suitable fuel: comminution, drying, classification, aggregation and wood densification.

There is a wide range of equipment available to prepare forest biomass as a fuel and there do not appear to be any major technological gaps among the various processes; however, whereas some processes, such as chippers, are very well developed and commercially available, other processes, such as chunkers, are still in the development stage and require further research.

2.5. Storage, drying and internal handling

The main purpose of storing wood fuel is to have fuel available when there is a market for it or it is needed for heating. Storage may also be desirable/necessary to reduce the moisture content to a desired level. Compared to some other energy carriers, wood fuels are relatively simple to store.

Studies (9) have been conducted into the storage of various forms of wood fuel (felled trees, unbarked logs, firewood, green particulate wood, fuel chips, dry particulate woodfuel and logging residues) with respect to drying and biological degradation.

2.6. Evaluation methods

A common systematic approach to machine and system evaluation which can be used to expedite international exchange of technology and experience has been developed (10).

The evaluation system was designed as a series of stand-alone components which can be selected to meet the needs of a particular project or study. Greater emphasis was placed on relatively low-intensity data collected over an extended period of time, with considerable freedom for the agency or group conducting the study to incorporate its own methods of intensive, short-term work studies.

Four broad areas of data collection were identified: the operating environment description of the operation, performance measurements and machine characteristics.

In order to facilitate the use of the evaluation package a simplified system for editing field data, constructing a data base, and then linking this with a broad range of parametric and non-parametric statistical analysis procedures have been developed.

2.7. Ecological consequences

One of the factors potentially limiting the harvesting of a greater part of the forest biomass is the long-term nutritional consequences of so doing.

Information available from empirical studies and existing computer simulation models has been reviewed (11). The main aim being to improve decision-making with respect to harvesting energy wood in relation to long-term site productivity. This is a complex problem and is still under co-operative investigation.

Information on damage to forest soils has been reviewed (12). Soil compaction is probably the worst effect. Correct tyre inflation pressure is generally considered to be the best and easiest way to prevent compaction. However, it is thought more likely that proper tyre construction and mode of loading will have a greater impact in reducing soil damage.

The recycling of wood and bark ash has been examined (13). It can be concluded that ash is most useful as a soil improvement agent and a fertilizer on agricultural land suffering from acidity and a lack of trace elements. For application on agricultural land existing equipment is adequate but for application in forests further development, possibly based on conventional forwarders, is necessary.

3. BIOENERGY AGREEMENT

The objectives of Task III are to develop improved methods and harvesting systems for producing wood for energy from various assortments (logging residues, stumps, small and cull trees) from conventional forests by cooperative R&D in the following research areas:

(a) Forest operations in conventional forests
(b) Preparation, drying, storing and internal handling
(c) Standardization of measurements

The Task will operate for three years (1 January 1986 - 31 December 1988) but it may be extended if so desired by two or more of the participants.

Participants contribute to the funding of the work both in-cash and in-kind. In-kind contributions consist of such funds and manpower necessary in order to execute the work.

3.1. Programme of Work

In the three year period the work of the Task will involve a systematic development of cooperative activities and information exchange in:

(a) Forest Operations in Conventional Forests. Methods for harvesting forest biomass for energy including whole tree harvesting of small trees, forest residues (separately and integrated with timber harvesting) and scrub wood land will be examined systematically in comparative trials. Systems analysis techniques will be used to develop new and improved harvesting systems which will minimise their ecological impact.

(b) Preparation, Drying, Storing and Internal Handling. Techniques for preparing wood fuel will be tested in comparative trials. These will be done in conjunction with studies of drying, storing and internal handing.

(c) Standardization of Measurements. Standardized methods for the measurement of wood fuels will be developed cooperatively.

Each activity is undertaken by a number of cooperators and led by the person best able (from the point of view of support from the national programme and expertise). The activities being undertaken in each of the research areas, the leader and expected results are shown in Table II.

4. REFERENCES

(1) NILSSON, P O; GAMBLES, R L; MITCHELL, C P; STEVENS, D; ZSUFFA, L. (1985). Programme Group Activities - Present status, collaboration and future directions. In Biomass Characteristics for Efficient Energy Conversion. Ed. D J Morgan et al. IEA/FE/PGB Report 1985:3, 1-18.
(2) JONES, K C. (1983). The status of forest energy plantation mechanization. IEA/FE Report NE 1983:4, pp49.
(3) PAPADOPOL, C S. (1986). Stand establishment of fast-growing species for energy plantations. IEA/FE/PGC Report. pp16.
(4) MATTSON, J A & CHRISTOPHERSON, N S. (1986). Harvesting system developments for biomass plantations. In Energy from Biomass and Wastes X. Proc Conf April 7-10, 1986. Washington DC. In press.
(5) POTTIE, M A & GUIMIER, D Y. (1986). Harvesting and transport of logging residuals and residues. FERIC Special Report No SR-33, pp100.
(6) BEARDSELL, M G; STUART, W B & MITCHELL, C P. (1985). Integrated harvesting systems to incorporate the recovery of logging residues

with the harvesting of conventional forest products: Harvesting simulation studies. Swed Univ Agri Sci, Dept Op Efficiency, Upps o Res Nr 17, pp78.
(7) KOFMAN, P. (1985). Felling and bunching small trees from thinnings with small scale equipment on gentle terrain. Danish Inst For Tech/ IEA CPC7 Final Report. pp116.
(8) POTTIE, M A & GUIMIER, D Y. (1985). Preparation of forest biomass for optimal conversion. FERIC Special Report No SR-32, pp112.
(9) GISLERUD, O. (1986). Storage and preparation of fuelwood. In Research in Forestry for Energy. Ed C P Mitchell et al. Swed Univ Agri Sci, Dept Op Efficiency, Upps o Res Nr 49, 307-314.
(10) STUART, W B. (1986). Standardized evaluation methods. In Research in Forestry for Energy. Ed C P Mitchell et al. Swed Univ Agri Sci, Dept Op Efficiency, Upps o Res Nr 49, 328-333.
(11) DYCK, W J; MESSINA, M G & HUNTER, I R. (1986). Current research on the nutritional consequences of intensive forest harvesting on site productivity. IEA/BIA Project CPC-10 Report No 3, pp171.
(12) BATARDY, J & ABEELS, P F J. (1985). Forest harvesting operations and soil damage. IEA/FE/PGC/JAC18 Report, pp74.
(13) HAKKILA, P. (1986). Recycling of wood and bark ash. A state-of-the-art review for Programme Group C under the IEA Forest Energy Agreement. Metsantutkimuslaitoksen Tiedonontoja 211, pp44.

Table I. Cooperative Projects in Programme Group C

Project	Cooperating Countries
Coordinated development of short rotation harvesters	Canada, Ireland, Sweden, United States
Preparation of forest biomass for optimal conversion	Canada, Denmark, Norway, Sweden, United States
Stand establishment	Canada, Sweden, United States
Standardised evaluation	Canada, New Zealand, Sweden, United States
Harvesting and transport of logging residues	Austria, Canada, Denmark, Ireland, Norway, Sweden, United States
Felling and bunching small trees in gentle terrain	Austria, Belgium, Canada, Denmark, Ireland, New Zealand, Norway, Sweden, United States
Integrated harvesting systems for conventional and energy products	Canada, New Zealand, Sweden, United States
Harvesting whole trees with processing in the forest to conventional and energy products	Canada, New Zealand, Sweden, United States
Nutritional consequences of intensive forest harvesting on site productivity	Canada, New Zealand, Sweden, United States
Chunkwood-technologies for comminuting and utilising forest biomass for energy	Denmark, Finland, Norway, Sweden, United States
Comparative trials of drying and storage of wood fuels	Canada, Ireland, Norway, Sweden, United States

Table II. Activities in Task III "Energy from Conventional Forests"

Area/Activity	Leader, Country	Result
FOREST OPERATIONS		
Harvesting whole trees with processing in the forest to conventional and energy products	Goulding New Zealand	Improved harvesting systems
Harvesting energy wood from early thinnings	Brenoe Denmark	Improved harvesting systems
Nutritional consequences of intensive forest harvesting on site productivity	Dyck New Zealand	Improved management
Application of systems analysis	Lonner Sweden	Improved operating systems
Development of standardised evaluation methods	Stuart USA	Standardised procedures
PREPARATION		
Microfractionation of marginal wood resources for energy utilisation	Brenoe Denmark	Improved technology
Chunkwood-technologies for comminuting and utilising forest biomass for energy	Danielsson Sweden	Improved technology
Comparative trials of drying and storage	Gislerud Norway	Improved storage and handling
MEASUREMENT		
Measurement and evaluation of wood fuels	Nylinder Sweden	Standardised procedures

WOOD HARVESTING SYSTEMS FOR ENERGY PURPOSES IN DEVELOPING COUNTRIES

R. Heinrich
Chief, Forest Logging and Transport Branch
Forest Industries Division
Food and Agriculture Organization of the United Nations

Summary

The paper highlights the importance of wood for energy, of which some two billion people are dependent. One means to improve the supply of wood for energy, of course, is to increase the efficiency of harvesting and transportation systems. The paper discusses the choice of technology in developing countries and various systems of basic and intermediate technology as a means of increasing the wood supply for energy needs with strong participation of the rural communities. Various examples of harvesting operations are described for forest plantations as well as natural forests, including tropical high forests and mangrove. Operations comprise manual felling by axes and handsaws, wood extraction by manual means, animal and machinery as well as short and long distance transport. With reference to wood extraction by machinery, the use of agricultural tractors with various forestry attachments such as winches, trolleys, cable cranes and trailers, is described. Finally, the paper refers to actions proposed by FAO to promote appropriate technology in forest harvesting operations with due consideration to improve the fuller utilization of the forest resource base and protecting the land by carrying out environmentally-sound forest operations, thus making more wood available for energy on a continuous basis to support the overall rural development.

1. IMPORTANCE OF WOOD FOR ENERGY

It is a well-known fact that wood in the form of fuelwood is a very important energy source for the rural communities and the poorer section of the population in urban areas, especially in developing countries. Still some 40-50 years ago, even in some parts of Europe, it was a widely-used energy which has been replaced to a large extent during the industrialization process. In early 1980, FAO estimated that although wood-based fuel accounted only to some 6 percent of the total world energy supply, some 2 billion people were dependent on wood for energy and some 100 million were unable to satisfy their minimum energy needs for cooking and heating.

When looking from the point of view of wood removed for energy uses one realizes that on a world scale some 60 percent of wood removed serve as fuelwood, and in developing countries this even increases closely to 90 percent.

Having this in mind, one can easily imagine that there is a great pressure on the forest resource base to satisfy the needs of wood for energy.

Although the annual growth of the world forests is estimated to correspond to an amount which is several hundred times the total world energy consumption, as indicated earlier, it is not sufficiently available for a large segment of the world population. Sometimes, it is too costly to serve for energy needs.

What could be done to increase the availability of wood for energy purposes:

(i) Increase the productivity of existing forest resources.
(ii) Establish new fast-growing fuelwood plantations.
(iii) Improve wood harvesting and transportation techniques and systems.
(iv) Improve conversion technologies.
(v) Substitute fuelwood with alternate energy sources.

This paper, as the title suggests, will identify some harvesting systems which could help to improve the supply of wood for energy purposes from the point of view of the traditional use, namely the home consumer, and for industrial consumption in developing countries.

2. CHOICE OF TECHNOLOGY IN DEVELOPING COUNTRIES

In natural forests, especially when many different species and dimensions are represented, wood harvesting can be a complex matter. This can be further complicated where access is difficult due to terrain and soil conditions. In some tropical forests, fuelwood is considered just as a byproduct and certainly requires a different approach in the use of technology than when using only the high-value timber species.

The utilization of lesser-known species and smaller dimensioned, less valuable trees, logs and tops of high value timber, will certainly contribute to a fuller utilization of the resource base and reduce the waste in selective wood harvesting operations.

In collecting, handling and transporting wood for energy needs, the local population could play a dominant role to guarantee a certain level of the raw material supply. This could be a desirable opportunity to provide local people with an employment opportunity from non-farm income and at the same time improve the productivity through silvicultural improvement utilization of the natural forest stand.

The technology of course needs to be simple so that, first of all, the local people can handle it, then can afford to use it and/or earn sufficient money.

In Malaysia and the Philippines, timber stand improvement programmes improved significantly the quality and quantity of the tree stand. In some cases, the volume of valuable tree species increased by 50 percent and at the same time it produced the welcome byproduct wood for energy.

3. BASIC AND INTERMEDIATE TECHNOLOGY AS A MEANS TO INCREASE WOOD FOR ENERGY NEEDS

In this type of operation the labour input is predominant. By use of good quality handtools and simple equipment, quite remarkable results can be achieved in improving the productivity of traditional labour-intensive logging methods. The handtools and equipment are relatively inexpensive and can be easily manufactured by using local skill and material. By the provision of the very modest tools and equipment the worker will be able to work in a smarter way rather than to use more power, thus increasing output with reduced energy input.

This will also lead not only to a more efficient operation, but give an improved safety in the work performed and finally result in increased earnings. Besides the use of improved tools and equipment, of course proper maintenance is equally important if they are to continue to serve for the purpose for which they were designed. The use of the basic logging technology promotes employment and is therefore of direct benefit to the worker. By getting acquainted with the technology this should be considered as a tool to further development, reaching finally a more sophisticated state of the art on equipment and systems.

The utilization of basic technology should not be misinterpreted as a primitive backward technology; on the contrary, it should include the latest scientific developments in technology and material.

Basic and intermediate technology is considered to be appropriate when it is fit for the human, economic and material resources which are available at a given time and situation.

The operations where basic technology can be used consist of all kinds of work including felling, bucking, extracting, transporting as well as lifting logs, comprising the whole range of activities from the stump to the landing/road site and/or consumer as well as processing plant.

4. APPLICATION OF BASIC AND INTERMEDIATE TECHNOLOGY

4.1 Felling by axes and hand saws

Although in almost all industrialized countries manual saws and axes have been replaced by chainsaws in felling operations, in developing countries the use of traditional bush knives (machete, bolos, etc.), axes and hand saws is still a valid option which, with high quality steel, can give good results. Manual felling methods prevail when cutting small to medium-sized trees in plantation forests, as well as small-sized non-commercial species in natural forests. For instance, in India, studies carried out by Sweden in felling operations when using hand saws showed that saws with good design and high quality steel gave a double work output with half of the energy input.

A similar case study was made on two-man peg tooth crosscut saws and one-man bow saw, which showed that the one-man bow saw was really twice as efficient as the two-man crosscut saw.

4.2 Manual wood extraction

As far as extraction, transport and lifting of logs are concerned, there are still plenty of examples where logs are moved manually, by means of animals and simple handtools and equipment which have been developed in various developing countries. In recent years in Tanzania, a simple sulky was developed for manual logging, which proved to be very efficient and reduced the work load of the forest workers considerably. Special skills for instance have been developed in mountain logging as well as in swamp logging operations. In mountain forestry operations it exists a long tradition of moving logs by means of gravity, either on the soil or where the terrain is covered by stones and bolders by means of constructing log or timber chutes. Even in some industrialized countries, for instance in Austria, some 40 percent of the annual wood harvest or 4.5 million m^3 are still moved downhill manually, primarily with the help of hookeroons or sometimes by skid pans and chutes, although labour costs are at least fifty times higher than those in most developing countries.

Some efficient and very cost-effective systems used are simple cable systems. The most pioneering ones are those where wood slides down by means of gravity in small bundles. There is no power required, the only essential input is the cable and wooden hooks. A somewhat higher technology is required when using a pendulum cable system. While the load is guided down by a winch drum breaking system, the unloaded carriage is pulled uphill on the cable at the same time to be ready for loading, once the load has reached the landing station at the valley.

The most recent developments are mobile short-range cable logging systems, which are attached to agricultural tractors.

A recent study carried out by FAO on mangrove harvesting in the peninsula of Malaysia revealed that charcoal and firewood billets can be efficiently transported by wheelbarrows. In the operation studies, the wheelbarrows are moved on timber tracks over an average distance of 150 m from the felling site to the river or canal. There the billets are loaded on to boats and transported to the charcoal kilns. The wheelbarrows are constructed locally and are equipped with a shoulder strap to lift and keep the wheelbarrow in balance. The average load of the wheelbarrow is around 300 kg.

4.3 Wood extraction by animal

In several countries, animal power is still used successfully and economically, although productivity is rather low and despite the fact that often specialized high-productive logging machinery has been introduced.

Especially in plantation forests, traditional logging by means of oxen has been and still is used successfully to a considerable extent in various countries. Some of the countries where successful log extraction and transport are carried out by oxen are Burma, Chile, India, Malawi, Mexico, Zimbabwe, etc. In some of the countries, buffalos and horses are used for wood extraction.

In Mexico and Chile, where comparative studies on manual, animal and mechanized log extraction were carried out, it became evident that under the then prevailing conditions, animal skidding, especially for short distances in favourable terrain conditions, was compatible with the log extraction system by agricultural tractor with forestry attachments. Similar results were obtained in Malawi where an oxen logging training centre was established in the mid-70's and nowadays extraction of logs is carried out successfully in pine plantations. The average skidding production per day per pair of oxen was 4.5 m^3 when skidding logs over a distance of 350 m. In 1979, extraction costs amounted to US$ 0.55 per m^3. When selecting manual and/or animal systems it can be a difficult and complex task to coordinate and guarantee a continued wood supply on a larger-scale basis due to the involvement of numerous people and animals.

The introduction of soft technology, which is labour-intensive and capital-intensive, has been more and more recognized both by the donor community and by the receiver countries. This shift in the development strategy is of course very much influenced by the fact that many developing countries are facing difficulties in having access to sufficient funds in foreign currencies to purchase machinery produced in industrialized countries. In addition to the shortage of funds, the rising costs of machinery has had a significant impact on the decision-making process of utilizing more basic technology.

In 1982, ILO undertook a study in the Philippines to promote appropriate forest technology which deals, among other operations, with basic logging techniques using improved handtools.

FAO has just recently commissioned a study to investigate the viability of using carabaos to harvest wood for fuelwood-based factories in the Philippines. A case study was undertaken in a seven years old giant ipil-ipil (*Leucaena leucocephala*) plantation. This is a fast-growing species which is not only a good source for fuelwood due to its caloric value, but also provides good material for fencing and the leaves can be used for feeding animals. Trees removed had a BHD of 17 cm, the spacing of planting was 3 m x 3 m and the annual increment was estimated to be some 24 m^3 per ha.

For felling, bucking and delimbing, bush knives, bolos and axes were used. Wood extraction was carried out by carabaos, using a wooden sledge to which the logs were tied and then transported from the felling site to the landing, where they were bucked to 1 m length and piled. There, logs were loaded onto trucks and transported to fuelwood energy plants. Based on six working hours per day and depending on the extraction distance, productivity per carabao ranged from 1-2 m^3 per day.

During the case study, wood was extracted on a distance of 900 m. The cost of the wood production, delivered at roadside, bucked to 1 m log length and piled, amounted to US$ 3.75/m^3. In addition, transport cost by truck to the consumer will need to be added.

In a number of countries, FAO has registered a growing interest in the involvement of local people in supplying wood to the roadside by using a combination of animals and light cable systems.

4.4 Wood extraction by machinery

Ground-skidding winches

Independent ground skidding winches are designed for extracting logs from difficult areas such as gulleys, creeks, ravines and broken terrain in small-scale operations. They are also designed for extracting small-sized timber from plantation forests. Depending on the make, their line-pull capacity ranges from 600 kp to 2 200 kp. They are able to cover extraction distances which range from 80 to 165 m. Cable diameters of 5 mm to 9 mm are generally recommended. Some of the winches are equipped with a power chainsaw engine, others with an engine of their own brand. The available hp for various winches ranges from 4.5 to 16 hp (DIN).

Their weight varies from 42 kg to 560 kg. The 42-kg winch can easily be carried by hand, whereas the others must be transported by others means to the working area.

The winch is generally tied to a tree; the logger then has to pull the cable to the log, choke the log and then, by means of radio control, winch it in. An exception is the Akja winch. This is built on a sledge. One end of the cable is fixed to a tree and the winch is moved to the log to be extracted; the log is put on the sledge and winched in to the landing on the sledge.

One manufacturer has produced a ground skidding winch with an engine on a small-single-axle trailer; winch and engine are linked by hydraulic hoses. The trailer can be moved by hand to the place where log extraction is to take place.

The sledge winch is generally used for prebunching of logs; however, it is also very useful for transporting cables and cable crane support material to installation sites in the forest area. Another application is for placing polyethylene chutes in thinnings. The sledge winch is equipped with a 4.8-kW chainsaw engine, with a line pull of 800 kp, a 110-m cable (6.5 mm 0) having a total weight (including cable) of 70 kg.

The Multi KBF lightweight ground skidding winch is equipped with a chainsaw engine (Jonsereds) of 4.2 kW, with a line pull of 1 000 kp and a drum capacity of 80 m with 6-mm cable, or 150 m with a 5-mm cable.

The Radiotir Alpin 1 200 winch is used with radio control and can be operated by a single man. It is equipped with a 6 kW engine, and has a line pull of 1 200 kp and a cable drum capacity of 165 m when 7 mm 0 cable is used or 125 m when 8 mm 0 cable is used.

Agricultural tractor

The agricultural tractor equipped with appropriate forestry attachments may in many cases do a sufficiently-good job. Numerous types of forestry attachments do exist for the various jobs and sizes and models of agricultural tractors. Forestry attachments such as single and double drum winches, logging trolleys, cable equipment, semi-trailers, trailers with or without mechanical cranes, are presently in use, in many countries, to supply wood to the roadside, nearby forest-based power plants, etc.

The limiting factors for the introduction of this machinery is often the type of soil and terrain, the size of trees and the accessibility. When utilizing the agricultural tractor for skidding in the forest stand the maximum gradient the machine can negotiate is 25 percent, when skidding logs downhill. Although this type of intermediate logging technology is widely used in the industrialized countries in plantation forests, it has not yet received the same popularity in developing countries.

Some of the reasons are that sometimes more advanced technology has more prestige and is more aggressively advertized, which has certainly an influence on the decision-making process.

One of the major advantages is that this type of equipment with additional forestry equipment can be used for forestry purposes without any major investment and will give the farmer an additional use of the tractor and income. In fact this is one of the reasons why in a number of European countries agricultural tractors outnumber by far the specialized logging machines such as skidders and forwarders. A similar trend to use more agricultural tractors in forestry work was observed in New Zealand and Australia.

When utilizing the agricultural tractor in forestry work it should have the following specifications in order to meet the job requirements and safety standards:

- four-wheel drive tractor with roll-over protective structure (either roll-over frame or safety cab);
- three-point linkage (except for those tractors which have forestry attachments directly mounted on the tractor);
- power take-off;
- bottom safety shield (a pan to protect the engine);
- power source 35-56 kW.

Tractor-attached winch

Tractor-attached winches are used for ground lead uphill extraction of logs on a distance of 30 to 50 m. Tractors are positioned at the forest roads with the winch pointing towards the valley side of the road where logs can be pulled up from the slope below the road. Winches can either have a single or double drum and are equipped with or without a logging plate. The logging plate is very useful for stabilizing the tractor, especially in preventing the tractor of skidding back while pulling in the logs. Various types and sizes of winches do exist, as well as of skidding plates. Some of the winches have drums which are located parallel to the tractor axles whereas others have them perpendicular. Tractor-attached winches generally have a line pull ranging from 1500 Kp to 5000 Kp. Recently in the logging project in Southern Viet Nam in pine plantations an agricultural tractor with an attached winch was introduced and time and work studies carried out. The studies revealed that in comparison with the traditionally-used tracked skidders the newly-introduced simple technology had a much higher production, especially when used for ground level logging. Studies in Mexico in pine plantations showed that the production per man/hour when skidding logs over an average distance of 80 m was 2.60 m^3.

Tractor-attached trolley

This tractor attachment is essentially a small trailer with two wheels, a built-in single drum winch and a skidding plate. The advantage of this logging trolley is that it can be used efficiently both for winching as well as for skidding on the forest floor as well as on the skid road. When pulling logs uphill the skidding plate acts as a safety protection structure whereas when skidding the load, it acts as support for the log and this reduces the friction on the floor. A further advantage of this system is that the pulling forces are exerted on the axle of the trolley and not on the rear axles of the tractors directly. The maximum line pull of the winch built in the trolley has 4000 Kp.

A few of these trolleys were tested in developing countries and they showed very encouraging results. Studies were carried out in Bhutan, Mexico and Sudan.

Tractor-attached cable equipment

Tractor-attached cable equipment is an ideal supplement in opening up forest resources and utilizing them properly, preventing damages from transport to the soil and terrain as well as to the remaining stand, and thus are environmentally-conformed harvesting systems on difficult and steep terrain. This type of equipment requires a good road network as its maximum working range is generally limited from 300 m to 500 m. This cable equipment is essentially a tower equipped with a main line drum, a skyline drum and rigging drums, cables as well as a carriage. It is operated as a skyline system and can be used uphill and downhill. Generally, the maximum payload to be transported by this system is not more than 1.5 tons; therefore it can only be applied in plantation forests which have smaller-sized trees, or in plantations with larger-sized trees, short log transport needs to be carried out. It can also be very efficiently used in thinnings as setting up time by a well-trained team does not require more than two working hours and therefore it does not require large quantities of logs per setting. With reference to its high mobility and the small volume of logs required per unit area, this equipment is considered to be highly recommendable to improve silvicultural work. Essentially this type of system should be used on steep and difficult terrain where ground skidding is not any more feasible or permitted due to environmental reasons.

In several developing countries FAO has assisted in pilot projects to study the use of this simple mobile cable system. Studies were for instance carried out in Bhutan, El Salvador and Mexico. The idea was to study low capital investment machinery, using environmentally-sound techniques on fragile soils and improving the efficiency in forest harvesting operations.

Traditional skyline cable equipment

In mountain plantation forests with larger-size trees and which have a low road net density the traditional skyline cable system can be considered as an appropriate method to harvest logs both selectively and in strip cuttings.

The simplest system in skyline operations which can be recommended for use in mountain logging, especially in plantation forests in developing countries, is the gravity cable system. The cable equipment is composed of a single drum winch stationed at the upper point of the setting, a skyline cable, a main line and a carriage with or without built-in stopping device. When utilizing this equipment, wood can be harvested generally on a cutting area of up to 5 ha, covering a span of up to 1000 m and a reach on both sides of the skyline of 25 m.

5. SHORT-DISTANCE WOOD TRANSPORT

Simple trailers often made in a local workshop are an ideal supplement for the agricultural tractor, when already employed in the extraction phase, for transporting wood over short distances. This type of machine has been essentially designed for farmers and small contractors to carry out the transport of logs from the road to the main log yards, to local forest industries in the vicinity of the forests or for their own needs.

Some of the trailers are single-axle trailers equipped with tilting devices which makes the unloading of shortwood especially easy. This type of trailers is generally used in transporting pulp or fire wood and it requires that logs are loaded parallel to the trailer axle. Such trailers can have a load capacity which ranges from 2 to 4 steres. Two axle trailers can be used both for shortwood and for log transport.

They are often equipped with mechanical cranes which are mounted onto the trailers and have a carrying capacity which is generally below one ton.

In a recent study carried out for FAO in the Philippines a farm tractor with trailer was studied to identify production and cost levels in transporting fuelwood. In a four-year old ipil-ipil forest trees were felled and bucked manually, piled at road side and loaded manually, average transport distance was 5 km, using a 60-kW agricultural tractor and trailer. The average volume transported amounted to 3.8 m^3. Depending on the distance, productivity ranged from 7-30 m^3 with costs of US$ 1.5 to 3 per m^3 of fuelwood delivered to the main road landing.

A further case study in Ethiopia, where an agricultural tractor and trailer with a loading boom was used to transport fuelwood and sawlogs in a seventeen-year old cypressus lusitanica plantation, revealed that wood delivered to the main landing amounted to US$ 7.3/m^3. The productivity was 2.5 m^3/hour.

6. LONG DISTANCE TRANSPORT

Transport costs are often the most important factor in determining the cost of fuelwood. In 1975 in Nepal the cost of collected fuelwood in the vicinity of the forest amounted to US$ 1 per m^3 whereas in Kathmandu the price was US$ 16.

In 1978, the price of fuelwood became as expensive as oil when it was transported on a distance of about 400 m, and charcoal on 800 m. In reality, however, it is often not economical to transport fuelwood by truck over distances longer than 100 km. Water transport is a traditional means of wood transport over longer distances in a number of developing countries. As far as water transport of fuelwood is concerned, FAO has recently carried out studies in Benin and other African countries, which concluded that under certain conditions floating of fuelwood would be feasible.

In a number of countries in the recent past, a considerable number of studies have been carried out on the use of fuelwood or charcoal to produce heat or generate electricity from wood fibre. Many forest industries were run on energy produced by wood waste, which has helped to reduce costs and dispose of the waste. Some of the developing countries with extensive forest residues and dependent on oil imports have even established dendro thermal power plants to save foreign exchange. One of the examples is a study undertaken by FAO for some Pacific Islands which suggests wood fired dendro thermal plants from old dying coconut trees which need to be removed and from waste wood.

7. PROMOTION OF APPROPRIATE TECHNOLOGY IN WOOD HARVESTING

Within the training and educational activities carried out by the Forest Logging and Transport Branch (FOIL) in the past, the promotion of basic logging technology has received considerable attention. A number of pilot studies were undertaken to introduce new simple methods which are suitable to fit with the local conditions.

Besides the case studies, FOIL organized training activities which focused partly on the introduction and transfer of basic technology in forest harvesting.

In connection with a logging training course carried out in 1981 in Sri Lanka, a handbook on basic logging technology was prepared, made possible by a special contribution from Sweden.

In 1982, FAO published the handbook on basic technology in forest operations as Forestry Paper No. 36. Recently, as a follow up to the handbook a design manual on basic wood harvesting technology has been prepared and will be published in 1987.

FOIL, together with the Austrian Government, has already organized four training courses on mountain forest roads and harvesting in which the transfer of basic logging technology has received appropriate attention. FAO Forestry Papers No. 14, 14 Rev.1 and 33 dealing with the subject of mountain logging were prepared in connection with the training courses.

In addition, film strips have been prepared, taking into consideration different levels of logging technology, giving special attention to handtools and simple machinery, to assist developing countries in providing more employment opportunities.

In 1984, through an André Mayer fellowship, a study on the utilization of animal power and use of intermediate technology in wood extraction and transport was carried out in Chile. As the result of the study, FAO Forestry Paper No. 49 "Wood extraction with oxen and farm tractors" was produced and widely distributed.

In future activities, the main emphasis will be put on training and supply of information on logging.

The Branch has initiated a survey of logging training needs in selected developing countries, which will be carried out during the biennium 1986/87. This project was made possible by a special contribution of the Finnish Government and is executed jointly by the Forestry Training Programme of Finland (FTP) and FOIL. Its primary objective is to gather sufficient information on wood harvesting and training presently carried out. On the basis of the information collected, an action programme should be developed and further in-depth studies be undertaken for the implementation of specific training activities.

Several training courses will be carried out on appropriate forest operations and wood harvesting. In June 1986, one course was held in Zimbabwe for SADCC countries and other English-speaking African countries where forestry and forest industries development play an important role in the overall development of the countries. A further course is planned to be held in the Philippines in November 1987. Further courses are planned on specific topics such as mountain wood harvesting, small-scale logging, wood transport infrastructure, etc.

In the near future, the Forest Logging and Transport Branch will continue with the case study programme on basic and intermediate wood harvesting technology, with a strong participation of the local population.

Under this programme, various case studies will be carried out so as to evaluate the suitability of different alternatives of wood harvesting techniques under certain conditions and disseminate the results.

One of the important programmes in the future is the promotion of a fuller use of the forest resource base in the tropics, and the reduction of wood losses, both in the forests as well as at the mill sites. This will include a series of studies on activities such as the utilization of lesser-known species, smaller-dimensioned trees from silvicultural operations as well as ways and means to improve harvesting systems in order to reduce harvesting and post-harvest losses.

As a first step, an interregional study will be undertaken to identify the originating points of the residues to determine, among other aspects, their characteristics and volume.

On the basis of these initial studies further future activities will be drawn up to provide solutions to the under-utilization and wasteful use of forest resources.

Under this programme various studies will be undertaken to investigate the potential use of special forest products such as coconut and rubber plantation wood.

The newly set up computerized equipment information system contains data on logging, transport and road construction machinery. It is hoped that during the forthcoming biennium the system could be further improved and expanded to build up a comprehensive data base on manufacturing companies and technical specifications of equipment in order to improve the servicing of member countries. Little information as of yet has been collected on basic technology in comparison with the more advanced and capital intensive machinery equipment, thus in the forthcoming biennium emphasis will be put on the collection of information on basic and intermediate logging transport and road construction equipment.

During the biennium, FOIl will develop a computerized forest harvesting production and cost information system which will enable the Branch to provide immediate data on productivity and costs of different logging methods for specific conditions. It will also serve as a source of information for prefeasibility and investment studies. The software developed will be later on available for member countries interested in this field. Production and cost data will be put in from FAO field projects and other sources of information. In addition the system can be used to determine wood production costs for a simulated given set up of input data, to identify raw material procurement costs, based on social, forest, terrain and environment conditions.

The Forest Logging and Transport Branch has produced a number of reports and publications which deal with various aspects and issues on wood harvesting. Publications are available either from the FAO Sales and Distribution Unit or from the Forest Logging and Transport Branch.

SESSION REPORT

A general paper on the subject of the session was presented by Mr Baldini, plus seven further papers on the following topics :

- two on harvesting machines and techniques (Curro-Verani and Morvan) ;
- one on harvesting and storage (Lisa) ;
- one on harvesting, storage and transport (Laufer) ;
- one on the energy obtained from the crop (Berti et al.) ;
- one on the techniques of using wood biomass in developing countries (Heinrich) ;
- one on the IEA R & D programme on forestry biomass harvesting methods (Mitchell).

The papers and subsequent discussions showed that the harvesting stage has received most attention hitherto. There are special machines and a great deal of experimental data for this stage. Present information indicates that semi-mechanized systems have a greater economic advantage, when the entire tree is skidded and chipped up along the forest tracks or in clearings within a radius of one kilometre.

Highly mechanized systems do not appear to be suitable, because of the existing structure of forestry ownership in the EEC countries, particularly in the private sector, where the average holding covers some 10 ha. It appears more appropriate, on both economic and energy grounds, to use a machine of limited dimensions and capacity or, for skidding, farm tractors suitably modified and combined with suitable appliances, such as winches, cranes, etc.

Further experiments need to be carried out on the use of biomass in the form of small branches. In addition, the need was stressed to restrict the amount of damage done by the machinery to the trees and the soil.

The drying and storage phases require more information and evaluation of various aspects, including the particle size of the chippings, drying techniques (including the energy balance considerations), degradation of the material during storage with a consequent reduction in the energy produced.

With regard to transport, which according to some sources accounts for 25-30 % of the total cost of harvesting biomass, more research is required on transporting the material both to the main collecting area and to the user, in order to determine at what point it becomes viable in terms of economics and energy to use vehicles and given loading methods.

Much appreciated were the paper on techniques of using biomass in developing countries and that on the IEA Programme on forestry biomass harvesting methods. Which showed, inter alia, the need for coordination among the various parties concerned, so as to optimize research and cooperation with the developing countries.

The papers pointed to the following recommendations :

The field of study and experimentation should be broadened to provide more information on aspects which have been neglected hitherto, and to cover the less important aspects, with a view to coordinating the whole process. With regard to further developments in mechanization, particular attention should be paid to evaluating the type of machinery used in relation to the costs of using the machines and the value of the biomass.

There should be more coordination with the other bodies which are active in this sector with a view to providing more information on the interaction between the different stages and optimizing research and cooperation with developing countries.

SESSION II - ENERGY CROPS

Chairman: A. STREHLER, *Technische Universität München*
Rapporteur: L. BRUNETTI, *Associazione Provinciale Olivicoltori Brindisi*

HARVESTING, DRYING AND STORAGE OF SHORT-ROTATION FORESTRY ENERGY CROPS

GERARD J. LYONS
Agricultural Institute, Carlow. IRELAND

1. INTRODUCTION

Motivation for development of biomass energy sources has, over the past decade, come both from the energy sector's vulnerability to fossil fuel imports and, more recently, from agriculture's search for non-food land uses and additional markets. At an early stage, short-rotation forestry (SRF) was identified as the most suitable biomass crop in Northern Europe, providing high yields of combustible dry-matter on relatively short (3- to 10-year) rotations. But while high conventional energy prices guaranteed the apparent economic feasibility of SRF during the 1970's, biomass energy is now an un-attractive option for most consumers at currently low fossil fuel price levels. Estimates of production cost for SRF wood chips are of the order 51 ECU per oven-dry tonne (in 1986 ECU), or 158 ECU per tonne oil equivalent (toe - of usable heat) (1), while heavy fuel oil and industrial coal can yield an equivalent amount of useful energy at 200 ECU/toe and 125 ECU/toe, respectively. The future economic potential for energy cropping may thus rely on either a. an increase in fossil fuel prices, b. subsidisation of energy crop production, or c. reduction of unit production costs, or a combination of these factors.

Acceptance of SRF biomass in commercial energy markets depends upon the producer's ability to consistently supply fuel, of an agreed specification (size, shape, moisture content, etc.), to consumers at a competitive price. The extent to which harvesting/transport costs influence the delivered fuel price is illustrated in Figure 1. These costs account for up to 73% of biomass production costs, and 75-80% of total energy inputs. Apart from the costs involved, methods employed in harvesting and post-harvest handling/processing of SRF biomass intrinsically determine the final fuel characteristics, and consequently affect its acceptability to energy consumers. The emphasis of SRF biomass research work in Europe over the past 10 years has been firmly placed on biological, genetic, and cultivation factors affecting crop productivity. Yet, any advances made in these areas will be meaningless in a competitive energy marketplace, unless strict compliance with users' fuel specifications can be ensured. With presently available techniques for harvesting and drying of SRF coppice, such specifications cannot be met; these areas thus represent the main drawbacks to forest energy commercialisation.

2. HARVESTING

The principal SRF species under investigation in North-European countries are Populus spp. and Salix spp. On a 4-year rotation, these produce multiple coppice sprouts (of basal diameter 25-40 mm) from each harvested stool. Branching characteristics and sprout proliferation are dependent upon species and clones grown. But branching is typically erect, with a small number of shoots in Populus clones, while Salix spp. produce a large number of smaller diameter sprouts in a curved, more

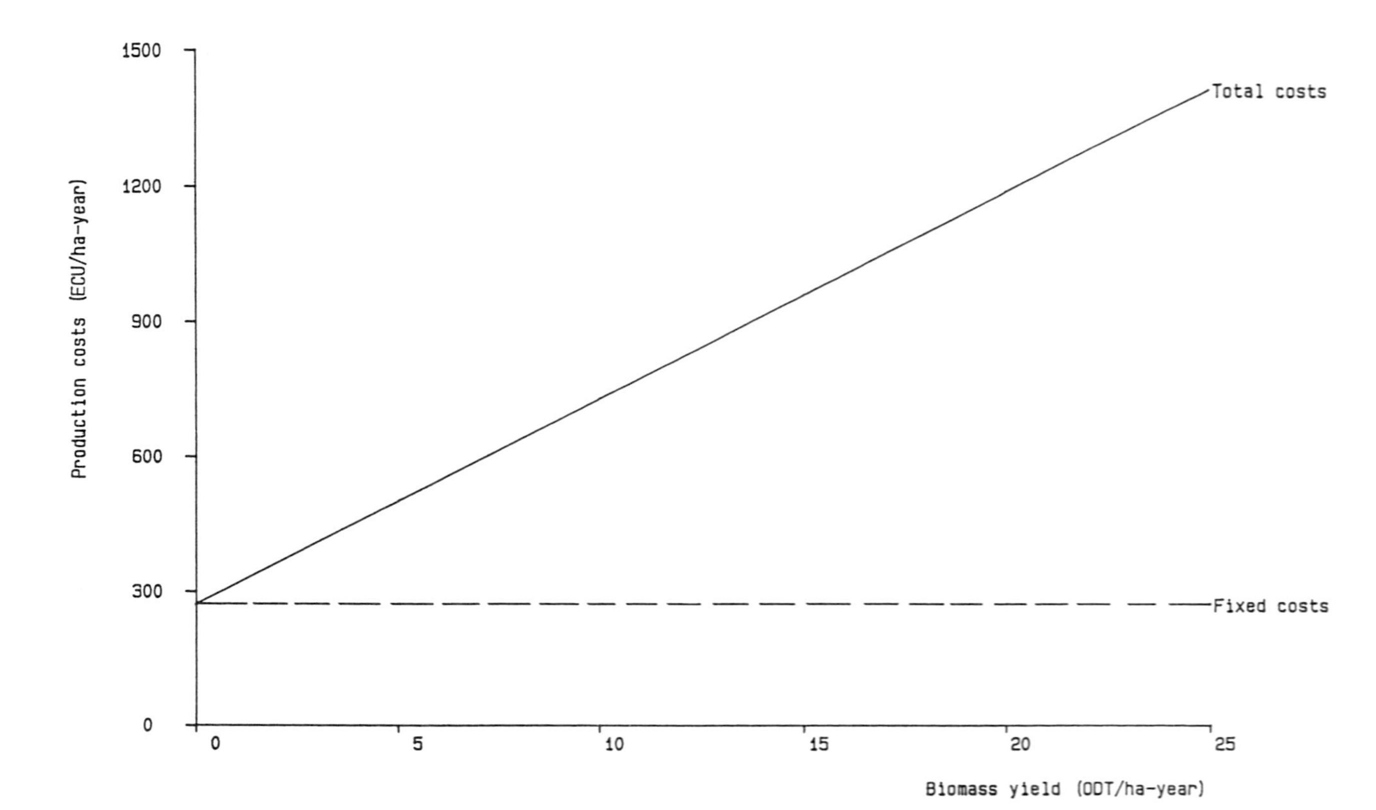

Figure 1: Production costs for SRF in Ireland (1986 ECU)

(Source: Based on Lyons, G.J. Ref. (7))

prostrate branching profile. Coppice yields recorded in Irish experimental plots are in the range 10-15 ODT/ha (oven-dry tonnes) annually, with planting densities of 10,000 trees/ha (and higher densities).

In most European countries SRF production studies have been limited to regions of poor agricultural potential. In France, however, high quality soils are being planted for provision of multiple products (sawlog, pulp/paper, and energy) on short (7-12 years) rotations. The sites most likely to become available for SRF production in Ireland are in soil categories 2 and 3. These consist of poorly drained lowland gley soils and heavy clay drumlin sites - category 2 - and cut-over peatland (category 3) remaining after industrial peat harvesting. On the category 2 sites, machine ground pressure must be less than 80 kPa and equipment is expected to operate on up to 20% gradients. While the peatland sites are uniformly flat, very low ground bearing capacity (10-12 kPa) necessitates the use of tracked harvesting machinery to avoid damage to the soil and root structure.

Several harvesting trials were conductred by the Irish Agricultural Research Institute using conventional felling equipment in densely populated stands of 3- and 4-year old coppice. Equipment tested included: chain saws, hedge cutter, feller-buncher, and tractor-powered disk and drum chippers. However, such discrete approaches to harvesting are tree-size dependent and require large diameter trees on high densities per hectare to operate efficiently and economically. Cost estimates for felling and chipping using such conventional systems were up to 31 ECU/ODT (in April 1986 ECU) in 1983 (1). On small farm-scale SRF plantations, the real cost of these conventional harvesting methods could be considerably less, if the opportunity cost of farm labour used in fuel harvesting was not accounted; that is, SRF harvesting for home fuel consumption could be conducted in the farmer's 'spare time', not infringing upon productive farm activities.

Two distinct harvesting strategies have been applied in development of purpose-designed SRF harvesters. These are: 1. Felling - collection - size reduction (to chips) - storage, and 2. Felling - bundling - storage of bundles. The final fuel product from these operations is pulpwood-type chips (usually 12-40 mm long and up to 15 mm thick). Such chips produced from SRF sprouts, using disk- or drum-type chippers, yield a high percentage of undersize particles and oversize twigs, which do not comply with fuel specifications desired for combustion and gasification conversion. Even the acceptable fuel chips are not well-suited to conventional combustion systems, as their close packing inhibits mixing of combustion air with the fuel. A noteworthy departure from a chipped product is the chunkwood concept developed by Arola (2), and now employed in a number of SRF prototype harvesters.

SRF harvester development work has been conducted in Ireland (that is, in the Republic of Ireland) by Bord na Mona (the Irish Peat Board) and the Irish Sugar Company. As part of the EEC/DGXII biomass energy programme, these companies designed, fabricated, and site-tested a prototype SRF harvester-chipper for large-scale biomass plantations on cut-over peatland sites. A fully-trailed harvester (of gross weight approximately 5 tonnes), it was drawn by a 98 kW tractor and fitted with tracks to provide an average ground pressure of less than 13 kPa. The prototype was designed to fell, collect, chip, and transport a payload of 2 tonnes to an end-of-row tipping site. The innovative chipping mechanism tested in the prototype, severed felled coppice sprouts at about 150 mm intervals, to provide fuel billets (10-40 mm diameter x 150 mm length) as an alternative to conventional pulpwood-type chips. While performance tests identified many conceptual and design problems, they also highlighted the special difficulties of harvesting/handling coppice material. These performance tests and the harvester development are described fully in papers by Keville (3,4).

3. STORAGE AND DRYING

One of the main deterrents to increased utilisation of wood as a fuel, is the characteristically high and variable moisture content of freshly-harvested biomass. Excessive moisture and fuel variability result in poor ignition, high volume of flue gases, reduced furnace capacity and low conversion efficiency; moisture levels below 30% (wet basis) are desirable for wood combustion and gasification. High moisture content also gives rise to serious problems in storage of chipped wood fuel; these include: self-heating, decomposition and dry matter losses, spontaneous combustion, and occurrence of allergy-causing microbial spores. A further effect of high moisture is its influence on biomass transportation and handling costs.

In order to sustain high biomass yield levels, harvesting of SRF plantations must be confined to the winter dormant season (November to March) of the species grown. However, to ensure continuity of supply to match the fuel demands of conversion plants, biomass stocks must be held in storage either by the consumer, in fuel yards or bunkers, or by the producer on the SRF plantation site, or both by the producer and consumer. Harvested biomass may be stored as either whole bundled coppice or whole-tree chips, chunks or billets, depending upon what harvesting strategy and equipment are employed. The form in which SRF fuel is presented for storage not only affects the cost of post-harvest handling and delivery, but it is critical to the mechanisms of natural drying and wood decay, which may enhance or reduce the final energy value of the biomass crop.

On harvest, SRF whole-tree biomass typically contains 45-55% moisture (wet basis) (5). Moisture reduces the heat energy potentially recoverable from SRF fuel combustion in two ways. Firstly, the initial gross calorific value (per unit weight of biomass harvested) of the wood is lowered by the presence of water, which does not contribute to the heat energy available. Secondly, combustion efficiency is reduced, because (i) heat is absorbed in the evaporation of water, and (ii) flame temperature and radiant heat transfer are lowered. As biomass yields are normally quoted on a dry weight basis, it is useful to illustrate the effect of fuel moisture (in the 'as-burned' fuel) on unit dry-matter (Figure 2). Thus for instance, if fuel moisture at harvest is 50% and biomass is allowed to dry naturally to 30% moisture, there is a usable heat gain of 1.6 GJ per tonne of dry-matter produced (6). If, on the other hand, fuel absorbs moisture during storage, a point will be reached where combustion can no longer be sustained by the dry matter present, and furnace 'black-outs' will occur.

There is ample industrial evidence, in North America and Scandinavia, of storage losses caused by wood decay, self-heating and spontaneous ignition in piles of pulpwood chips. However, the pulp and paper industry is not concerned with reducing the moisture level in pulpwood chips. On-going research at the Agricultural Institute, Oak Park, is investigating both the potential for natural drying of SRF biomass, and the mechanisms (wood respiration, fungal/bacterial colonisation, etc.)

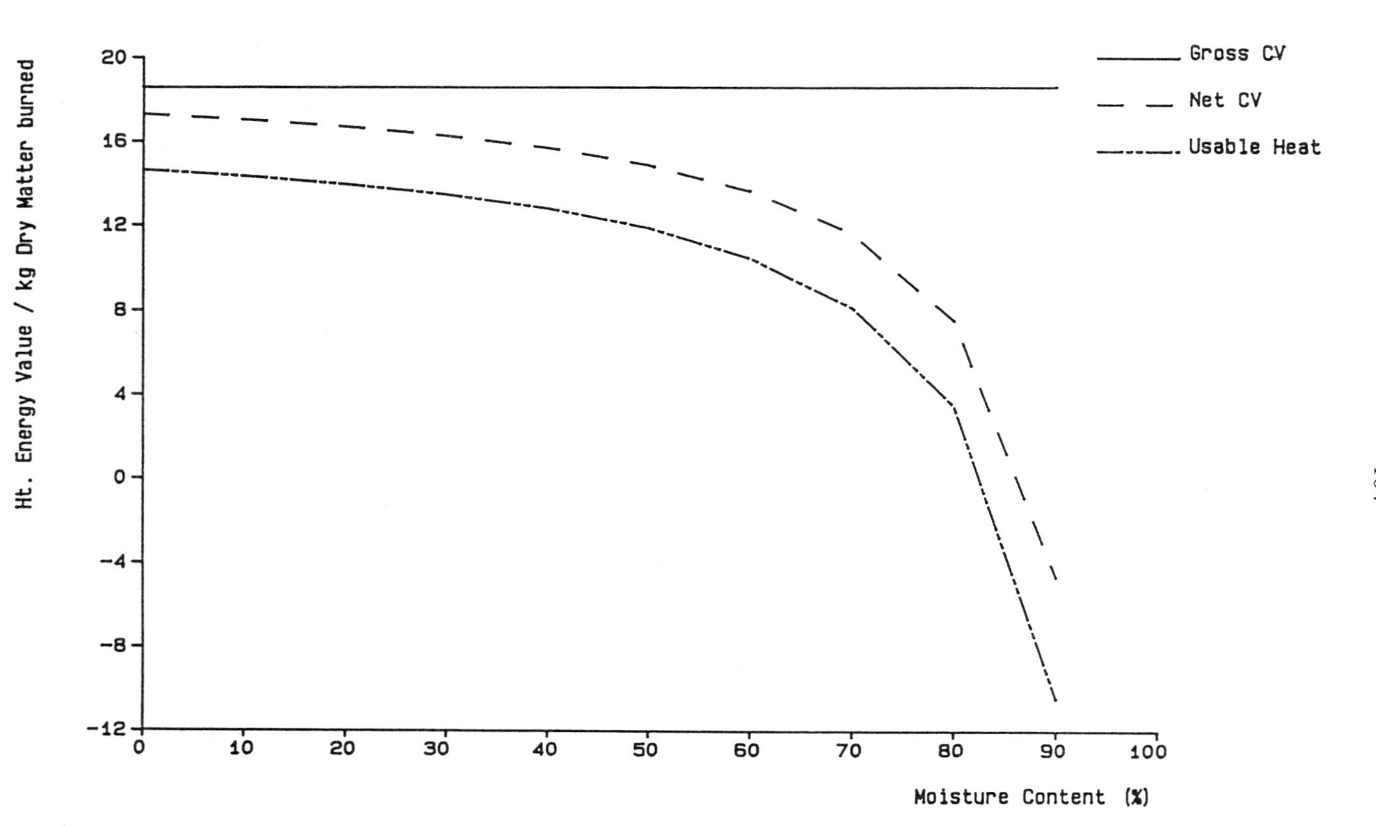

Figure 2: Variation of heat energy recoverable per unit dry matter (MJ/kg) of fuelwood with fuel moisture content

(Source: Lyons, G.J. et al. Biomass 8(1985) pp. 283-300) (6)

affecting fuel quality and value during storage of pulverised fuelwood. Theoretically, wood fuel should eventually approach an equilibrium moisture content (EMC) when exposed to given environmental conditions. Figure 3 shows the monthly variation of this theoretical EMC for Irish conditions, plotted with typical values for air temperature and humidity and potential evapo-transpiration. It would appear from this data that there is considerable scope for natural drying, even in the cool moist Irish climate. But natural drying relies on intimate mixing of ventilation (i.e. naturally circulating) air with fuel particles, and this is greatly inhibited in storage piles constructed from conventional whole-tree chips.

Figure 4 presents the measured moisture variation in 2 m-high piles of Fraxinus spp. coppice fuelwood chips, over a 90-day storage period in summer and early autumn. While average moisture content (MC) declined steadily over the first 30 days, re-wetting of the outer chip layers in early autumn again raised the average MC; it was noted, however, that chip pile interiors remained relatively dry even in winter, protected from re-wetting by the 'roof' effect of outer layers. Laboratory studies, designed to simulate conditions in large-scale piles of wood chips, have shown that respiration of freshly-harvested fuelwood promotes a rapid temperature rise in the first weeks of storage; this is later followed by bacterial and fungal colonisation. Measured dry-matter losses in these experiments were of the order 0.9-1.8% per month in storage (7).

As indicated in the discussion on harvesting, pulpwood-type chips are poorly suited to most combustion and gasification systems, because of their tendency to layer and form densely-packed fuel beds. This criticism also applies to storage of chipped fuelwood, and larger irregularly-shaped cuboidal or cylindrical particles, such as chunks or billets, would greatly enhance the potential for natural drying and should reduce respiration losses in storage. Different harvesting strategies could also contribute to better storage and drying conditions. Samples taken from bundles of whole-coppice Salix spp. and Willow spp. showed a 30% decrease in fuel moisture from date of harvest, in January 1986, to time of sampling, 9 months later in September 1986. Final moisture values were in the range 23-28%, which is an acceptable MC for most applications. This research work is continuing at Oak Park, Carlow, with support from the EEC/DGXII Non-nuclear energy R & D programme.

REFERENCES

(1) LYONS, G.J. Economics of short-rotation forestry energy plantation. In: Neenan, M. and G.J. Lyons, eds. The production of energy from short-rotation forestry. Report to the CEC on contract no. ESE-R-036-EIR(H). Commission of the European Communities EUR9959EN, 1985. pp 133-155.

(2) AROLA, R.A. ET AL. A new machine for producing chunkwood. USDA Forest Service. Res. paper NC-211, 1982.

(3) KEVILLE, B.J. Short-rotation forestry harvester chipper. In: Energy from Biomass, Proceedings of the EC Contractors' Meeting held in Brussels, 5-7 May 1982. Series E Vol. 3. D. Reidel Publ. Co. 1982.

(4) KEVILLE, B.J. AND E.J. DEVENISH. Short-rotation forestry harvester chipper. In: Energy from Biomass, Proceedings of the Workshop and EC Contractors' Meeting held in Capri, 7-8 June 1983. Series E Vol. 5. D. Reidel Publ. Co. 1983. pp 128-135.

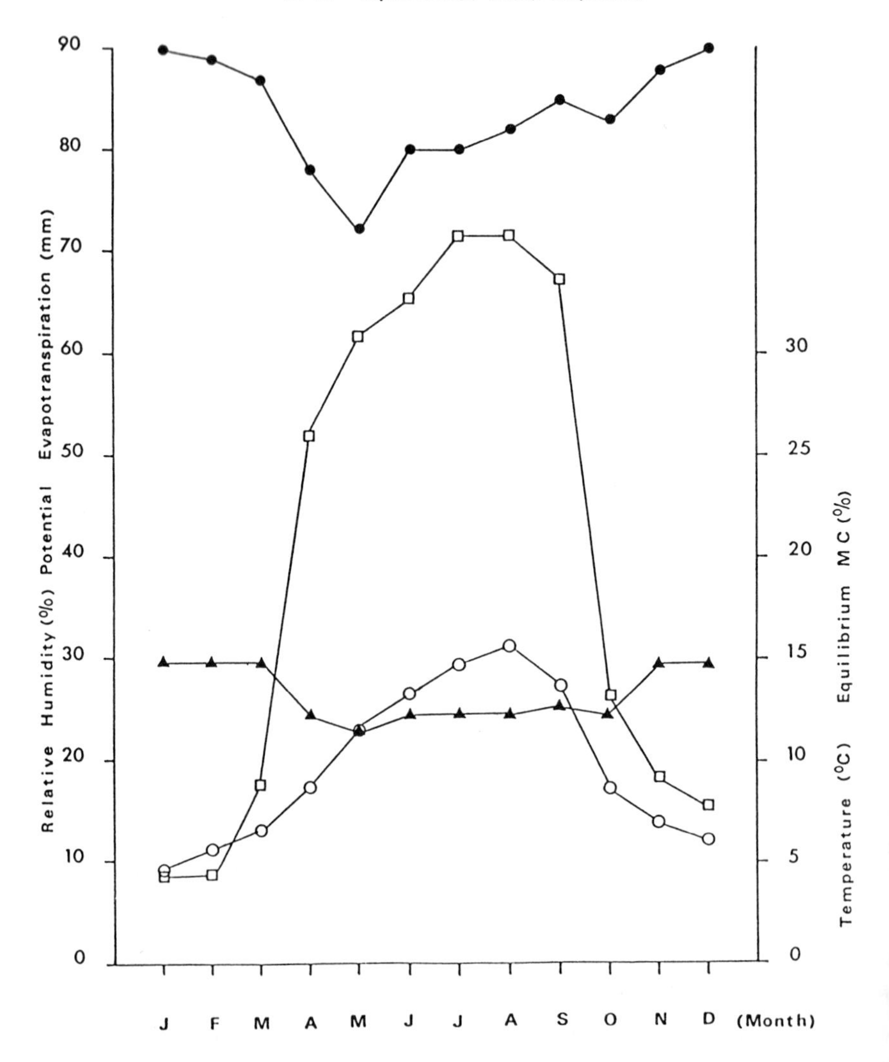

Figure 3: Seasonal variation of relative humidity, ambient air temperature, potential evapotranspiration, and wood equilibrium moisture content. Mean of 3 years' values, for Oak Park, Carlow.

(Source: Lyons, G.J. Ref. (7))

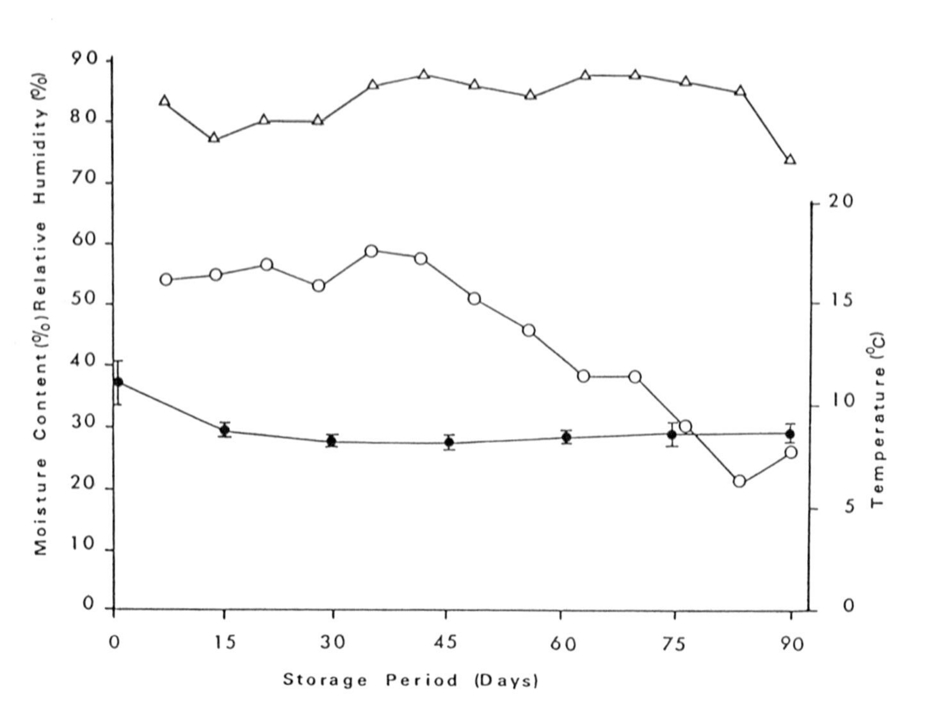

Figure 4: Summer-autumn storage. Average moisture content of whole-tree chips in experimental storage piles. Brackets denote standard deviation of seven samples. Mean weekly (7-day) relative humidity and ambient air temperature also shown.

(Source: Lyons, G.J. Ref. (7))

(5) LYONS, G.J., H.P. POLLOCK AND A. HEGARTY. Fuel properties of short-rotation hardwood coppice sprouts. Journal of the Institute of Energy. Sept. 1986. pp 138-141.

(6) LYONS, G.J., F. LUNNY AND H.P. POLLOCK. A procedure for estimating the value of forest fuels. Biomass. Vol. 8, No. 4. 1985. pp 283-300.

(7) LYONS, G.J. An assessment of technical, economic, and energetic factors affecting the development of short-rotation forestry as an energy source. Ph.D. dissertation, National University of Ireland, May 1984. pp 504.

INNOVATION IN THE MECHANIZATION OF BIOMASS HARVESTING

A. RIEDACKER
Forestry and Forest-Based
Industries Department of the Agence
Française pour la Maîtrise de l'Energie, Paris

Summary

The success of energy from biomass still depends on whether machines for harvesting biomass can be developed. However, present energy prices and recent experience call for some rethinking on the question of what form innovation incentives should take. Forms must be found which involve fewer risks for manufacturers who wish to continue innovating. Cooperation between technical research centres and manufacturers would seem to be one of the factors of success. Another is the amount and type of aid granted to firms.

1. INTRODUCTION

The price of timber biomass, whether standing or felled, is very often low, even nil. And yet at the furnace door these same products are sometimes so expensive that they cannot compete with other forms of energy. Thus all the participants in this seminar will doubtless acknowledge that it is essential to devise more efficient methods of harvesting biomass, and particularly machines.

If we are to achieve this end, we must identify how and where such machines can be designed and produced.

In the wake of the second oil crisis, the COMES (Commissariat à l'Energie Solaire), whose projects have been taken over by the Agence Française pour la Maîtrise de l'Energie, issued an invitation to tender to various manufacturers and technical research centres. This paper sets out to give a brief analysis of the projects which have been carried out and on the basis of them, to make recommendations for the future.

2. WHO SHOULD BE ASKED TO DEVELOP BIOMASS HARVESTING EQUIPMENT ?

Several types of candidates may apply when new machines are to be invented : technical research centres, industrial firms of various sizes, or even small craft firms. None of these applications should be rejected out of hand at this sounding-out stage, in which the net should always be spread as widely as possible. Experience shows, however, that technical research centres often produce very interesting work at the research stage, but they often find it difficult, when working alone, to follow through to the production of working machines.

Of the various projects which have received funding, only those carried out by firms - e.g. the CIMAF Scorpion project - have had any practical success. The others have had to be abandoned before completion or have proved industrially impracticable.

It would seem therefore that, if a project is to be successful, a fundamental requirement is for an industrial partner to be involved from the outset, at the design stage, and while the project is being carried out. This is a significant factor, but it is not always enough to guarantee success. The failure of the Nicholson harvester produced by an American manufacturer proves this, if ever proof was needed. This failure was not too damaging for the firm, however, since it appears it had considerable public funding.

It must be said that failure is inevitable where innovation is concerned ; the only hope is that the failure rate can be reduced by combining from the outset as many success factors as possible.

3. WHICH INDUSTRIAL PARTNERS ?

Only ten years ago there were far more firms producing, for example, forestry machines in the world or in Europe than there are today. Major concerns like Volvo and Renault have pulled out of this field. Quite a number of firms have gone into liquidation or have been taken over by the remaining concerns. This goes to show how fragile this industrial tissue is and how several years of effort may suddenly be jeopardized if a company goes into liquidation or is restructured.

Furthermore, in a generally depressed economic climate, inciting certain firms to innovate may bring about their collapse.
All these difficulties may well explain why only two of the biomass projects submitted to the EEC were for biomass harvesters.

It is therefore necessary to come up with new forms of innovation aid, failing which all appeals to mobilize innovative potential will probably remain unanswered.

4. COOPERATION BETWEEN FIRMS AND TECHNICAL RESEARCH CENTRES : A PROMISING OPTION

The know-how and the theoretical knowledge acquired in laboratories may indeed serve as a guide to innovation in industrial firms. In return, the involvement of the latter can prevent projects from being diverted towards machines which are impossible to manufacture or to sell.

5. HOW MUCH AID ?

In addition to this cooperation, and in view of the current energy market, it is essential to increase the aid granted to manufacturers if research in this field is to continue. Developing new machines is, in fact, very costly and, since there is no supporting market, is not an attractive proposition. It is likely, in fact, that very few realistic manufacturers could be persuaded to start designing new machines in the present climate with aid amounting to only 50 %.

But must we give up all effective research in this field ? We do not think so. Let us therefore try to make some suggestions. With things as they are, would it not be desirable to support innovation through loans covering up to 100 % of costs ?

A small proportion of this loan (10 to 30 %) could be repaid on the basis of sales of the whole range of machinery produced by the firm in question (or on its profits), while the greater proportion would not become repayable until the new biomass harvesters were sold, i.e. when the market became more favourable again.

Another means of motivating manufacturers would be a minimum sales guarantee once the new machines were successfully produced. This would, however, be more difficult to apply than the loan scheme referred to above, unless European pilot projects requiring biomass harvesters were launched.

Lastly, it would be desirable in certain areas if two or more countries were to pool their efforts, as in the case of biomass harvesters for the Mediterranean region. Would it not be a good idea in such cases to set up research with the above level of funding provided jointly by the EEC and specialized agencies in various countries ?

Unless there are reforms in this field, it will not be surprising if there is no advance at all in mechanized biomass harvesting until the next oil crisis hits us.

THE SCORPION : A FORESTRY TRACTOR FOR HARVESTING BIOMASS

Emile VAN LANDEGHERM - Gilbert GASQUET
Société CIMAF, Esternay, France

Summary

The Scorpion is a six-wheeled forestry tractor with hydrostatic transmission used, in particular, for harvesting biomass with the aid of the TRV and TRH harvesting heads. Fitted with the TRV it becomes the first ever combine harvester for forestry, capable of felling and chipping trees 15 cm to 20 cm in diameter. Fitted with the TRH it can harvest smaller-type scrub. It can also be used as a non-harvesting rotary slasher. The profitability of harvesting depends upon the degree of site incline, the quality of the biomass harvestable per hectare and its market value. The Scorpion's major attraction is its versatility, especially at a time of fluctuating energy prices.

1. INTRODUCTION

The Scorpion is a forestry tractor designed to carry out various types of work, in particular harvesting of biomass. It was developed by the Société CIMAF in collaboration with the Agence Française pour la Maîtrise de l'Energie, the European Economic Community, the Champagne-Ardenne, Provence, Alpes, Côte d'Azur and Corsica regions, Gard département, the Ministries of the Environment and Agriculture, and the Secrétariat d'Etat aux Risques Naturels et Technologiques Majeurs.

2. DESCRIPTION OF THE BASIC SCORPION UNIT

The basic unit is a forestry tractor - capable of carrying various tools - with the following main features :

- six side-driven wheels ;
- double-walled monobloc chassis serving as oil reservoir ;
- hydrostatic transmission ;
- six-cylinder 216 hp Mercedes engine, turbocharge option ;
- hydraulic pump rating 150 hp, in both directions ;
- continuous advancing speed of up to 12 km/h ;
- ability to manoeuvre on inclines of up to 100 % ;
- turning circle : own length ;
- length 4.5 m, width 2.10 m, height 3 m ;
- total laden weight 9 tonnes ;
- weathertight, air-conditioned cab ;
- all parts accessible ;
- very small ground-contact area ; option of fitting dual wheels or caterpillar tracks ;
- wheel measurements : 400 x 22.5 - 500 x 22.5 - 600 x 22.5 ;
- variant : Scorpion with caterpillar tracks (US type).

3. TRACTOR ACCESSORIES

The tractor can be equipped with several tools, thus making it a multipurpose vehicle :

A) Rotary slasher

Fitted with a conventional rotary slasher the Scorpion can carry out most scrub clearance work (not involving biomass harvesting), operate under electricity cables or create firebreaks in Mediterranean forests (Photo No 1). We shall compare the rotary slashing option with that of biomass recovery later on.

All existing rotary slashers can be adapted to fit the Scorpion. Cutting diameter 0.25 m max., cutting length from 1.80 to 2.50 m.

The GIROCIMAF rotary slasher, built by CIMAF, is a vertical-axis stripper equipped with two moving blades mounted on a disc.

B) The TRV-CIMAF vertical recovery head

This head (Photo No 2) is manufactured by CIMAF and consists of two discs equipped with two blades mounted on two variable-speed rotating cones (forewards and backwards). A drum located behind the cones guides the vegetation to a cutter producing chippings for immediate use in wood burners. These are collected in a 4 - 7 m^3 hopper - located on top of the Scorpion - which can discharge at a height of up to 4 m (Photo No 3).

Scorpion and recovery head : - overall length : 7 m
- overall width : 2.10 m
- total weight : 14 tonnes

This patented recovery head can be used on coppice between 15 cm and 20 cm in diameter, and is the first ever "combine harvester for forestry".

Fitted with this head the Scorpion can harvest scrub in the Mediterranean region, coppice in coppice-with-standards (Photos Nos 2 & 4) and trees in initial thinnings (Photos Nos 5 & 6).

The chips thus harvested can be discharged into a container or at the edge of the plot, and in the Mediterranean area they dry naturally when piled into heaps.

C) The CIMAF-TRH horizontal recovery head

Harvesting slash or scrub not containing large stems simply entails fitting the Scorpion with a TRH recovery head ; this has the advantage of being much lighter than the TRV.

Description

The cutter consists of a rotor driven by a hydraulic engine and equipped with rows of moving blades.

The recovery unit consists of :

- an extractor screw with special thread ;
- an expulsion turbine, with interchangeable blades, connected to a discharge tube.

The screw and turbine are driven by the same hydraulic engine.

The TRH head can be fixed to any vehicle with a hydraulic power rating of 90 to 150 hp.

The chips obtained can be :

- spread evenly on the ground, or
- collected in a storage hopper.

When the head is fitted to the Scorpion the hopper is an optional extra (Fig. 1).

It should be possible to adapt this recovery head to fit tractors other than the Scorpion. Further studies are necessary for this, however.

D) Other tools for use with the Scorpion :

- CIMAF-B.T. extendable boom

This can be fitted with a number of tools : cutting nippers, grab, lopper, saw, skidding cable, etc. All existing cranes can be adapted to fit the Scorpion.

- CIMAF-TAR borer

This borer has a very high level of performance : 140 holes per hour.

A special device avoids compaction of the drillhole sides, thus allowing seedlings to establish quickly.

The Scorpion can work on steep inclines inaccessible to conventional vehicles.

Existing borers can be adapted to fit the Scorpion.

4. ECONOMIC ASSESSMENT

Long-term tests in Corsica (see the Amandier report), and the Var, Mediterranean and Champagne-Ardenne regions have made it possible to gauge the economics of using the various versions of the Scorpion.

Fig. 2, based on sites in the Var, shows the daily area covered by a Scorpion working on various inclines solely as a rotary slasher or with a vertical recovery head. It is quite clear that recovery slows down the tractor's workpace in all cases. When working as a rotary slasher the Scorpion can cover between 1 and 4 ha/day on inclines of under 50 %, whereas for the harvester version this varies between 0.4 and 1.2 ha/day.

Thus, if scrub exploitation is not possible, or if inclines exceed 30 %, slashing only is preferable.

If the incline is under 30 % and there is a market for the chips, the result clearly depends on the quantity of product harvestable per day (expressed here in tonnes of fresh matter) and the market price per tonne for chips (Fig. 3).

In Zone A, with a low daily harvest rate - either because the vegetation is sparse or because the inclines are too steep - it is preferable not to harvest the biomass. In Zone C utilization of the biomass not only pays for the harvest but also generates a small profit. In Zone B, situated between the two others, biomass harvesting can lower, but not completely cover, the scrub clearance costs.

Fig. 4, based on an average clearance cost per hectare of FF 12 000, shows how the price per therm of roadside chips varies according to daily biomass harvest rates. Thus, for a site with a clearance cost of FF 12 000 per hectare, a harvest rate of only 10 tonnes/day means the roadside chip price must exceed FF 0.40 per therm if scrub utilization is to cover the total cost of clearance. If the market price of the chips is only FF 0.20 per therm, scrub clearance will cost about FF 7 000/ha.

On the other hand, if the biomass harvestable amounts to 40 t/day, i.e. in places where vegetation is denser and the incline not so steep, a selling price of FF 0.11 per therm for roadside chips would cover clearance costs.

Selling wood chips at FF 0.10 per therm entails per hectare clearance costs of :

FF 1 000/ha at a harvest rate of 40t/day ;
FF 3 500/ha at a harvest rate of 30t/day :
FF 6 000/ha at a harvest rate of 20t/day.

These variations underline the importance of :

- the market price for roadside chips, i.e. the influence of the price of rival energy sources on the competitiveness of biomass harvested using the Scorpion ;
- the type of site, in particular the amount of biomass harvestable per hectare and the degree of incline ;
- the importance attributed by the forester to the various kinds of operation, because although harvesting might greatly reduce the cost of operations, it will cover the full cost only under very favourable conditions.

If the forester's aim is simply to lower the cost of his forestry operations, he should have no great difficulty in finding suitable sites.

However, if he wants to carry out such operations without spending money, he needs to be more selective in choosing sites.

The Scorpion's major attraction is that it is not restricted solely to harvesting biomass. Initially it can be used with the other tools, doing without the harvesting head until the market for wood chips picks up.

Figure 1

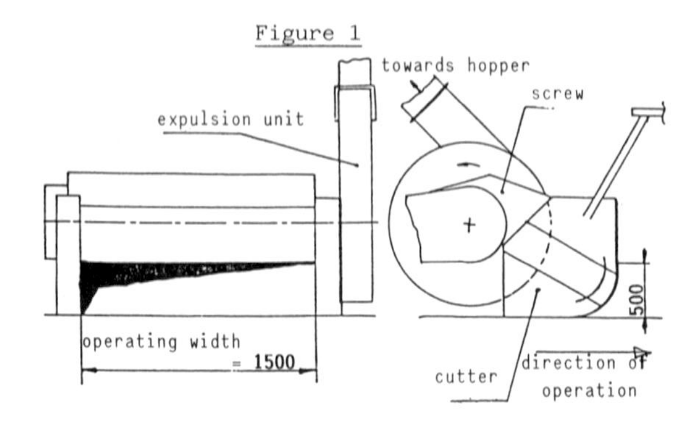

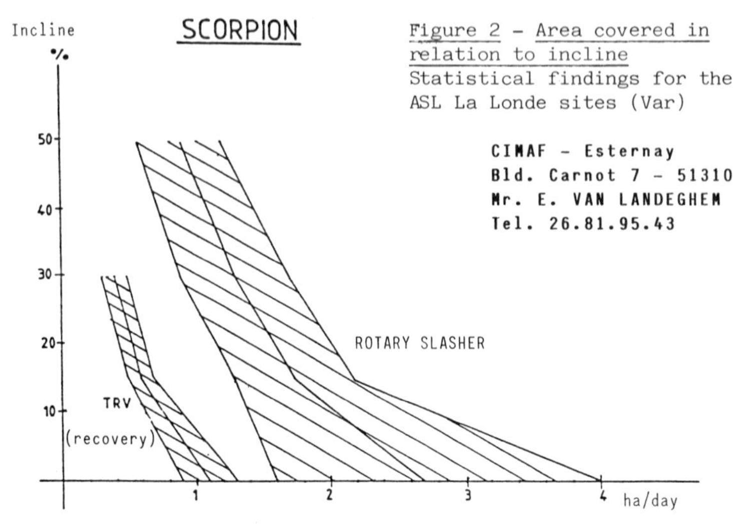

Figure 2 - Area covered in relation to incline
Statistical findings for the ASL La Londe sites (Var)

CIMAF - Esternay
Bld. Carnot 7 - 51310
Mr. E. VAN LANDEGHEM
Tel. 26.81.95.43

Figure 3 - Variations in profitability threshold related to market price for chips

Comparison with conventional rotary slashing

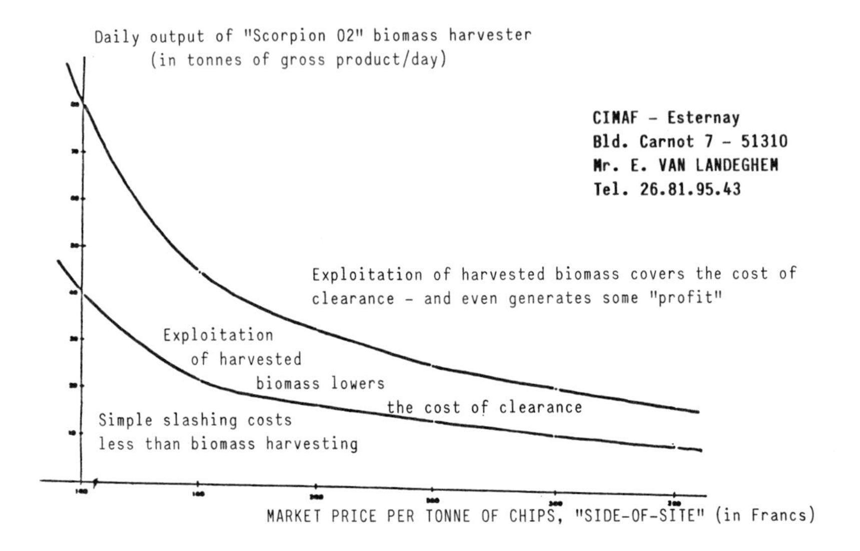

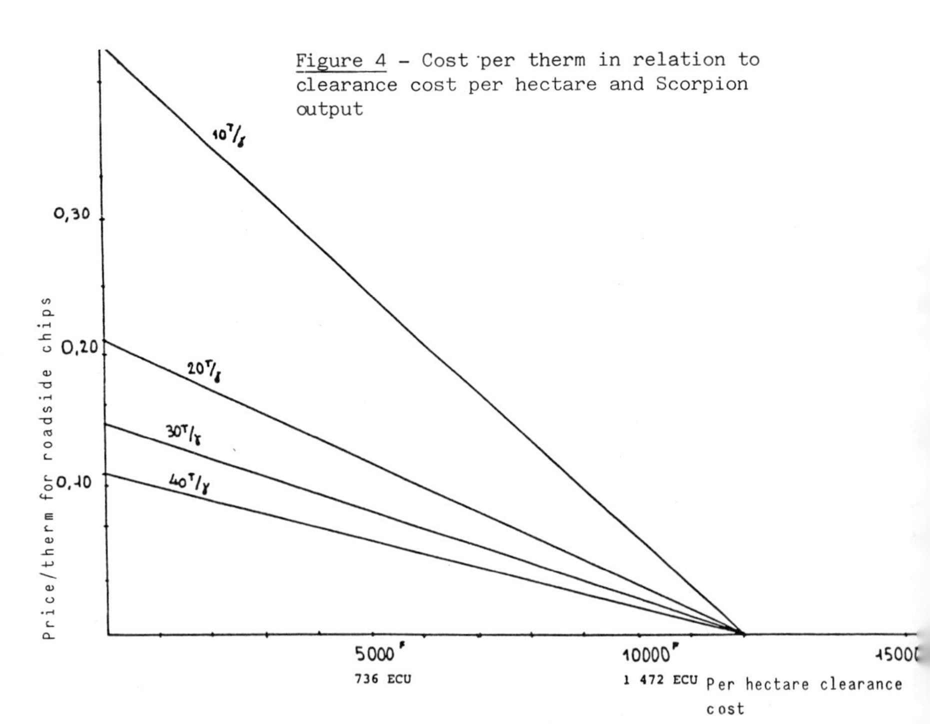

Figure 4 - Cost per therm in relation to clearance cost per hectare and Scorpion output

Photo 1 - A Scorpion fitted with a rotary slasher, in the Mediterranean region.

Photo 2 - A Scorpion fitted with a TRV vertical harvesting head and a chip transport hopper, in coppice-with-standards.

Photo 3 - Chips being discharged into a container.

Photo 4 - Coppice-with-standards after Scorpion has harvested coppice.

Photo 5 - A Scorpion harvesting stems of initial thinnings in a softwood stand (front view).

Photo 6 - Tree harvest line in a softwood stand, after the Scorpion has passed by.

BIOMASS HARVESTING TECHNOLOGY AND ENERGY BALANCE

Prof. Dr. ir. P.F.J. ABEELS
Université Catholique de Louvain
Louvain-la-Neuve, Belgium

Summary

The harvesting of the available biomass in agriculture and forestry is a kind of bottleneck carrying many consequences for energy recovery. Most of the time biomass production and utilization are analyzed neglecting necessary harvesting techniques. Since 1984-85, some distinguished organizations have studied the harvesting conditions of biomass in detail. This paper reviews those contributions technically and for economic and energy balance. The range of raw materials products, equipments and processes does not allow an easy digest. Fundamentals and some examples are given in order to put much emphasis on the necessity to deepen the knowledges in the area and to innovate if possible. Biomass is a renewable production that stores energy in the world. Much as energy is the condition of the progress, we must put the needed efforts to use it in the best way possible.

NOTION OF BIOMASS

Biomass implies "all forms of plant and animal material, grown on land in or on the water, and substances from biological growth" (Hall, 1981). This paper involves only cultivated plant and forest products including the main residues they bring along.

"The economic feasibility of utilizing residues is sensitive to the type of residue, the requirements for collection, processing, transportation and storage, the technology employed to perform these activities, the type and scale of the conversion facility, costs to mitigate adverse impacts and many other factors (Jenkins & Summer, 1986).

SIGNIFICANCE OF BIOMASS FOR ENERGY

The plants deliver commercial and marketable products, residuals and residues. When portions are of industrial interest or used directly in mills they must be considered as marketable. The residuals are left in the field after harvesting. The residues are parts removed and/or left from the harvested products at the processing sites or mills.

TABLE 1 - TYPES OF BIOMASS

Type of biomass	Agriculture	Forest
Conventional and merchantable	Grains, leaves, fuits, roots, ...	Stems
Residuals	Stalks, roots, prunings	Tops, branches, foliage, stumps, roots
Residues	Cobs, stalks, straw, hull, shells, etc...	Tops, branches, butts, slabs, bark, ...

Biomass has therefore different significations for the harvesting and processing technologies.

TABLE 2 - PROPORTIONS OF THE COMPONENTS IN BIOMASS

Cereals, corn		Conifers			Leafwood	
Portion	%	Portion	%		%	
Grain	40	Tops	2-10		5-15	
Cobs	7-8	Branches	5-15		10-20	
Husk, shells	2-3	Leaves	2-10		5-10	
Straw		Stem (+bark)	60-70		50-60	
Stalks	40		-----		-----	
		Total		75-85		70-85
Roots	10	Stumps	5-10		5-20	
		Roots	5-15		10-20	
			----		-----	
		Total		15-25		15-30

These components of the biomass involve selection procedures. For example, if we consider a conifer tree it delivers :

Bark free log to 10 cm top	58-65 %
Stump + roots	16-18 %
Stembark to 10 cm top	12-14 %
Top and branches with bark	8-10 %
Needles	4-5 %

and, in the bark free log the following proportions of components are recognized at the mill :

Lumber recovery from debarked log	45-68 %
Chips	20-30 %
Saw dust	20-25 %

If lumber must be considered as marketable, chips are marketable too. If not, they are left as residuals, or residues. Finally, at least 60 to 70 % of a full tree with stumps and roots are available for non lumber products. The next transformations of lumber will again leave residues.

The availability of the biomass is dependent on the time in the year. There are always growing and harvesting periods. An all year round utilization of raw biomass implies a storage. It introduces risks of degradations due to the biological nature of the products. The more of the biomass available to feed living organisms the more bio-degradation can occur. Some precautions are necessary.

TABLE 3 - AVAILABILITY OF BIOMASS FOR ENERGY

Type — Months in the year (temperate climate)

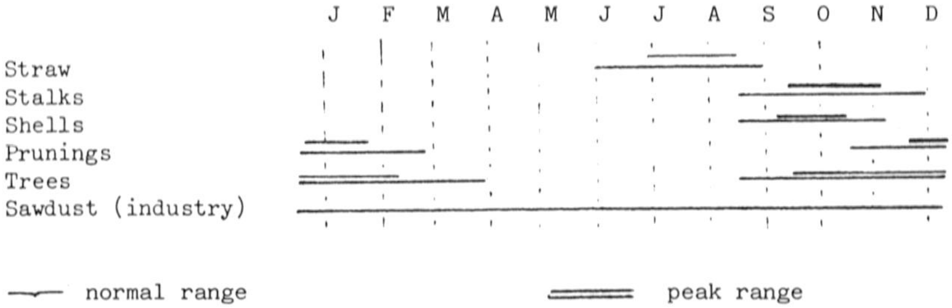

J F M A M J J A S O N D

Straw
Stalks
Shells
Prunings
Trees
Sawdust (industry)

—— normal range ═══ peak range

Due to storage necessities, biomass must be handled at an appropriate moisture content in order to avoid storage problems like spontaneous combustion or excessive degradation.

Considering agricultural and forest residues, the main heating values are low by comparison with fossil fuels.

TABLE 4 - HEATING VALUES OF PRODUCTS

Agricultural crops residues :	
Barley straw	17.31 MJ/Kg
Wheat straw	17.51
Corn stalks	17.65
Walnut shells	20.18
Grape prunings	19.35
Forest biomass for energy :	
Conifers :	
Tops, branches	18.8-19.8 MJ/Kg
Stumps, roots	18.8-19.2
Leafwood :	
Tops, branches	19.5
Sawdust	18.8
Fossil fuels :	
Peat	16.4-22.08 MJ/Kg
Coal	33.5-35.40
Fuel	44.16
Crude	41.83

From these values the following statements are possible :

1. Energy availability from biomass is different from fossil fuels
2. Material characteristics are different
3. Energy valorization of biomass is different from that of fossil fuels.

Notwithstanding these heating values we must consider that other ways for the production of energy from biomass must be suggested. The following data underline such possibilities when biomass is transformed into dust.

TABLE 5 - EXPLOSIVE ENERGY CONTENT OF BIOMASS

Nature of dust	Min. explosive contents in air g/m^3	Max. pressure of explosion kPa	Max. speed of increase in pressure kPa/sec
Cereal	40	871	28371
Sawdust	20	800	60795
Cocoa	35-65	466-871	8511-29384
Hay	200	445	4255
Corn	55	800	42556
Rice	45	740	19251
Peat	45	740	31410
Coal	55	588	16212
Asphalt	25	628	32424

Beside conventional combustion we must recognize that the control and the management of the explosive potential of biomass dusts represents a potential source of energy recovery.

HARVESTING TECHNOLOGIES FOR BIOMASS

1. The first and most typical technology applied to biomass is firewood production at the individual domestic level. Nowadays the feller has left the axe for the motor chain saw. Felling is followed by cross-cutting executed also with a chain saw in most cases but sometimes by circular saws on an appropriate bench. Afterwards, the billets are split and piled manually. Splitting is mechanized too. The firewood process involves :

   a. Felling, delimbing and cross-cutting in the forest ; productivity — 1.5-2 h/steres
   b. Splitting — 1.5-2 h/steres
   c. Manual piling in the forest — 0.3-0.5 h/steres

   The disposal at the farm or at the house implies other handlings and transportation.

2. Industrial production of firewood for sale to private or enterprise customers can be examined like FOLKEMA did it in the FERIC manual nr. 6. Two levels are examined. The first one implies a total annual production of 150 to 1500 cords/year. The process involves a felling team and a processing plant. Felling happens together with the normal harvesting of the forest. The processing team operates a pole -, a pulpwood frame - or a platform trailer truck with crane together with mechanical cross-cutter and hydraulic splitter. Other vans or trucks will deliver the firewood to customers. Large companies handle over 1500 cords/year. As a whole, the procedure is the same as for the small units but equipments are more numerous and larger.
   Firewood production is sensitive to the market prices of the domestic fuels.

3. The second procedure of harvesting biomass for energy concerns straw and stalks. The cutting of the cereals or corn requires a cutting bar or disc. The straw or stalks left after combine-harvesting are picked up, pressed and baled before handling and transportation. Few farms are equipped for burning straw and stalks due to the much greater interest in these products for farm operation.

4. The industrial use of biomass for energy must be correlated with the supply of chips of adequate shape and correct moisture content. The chipping happens in the forest stands or at the mills. The first operational modes were organized in the same way as for the firewood processing. Then instead of splitting the billets or the logs, the chipper devours the raw material. New developments are marketed now. They offer a chipping-combine for operations in the stand where the felled trees are directly fed to the grinding tool with storage of the chips in an integrated or accessory container to the machine. When the container is full, it is discharged at a short distance and tipped up or ensiled after which the chips are reloaded and transported to the mill.

THE DESIGN OF HARVESTING EQUIPMENTS

First, the collection and processing of the raw material involves different techniques among which the technologies of cutting are the most frequently utilized. The felling of the tree, the delimbing, the cross-cutting and the chunking or the chipping use appropriate cutting processes together with infeed and removal devices. The tools must be fitted on mobile or stationary machines and replaced at appropriate locations. Tools and working conditions have the most important role in the energy requirements. The main comminution principles are reviewed by Pottie and Guimier in the FERIC special report no. SR 2, 1985 (IEA Cooperative Project CPC3). The fundamentals of cutting are described by Peter Koch in Wood Machining Processes.

The suggested review of the comminution means are :

a. the disc chipper
b. the drum chipper
c. the hammer hog
d. the shredders
e. the jaw crushers
f. the chunkers with either a spiral cutter, an involuted disc (single or double), a wing blade or a saw auger.

Cross-cutting is obtained by shearing, by circular sawing or by a chain saw. The splitting is due to the shear action of a knife, of a spiral screw or of crushing rolls when logs and billets are pressed and pushed through. Several different infeed and outlet devices are combined with these tools. Bands or belts with rollers are used for the infeed systems while bands and augers are placed at the outlets.

Transportation over short and long distances requires vans and containers in accordance with the desired productivity, the transfer necessities and the highways regulations.

When chipping is combined with densification the raw chips must be at a convenient moisture content. The interest of densification processes is related to the shape and the standard size of the product for the feeding of incinerators and furnaces.

HARVESTING IN INTENSIVE PRODUCTION OF BIOMASS FOR ENERGY

Short rotation intensive culture mainly of woody biomass has developed special harvesting technologies. Most of the marketed equipments are of the combine type i.e. an integration of felling, comminution and forwarding devices. The same fundamentals are at the origin of designs for other recovery systems in biomass production. These energy plantations involve few tree species and are comparable with the harvesting of agriculture crops like stalks or standing straw.

MICROECONOMY OF BIOMASS HARVESTING

The microeconomy of the harvesting of biomass involves the costs for the disposal of certain types and quantities of material for energy production purposes. The paper now examines different typical cases.

a. The manual harvesting of firewood

The felling and/or the manufacture of firewood like splitted billets at the individual domestic or farm level is evaluated in the following data:

- machine costs for :	felling	0.10-0.15 ECU/stere
	manufacturing splitted billets	0.25-0.30
	transportation (3-5 km max)	0.60-1.40
	Total	0.95-1.85

- manpower is generally not accounted for in individual domestic uses. If taken into account for all operations and for conditions such as 3.5-4.55 h/stere and a salary of 7 ECU/h, it represents 24.5-31.9 ECU/stere.

- total (0.95 to 1.85) + (24.5 to 31.9) = 25.5 to 33.8 ECU/stere, or an average of 30/500 = 0.06 ECU/kg air dryed or 19 MJ i.e. 0.003 ECU/MJ, let us say : 0.005 ECU/MJ (price of wood not included).

b. The harvesting of straw or stalks

Cutting is operated by grain combines and corn harvesters. They leave felled straw or stalks on the field. The picking-up, baling and delivering costs are as follows :

tractor + operator and pick-up-baler	0.011 ECU/Kg
tractor + operator and self-loading trailer	0.018
transport field to farm	0.003
Total	0.032 ECU/Kg

Such a total brings the costs at 0.002 ECU/MJ or let us say 0.004 ECU/MJ for general conditions.

c. Various forest biomass harvesting methods are possible depending on the forest conditions and silvicultural mode of treatment. Again Pottie and Guimier have published a condensed report from which values are extracted.

- First condition : harvesting from thinnings by fully mechanized full tree feller-buncher operating in conjunction with a grapple skidder for row
  thinning = 10.4 - 19.6 ECU/t or 5 to 10.10^{-4} ECU/MJ
  row + selective = 17.1 - 26.2 ECU/t or 9 to 13.10^{-4} ECU/MJ
  selective = 10.4 - 29.9 ECU/t or 5 to 16.10^{-4} ECU/MJ

- Second condition : semi-mechanized tree section thinning with motor manual felling and boom grapple saw forwarder costs are :
  31.7 - 36.9 ECU/t or 17 - 19.10^{-4} ECU/MJ

The same method, but with feller-buncher and grapple saw forwarder changes the costs to : 30.8 - 35.8 ECU/t or 16 - 18.10^{-4}ECU/MJ.

- Third condition : integrated forest biomass harvesting with semi-mechanized whole tree chipping in a manually fed chipper-forwarder
  - in row thinnings : 23-42 ECU/t or $12\text{-}22.10^{-4}$ECU/MJ
  - in row + selective : 27-56 ECU/t or $14\text{-}29.10^{-4}$ECU/MJ

  The same method but with crane fed chipper-forwarder will cost
  - in row thinnings : 17,5-34,2 ECU/t or $9\text{-}18.10^{-4}$ECU/MJ
  - in row + selective : 14,2-34,2 ECU/t or $7\text{-}18.10^{-4}$ECU/MJ
  - in selective : 19,2-24,2 ECU/t or $10\text{-}13.10^{-4}$ECU/MJ

- Fourth condition : semi-mechanized whole tree chipping after manual felling and hauling to skid road where chipping is done by a grapple fed chipper forwarder, selective thinning with opening of strip access ways in the stand, then the costs are :
  50-100 ECU/t or $26\text{-}52.10^{-4}$ECU/MJ

- Fifth condition or fully mechanized whole tree chipping by a feller - chipper - forwarder, where the range of costs are :
  64-128 ECU/t or $34\text{-}67.10^{-4}$ECU/MJ

The costs of complete tree harvesting are in the range of :
23 to 45 ECU/t or $12\text{-}24.10^{-4}$ECU/MJ

The above review shows that the costs for the supply of chips for energy at the road side is assumed to be within a range of 0.0005 to 0.0067 ECU/MJ while straw is delivered at the farm for 0.004 ECU/MJ and classical firewood for 0.005 ECU/MJ.

MACROECONOMY OF BIOMASS HARVESTING

The fundamental data about the energy value of the equipment used in the processing of biomass harvesting are not found in the literature. Different assessments are necessary about the manufacture of machinery and also for the running of the equipment. These determinations are established systematically with the information available everywhere about industrial energy aspects.

The energy inputs for the manual harvesting of firewood can be related to the following quantities for the processes involved :

felling, delimbing, cross-cutting	4.52 MJ/stere
splitting	4.20 MJ/stere
manual piling in the forest	1.05 MJ/stere

or a total of 9.77 MJ/stere. Manual transportation and handlings must be added for a total of 4.2 MJ/stere minimum. Road transport represents about 17.8 MJ/stere for an average distance of 5 to 10 km between forest and delivery area. The full process needs about 31.77 MJ/stere or 0.064 MJ/kg of wood.

When felling is followed up by mechanical hauling and splitting phases, the total consumption energy rises to 0.141 MJ/kg. Thus less energy is used to harvest and deliver than is found as the potential of the delivered product : 19 MJ/kg.

The harvesting of products like straw or stalks is considered following the data given by Jenkins and Knutson (1984).

The inputs of fossil fuel energy for collecting and preparing 1 t of rice straw are established as follows : windrowing : 125.5 MJ/t, baling = 149.9 MJ/t, roadside piling : 117.2 MJ/t, loading : 59.6 MJ/t, transport =

286.3 MJ/t, storage = 59.6 MJ/t, comminution : 450.7 MJ/t or a total of 1243.8 MJ/t. So 1.2 MJ/kg is necessary to give a potential of 16 MJ/kg in a power plant for example. The fuel for the operations is diesel (74 %) while 26 % was embodied in lubricating, manufacturing equipment, repairs and labour.

The semi-industrial and the full industrial biomass harvesting systems have received data from Pottie and Guimier (FERIC) but the reported values must be completed because the manufacturing of the machines and the manpower involved are not included. The following energy need determinations have been attempted for the different harvesting methods.

The semi-mechanized thinning delivering tree sections by manual felling plus bucking and loader-forwarder needs 0.068 MJ/kg. The same process fully-mechanized requires 0.085 MJ/kg. Splitting and/or chipping will happen after transport at the mills or plants.

The fully mechanized thinning with a tree feller-buncher and a full tree grapple skidder requires the following average energy quantities when operated in row thinnings : 0.044 MJ/kg, while in mixed row + selective thinning it needs 0.067 MJ/kg and in selective thinning the records mention only 0.051 MJ/kg.

Integrated forest harvesting with semi-mechanized whole tree chipping provides prepared biomass for conversion or burning. The main forms of products delivered are chips or chunks. Such integrated processings have the following approximate energy requirements :

a. When the chipper is manually fed :
   in row thinnings = 0.044 MJ/kg, in mixed thinnings = 0.055 MJ/kg

b. When the chipper is crane fed :
   in row thinnings = 0.057 MJ/kg, in mixed thinnings = 0.049 MJ/kg
   in selective cuts = 0.052 MJ/kg.

The semi-mechanized whole tree chipping at the road side of the harvested stand by motor manual felling and hauling by winch will require energy in a range of 0.035 to 0.045 MJ/kg because chipping is rationalized in such operations.

The fully mechanized whole tree chipping by a feller-chipper-forwarder will require from 0.023 to 0.046 MJ/kg to store chips at the road side and have them ready for transport to the mills.

CONCLUSIONS

The energy crisis of the seventies has generated high interest for renewable sources of energy among non-marketable agricultural and forest by-products, residuals or residues. The main obstacles to the use of biomass in modern heating systems must be related to technical, economical and social or ergonomical factors.

The easiest and most convenient energy sources like fossil fuels and electricity are multivalent, have high energy values and are suited for a wide distribution.

The countries with an economy highly related to wood production and transformation have mechanized the harvesting of marketable trees since the fifties and sixties. For those countries it was only one step more in the mechanization process when they improved biomass harvesting. Those countries have a strong industry related to forestry. Other parts of Europe have opposite conditions where a real gap exists between the forestry and wood industries, moreover the harvesting contractors must fill this gap. Farm crop residuals have been always harvested. Most of them have utilized conventional recovery possibilities before using residuals as energy sources. Of course, forest and agricultural residuals are still used in the classical technologies for energy recovering. Some adaptations have been realized but no major changes are evident. Therefore, a fundamental question can be raised about other energy recovering possibilities like controlled dust explosions, because dusts are important technological residues in farm and forest product manufacture.

The establishing of an energy balance is an excellent opportunity, for a statement about the macro and micro-economic aspects involved in the harvesting of biomass. Only two kinds of raw materials are considered in the paper : straw and wood.

The macro-economic study of the harvesting alone, i.e. without preparation necessities for the delivery of a product at low moisture content or highly densified (compacted) and without transportation over long distances, shows a rather low to very low energy consumption. Particularly when we consider the portion of high value fuels in the processings.

We must admit that the harvesting of biomass is energetically interesting and that it offers financial and social possibilities for investments, employment and improvement.

The micro-economic balance of biomass harvesting must compete with the prices for other industrial or domestic fuels at equivalent energy values and ergonomic convenience. Individual domestic supply will react immediately to a comparison in prices for fuel oil and firewood due to the slenderness of the range and due to the physical constraints associated with felling, cross-cutting and handling (2.514-4.190 MJ per working hour). The industrial harvesting operations for the supply of plants with a convenient form of energy implicate large organizations starting in the forest and ending at the mills. Most of the big heating or electric plants ask for more than 25-50 MW which requires large, continuous and secure supplies. As seen the fully mechanized whole tree chipping by a feller-chipper-forwarder is most appropriate and does not require high energy quantities per volume unit of heating. Of course yields must be improved in the plants and further research is recommended. The main interest in the energy from biomass for supply purposes must be taken into account for the future due to the large, numerous and various implications for many technico-economic aspects. Therefore, the political determination of using biomass for the supply of energy at a regional level can be a source of real economic importance.

REFERENCES

1. CARILLON, R. (1979). L'analyse énergétique de l'acte agricole. Etude du CNEEMA, France, n° 458, 58 p.
2. JONES, K.C. (1981). A review of comminution energy requirements. FERIC ENFOR Report n° P-28, 44 p.
3. JONES, K.C. (1979). Energy requirements to reduce forest biomass to useable fuel. Interim Report n° 1, FERIC, Ottawa, 80 p.
4. HALL, C.W. (1981). Biomass as an alternative fuel. Ed. Government Institutes Inc., Rockville, MD.
5. FOLKEMA, M.P. (1984). Manuel sur la production à grande échelle et la commercialisation du bois de chauffage. Manuel FERIC n° 6, Ottawa, 55 p.
6. X (1984). Energy conservation with tractors and agricultural machines. FAO, European Coop. Network on rural energy, bulletin n° 4, 76 p.
7. POTTIE, M.A., GUIMIER, D.Y. (1985). Preparation of forest biomass for optimal conversion. International Energy Agency, FERIC Special Report n° SR-32, Ottawa, 112 p.

8. X (1985). Waldschonende Holzernte. Kuratorium für Waldarbeit und Forsttechnik, 9-KWF Tagung, Ruhpolding, Gross Umstadt, 390 p.
9. ABEELS, P. (1985). Personal technical notes. Unpublished 1979-1985.
10. POTTIE, M.A. and GUIMIER, D.Y. (1986). Harvesting and transport of logging residuals and residues. International Energy Agency, FERIC Special Report n° SR-33, Ottawa, 100 p.
11. GOLOB Th. B. (1986). Analysis of short rotation forest operations. National Research Council Canada, Division of Energy, Bioenergy Program, NRCC n° 26014, Ottawa.
12. CURTIN, D.T. and BARNETT, P.E. (1986). Development of forest harvesting technology : application in short rotation intensive culture woody biomass. Technical Note 858 Tennessee Valley Authority, USA, 90 p.
13. JENKINS, B.M. and SUMMER, H.R. (1986). Harvesting and handling agricultural residues for energy. Transactions of the ASAE, vol. 29 (3), May, June 1986, pp. 824-836.

SWEET SORGHUM : FROM HARVESTING TO STORAGE

A. BELLETTI - A. BIOTEC - FORLI', Italy

Abstract

The paper describes three prototypes of sweet sorghum harvester :

- One from United States (Figs. 4-5), that is able to cut off panicle ; at the same time, it cuts the stems and drops them on the ground without cutting off the leaves.
- One from CLAAS OHG (West Germany, Figs. 6-7), that cuts the stem, chops it into 20-30 cm pieces, blows away the leaves by airstream and separates, but does not recover, the panicle.
- One built as part of a research project supported by the Region of Emilia-Romagna (Figs. 8-9), that is able to collect the stems, recovering the panicles separately ; it bundles them on a truck, without cutting off the leaves.

Storage tests on whole stems, with and without leaves, have been performed, and on stems chopped, very small and to 20-30 cm lengths.

The conclusion is that on no account can the product be stored fresh, but it must be processed within a short time of the harvest.

However, at present storage in the form of concentrated juices is under study.

Among the different harvesting methods, the best is the one that uses 20-30 cm long pieces and gets rid of panicles and leaves.

INTRODUCTION

An experimental test programme to study the farming of sweet sorghum has been underway since 1980 with financing from the Region of Emilia-Romagna.

A number of research projects have been drawn up since 1985 which are aimed at the study of the various stages of the 'Sorghum Story' as shown below :

- production of biomass (agronomic tests and genetic improvement) ;
- harvesting ;
- removal ;
- storage ;
- biotransformation ;
- use of by-products ;
- economic and marketing evaluation.

Each stage represents a research project financed either by the Region of Emilia-Romagna or the EEC or else still awaiting financing.

The mechanization of harvesting, which is the subject of this report, is a difficult chapter in the sorghum story, and has been under study since 1985. Satisfactory results have still not been achieved.

Over the last two years work has also included the study of storage, which is a problem closely linked to the harvest system and type of removal equipment used.

THE PLANT

Sweet sorghum (Latin name : sorghum bicolor) is a herbaceous annual with a stem rich in fermentescible sugars. In Italy it can reach heights of 2 - 3 m.

The culm, which at its base can be anything between 10 and 25 mm in diameter, has at its summit a panicle of between 10 and 25 cm in length which, if recovered and threshed, can produce significant amounts of grain. The latter can, however, cause problems in the industrial processing of the stem unless it is removed.

The biomass produced can be anywhere between 50 t and 70 t/Ha. The transformation industries require the incoming product to be stripped of leaves and panicle.

Fig. 1 : View of a field of sweet sorghum ready for harvesting

Fig. 2 : Detail of stem base. Diameters range from less than 10 mm to more than 25 mm

Fig. 3 : The panicle at the stem summit may be anything between 10 and 25 mm in length

MECHANICAL HARVESTERS FOR SWEET SORGHUM : CURRENT PROTOTYPES

The only mechanical harvesters for sweet sorghum are one-off prototypes because the problem still has not been adequately tackled (and solved) at world level. A. BIOTEC has studied three machines which appear to be the only machines available - one American, one German and one Italian.

a) The American harvester shown in Figs. 4 and 5 can cut and recover the panicle separately. It then scythes the stem, dropping it unstripped onto the ground.

Fig. 4

Fig. 5

This harvester is used for sweet sorghum in the southern States of the USA where the tropical climate allows the leaves to dry quickly after the stems have been cut. The leaves are then burnt off with the aid of flame guns.

In Italy the crop is harvested in September/October and the weather would not allow a similar practice. Leaves will thus have to be disposed of by mechanical means.

b) The German harvester (a Zuckerrohrvollernter CC 1400 built by Claas OHG) is shown in Figs. 6 and 7. It scythes the stem, cuts it into 20 - 25 cm lengths, removes the leaves by airstream and removes, but does not recover, the panicle.
   The short lengths of stem are then loaded directly onto a truck moving alongside the harvester.

Fig. 6

Fig. 7

The crop if cut and not processed immediately afterwards, quickly loses much of the quantity of fermentescible sugars it contains owing to fermentation which, in the cut stems, takes place much more rapidly than in whole stems.

c) The Italian prototype was built in 1985 with funds from the Region of Emilia-Romagna. It was an attempt to reconstruct mechanically the practice of the period 1939 - 1945 when in the region sweet sorghum was harvested, stripped of its leaves, bundled, loaded onto trucks and stacked carefully in the factories in a way that allowed air to circulate, and all that by hand.
   Under such conditions the crop could be stored a number of months without excessive losses.
   The same results were obtained during tests carried out in 1982 and 1983 in which harvesting, stripping and stacking were done by hand.
   The Italian prototype shown in Figs. 8 and 9 is able to harvest the stems (with separate panicle recovery), bind them in bundles and load them onto a truck moving alongside. However, it cannot strip the stems. A special machine would be required for this operation.

Fig. 8

Fig. 9

Since the harvester had to perform a large number of operations in the field, it could not cover more than 4 - 5 hectares per day. The machine also needed four operatives - two on the machine itself and two on the truck to stack the bundles.

Overall, the resultant operating costs are very high. Stacking at the factory, which involved simply unloading the truck, offered no guarantee that the crop would keep well since the haphazard unloading of the bundles prevented free circulation of air within the pile. Any leaves still present only further accelerated the degradation of the product.

The possibility of storing the crop as it was for a number of months would have offered the major advantage of allowing the industrial processing equipment to be built on a smaller scale. However, if such an option was open when labour was available and cheap, it is not today because manual operations cannot be simulated by machines.

THE CLAAS PROTOTYPE

In 1986 the CLAAS harvester, a model CC 1400 'Zuckerrohrvollernter' (Figs. 6 and 7), was shipped to Italy and tested on experimental fields.

The machine was designed to harvest sugar cane and therefore it operated efficiently on large areas, being capable of handling up to 50 t/h of stems.

The product reaches the factory in cut lengths of 20 - 30 cm with approximately 10 - 15 % leaf content and some panicle (from small plants). There are, however, two major drawbacks : its size is such that it cannot be moved on public highways without a police escort and its price represents considerable initial outlay on a machine that is in use for only a few weeks of the year.

The harvested crop was excellent because sugar losses were low compared with a finer cut (maize cutter). Weight for weight the crop is less bulky (compared with whole cane) and improved leaf and panicle separation is possible.

The product cut in this way can be stored for a short period only as storage tests showed and immediate processing is therefore needed. Consequently, industrial-scale processing methods are called for with a daily throughput rating high enough to cope with a harvest concentrated in a period of 50-60 days maximum.

Now that the basic requirements have been determined, a new prototype harvester is needed which should be smaller overall to make allowance for the smaller size of fields under crop in Italy.

STORAGE TESTS

Tests were carried out on stripped and unstripped stems. The presence of the leaves was allowed for by assuming the hypothetical use of a maize cutter for harvesting the crop with the associated option of later removing the leaves at the refinery. The negative effect on transport costs and on the processing stages caused by presence of leaves was disregarded in the tests.

Tests were conducted on the following types of product :

- whole stems (with leaves) stacked in bound bundles ;
- whole cut plant, i.e. with a maize cutter, and including leaves and panicle ;
- whole stems, stripped of leaves, bound in bundles and stacked ;
- stems stripped of leaves, cut in 25-30 cm lengths and stacked.

In view of the high level of fermentescible carbohydrate loss the tests showed that owing to its high humidity level (75-80 %) the fresh product cannot be stored as it is but must be processed within 24-48 hours of the harvest irrespective of whether as whole stems or 25-30 cm lengths.

Cutting up the products into smaller pieces should be avoided due to the unacceptable degree of sugar loss - 15-20 % in the space of 2-4 hours - i.e. the time taken to harvest and transport the crop. After six hours storage in stacks at the factory 25 % loss was registered and sucrose had disappeared alltogether.

CONCLUSIONS

Unless future technology produces substantial innovation, at the current state of the art sweet sorghum will have to be harvested in 20-30 cm lengths with leave and panicle stripping in the field. The cut product will have to be processed within 24 hours.

The product cannot be stored fresh as has been shown but storage as concentrated juice may be possible. However, such a process is not without its problems because concentrating the product will have an energy requirement which will vary with the type of extraction plant used - crushers, continuous diffusion or a combination of both.

The choice of extraction plant with a different energy requirement for concentrating the juice for storage is still pending and central to the calculation of the energy balance. A. BIOTEC is currently drafting a research project to study the latter.

AQUATIC CROPS : PROBLEMS AND PERSPECTIVES CONCERNING HARVESTING TECHNOLOGIES

V. BOMBELLI
C.E.C. Consultant
Milano

Summary

Vegetable aquatic biomasses are a potential source of energy, proteins and raw material that have not yet been fully exploited. Spontaneous biomasses present in the natural environments and cultivated species can be harvested. The development and optimizing of harvesting systems - storage - of both the spontaneous and the cultivated biomasses is of fundamental importance for the containment of production costs and the improvement of the energy balances for these types of culture. The harvesting technology is at times a decisive element in the choice of the species of cultivation, the potential uses of the biomass produced being equal. The following is a brief analysis of the problems and prospects connected to the harvesting technology. For simplicity it has been decided to group together 5 biomasses having a similar behaviour as regards the harvesting-storage system. More specifically they have been divided into : microalgae, macroalgae, rooted and non-rooted macrophytes, marine forest. For each group the state of the art of existing systems will be focused on. Examples will be given of some plants (pilot and real scale). No analysis will be made here of the results, possible end-uses and integrations (energy, industry, agriculture, etc.).

INTRODUCTION

The cultivation and/or exploitation of aquatic biomass has essentially three objectives : Biomass production, environmental reclamation, water depuration. The end uses of the harvested biomass can be energy, proteins and raw material production. The potential uses offered by aquatic biomass are showed in fig. I.

Major areas of concern for the development of large-scale culture systems are production costs, economic harvesting, and control of culture (as related to harvestability, value of biomass produced, predator infestation, poisoning by various pollutants etc.). The following is a brief analysis of the problems and prospects connected to harvesting technology, both for spontaneous biomass present in natural or artificial environments (lagoons, ponds etc.) and also of mass cultured biomass.

Similar biomasses have been bracketed together in the following groups : 1 microalgae, 2 macroalgae, 3 rooted macrophytes, 4 non rooted macrophytes, 5 Marine forest. For each group the state of the art of existing harvesting systems will be highlighted based upon the different culture processes.

1. MICROALGAE

Mass culture of microalgae is mainly practised for the mass production of the most interesting species in monoculture systems, and for water reclamation (eg high rate oxidation ponds). Systems of water reclamation have been developed on a commercial scale particularly in the U.S.A. and Israel and, among the EEC countries, in France. According to De Pauw (2) waste water treatment of microalgae will only be effective and economically beneficial if the algal biomass and other suspended solids (including bacteria) are sufficiently removed and/or recovered for further utilization. Monoculture systems of mass producing microalgae have been developed on a pilot scale, and in a few cases, on a commercial scale in Italy, France, Germany and Belgium among the E.E.C. countries, and in several other countries including developing countries. The main objective in this case is the production of biomass which can be of interest for further exploitation.

Separation processes are different and include "direct systems" (e.g. using natural removal of algae with the aid of aquatic biota) and "indirect systems" (using artificial or concentration processes). The harvesting process is very important and has a big influence on such steps as drying (the final algal concentration after separation has a considerable bearing on drying costs).

The main available "indirect" microalgae harvesting processes are usually based upon weight, or size differences

a (size) : filtering, screening techniques

b (weight) : sedimentation, flotation, centrifugation, chemoflocculation.

a) Filtering, screening techniques

These techniques include the relatively low cost and simple filtration techniques which can be used to harvest large microalgae such as Spirulina sp, Coelastrum sp. The possibility of utilizing these techniques has been a decisive element in the choice of microspecies for possible cultivation on a commercial scale (e.g. see fig. II Spirulina large scale cultivation in Israel).

b) Sedimentation, flotation, centrifugation

Sedimentation and/or flotation can be made more effective by using systems like centrifugation and chemical flocculation. The disadvantage of these systems (especially used in oxidation ponds) is their high cost energy requirement and, in the case of chemoflocculation, the poor quality of the produced biomass for bioconversion. For those reasons the strategy in chemoflocculation is now concentrating on the recovery of flocculant (3) and the use of more efficient non toxic and less expensive flocculants. A more promising and economical method utilises microstraining techniques through species control which can be achieved by varying such operational parameters as detention time, pH, mixing, pond depth etc.

2. MACROALGAE

Macroalgae can be exploited both by removing the spontaneous biomass present in coastal lagoons and areas, and/or by harvesting cultured crops. According to the final report of EEC project Cost 48 "Marine primary biomass provision and conversion" (4) biomass can be obtained from :

a) Attached (Benthic) marine macroalgae
   . biomass from natural population
   . biomass from attached cultivated seaweeds
b) Non attached (free floating) benthic marine macroalgae including free floating cultures.

Most of today's utilization of marine macroalgae is based on natural population : the total biomass harvested from the sea (tons fresh weight) is :

	Europe	the world
Red Algae	72,000	807,000
Brown Algae	328,000	1,283,000

About half of the total world output of a least 2 Million tons fresh weight per year is provided by cultivation. All of Europe's harvest is taken from natural seaweed beds.

The removing processes are not always the same but depend on whether the species are attached or floating, spontaneous or cultivated, and on the harvesting period. According to the Report of an ad hoc Panel on utilization of aquatic weeds (5) "...plants can be lifted from the water by hand, crane, mechanical conveyor or pump. Mobile harvesters are usually expensive machines that sever, lift, and carry plants to the shore. Most are intended for harvesting submerged plants, though some have been designed or adapted to harvest floating plants or the mowed tops of submerged plants".

The major problems of harvesting systems of natural populations, in particular, biomass from attached (benthic, submerged) marine macroalgae, are :

a. Use and development of cutting techniques and methods not dangerous to :
   - the natural ecosystem in which macroalgae are growing (e.g. not recycling nutrients and/or pollutants from the bottom of the often eutrophic natural environment)
   - the capacity of reinstatement of the vegetation after harvest (e.g. problems related to the quantity of biomass removed without damaging the natural renewability of the crop).

b. Concerning harvesting in shallow water, the development of harvesting machines of optimum size and weight (e.g. problems related to playload capacity).

c. Handling : transport of macroalgae is difficult because of their high water content - choppers are often incorporated into harvesting machine, because chopping makes the plants much easier to handle and reduces their bulk thus simplifying transportation, processing and storage.

d. Transport to the shore : several methods are used, each suited to the particular environment in which biomass is growing. Some of these entail problems of accessibility to the shore (e.g. grabs, pumps, conveyor belts).

For particular species, simple methods of biomass removal (e.g. by hand or rake from small boats) seem to be the most suited both for environmental quality and for the subsequent conversion of the biomass removed.

In order to optimize exploitation from natural lagoons a very important point is the setting up of a procedure for managing the algal biomass resources, which may include remote sensing and in situ measurements. An example of integrated macroalgae exploitation is the project "Valorization of Gracilaria sp to agar and proteins in the lagoon of Orbetello by harvesting existing biomass and by mass culturing", coordinated by the undersigned and inserted in the EEC COST 48 Programme. The first step of this project, in part funded by E.N.E.A., was to quantify the biomass of the submerged vegetation, in order to assess the possibility of setting up harvesting for industrial purposes and adopt in situ measurements (phytosociological methods) correlated to remote sensing images. It made use of pilot harvesting and testing, on pilot scale of Gracilaria cultivation.

As regards cultivated biomass there are very few large scale experiences in the EEC countries, and for this reason and because it is promising, the development of culture systems is one of the main research topics outlined by the EEC Cost 48 Programme, together with harvesting techniques.

3. ROOTED MACROPYTES and

4. NON ROOTED MACROPHYTES

General

Management of these biomass includes exploitation of promising wetland crops, such as Phragmites (reed), Scirpus sp, Lemna sp etc., which are also cultivated in systems for mass production and/or water reclamation. The harvesting methods differ according to the species and the configuration of the production site (e.g. natural, artificial, channels, pools, marshes, shallow swamps etc.).

Some general considerations, which were reported in the previous points, are of concern, such as the opportuness of setting up a procedure for optimizing exploitation of resources, cutters, playload capacity, handling and transport.

Harvesting methods were first developed for commercial exploitation or for reasons of free water circulation, control of eutrophication, insects, and fishing in the EEC countries, and for Industrial purposes, especially in East Europe (e.g. Romania) and in the Far East (e.g. China).

In general, the most common systems thus far utilized in the E.C. Countries for removing biomass are (6) :

a. simple small boats for harvesting (width $\simeq$ 1.3 m., length 4 m.) with cutting systems in the front utilized to clear shallow channels ;
b. self propelled-floating harvester (width $\simeq$ 2.3 mt., length 4 m.) with a cutting system and mechanical conveyor. Utilized to clear lakes, pools etc (usable to remove floating macrophytes such as Lemna sp) ;
c. b + cutting rake and collecting bin utilized for the collection of the cut biomasses ;
d. systems integrated with the agricultural machines (cranes etc.) utilized for harvesting from the shore (problems of accessibility etc.).

"Direct Systems" of biomass removal with other biota can also be utilized for the control of culture density.

The most suitable mechanical removing processes can be developed, with adaptations, from the systems already utilized to harvest natural biomass, taking due account of economical consideration (value of biomass produced, opportunity of association with different common owners etc.).

In particular :

Rooted Macrophytes

Harvesting biomass from E.C. swamps could be accomplished by harvesting machines similar to those already utilized for harvesting macrophytes for industrial energy purposes in other countries. At the moment there is no large scale experience of harvesting mass produced crops in the E.C., in spite of experiences of cultivation in waste water in Germany, Holland, France, Belgium, Italy (e.g. fig. III culture on a pilot scale of rooted macrophytes in sewage water near Milano).

The trials carried out in other countries are not completely applicable to E.C. countries, one of the reasons being different climatic conditions (e.g. in Swedish energy reed plantations, the harvesting is usually done in winter time with tractors from frozen water surfaces (See fig. IV energy crop cultivation of reed in Sweden (7)).

Harvesting in winter when the material is dry (15-25 % of water for emergent macrophytes) can reduce harvesting and transportation costs, and integration into conventional agricultural systems may be interesting for further processes such as harvesting, storage etc.

Non Rooted macrophytes

These species (for simplicity we include rooted floating macrophytes) are promising because the removing system for certain species is simple and makes subsequent harvesting easier. The problems of removing biomass are similar to ones encountered in floating macroalgae.

The experience of harvesting processes carried out especially in the U.S.A., on water hyacinth (8) (e.g. combinations of harvester, chopper, or conveyor, or pusher boat etc.) can be adapted to our interesting indigenous species.

Finally, there exist interesting experiences of direct removal of biomass by the use of aquatic biota (e.g. systems of integrated aquaculture with phytophagus fish or integrated breeding systems with animals such as ducks).

5. MARINE FOREST

By marine forest we refer to the presence in several Mediterranean coastal areas of forests of phanerogams such as Posidonia sp, Zoostera sp, Cymodocea sp and algae such as Caulerpa sp from a few meters in depth down to 20-30 m.

Direct exploitation of marine forest is still to be very cautiously considered both from an ecological (e.g. harvesting systems have not yet been developed on an industrial scale) and production viewpoint (the production capacity of such systems has not yet been fully ascertained) (9). This resource is included in aquatic crops, with particular emphasis on biomass deposits of Posidonia accumulated by the action of wind and tides on the beaches and in preferential points (fig. V (10)) which form layers having a breadth of some meters (along the coasts of Provence, Corsica, Sardinia, Sicily, Greece etc.).

Such biomass, which also gives rise to environmental problems, can be usefully harvested for subsequent conversion using conventional agriculture machinery, suitably adapted for the purpose, once the accessibility difficulties have been overcome.

REFERENCES

(1) FUSO NERINI, F., BOMBELLI, V. (1981). Macrophytes cultivations. Technical economical analysis. I° International workshop on aquatic macrophytes. Illmitz, 3-10 May, Austria.

(2) DE PAUW, VAN VAERENBERGH, E. (1981). Microalgal waste water treatment system : potential and limits. International Congress on phytodepuration and use of the produced biomasses. 15-16 May, Parma I.

(3) SHELEF, G., ORON, G. and MORAINE, G. (1980). Algal mass production as an integral part of a waste water treatment and reclamation systems, p. 163-189. In Algae Biomass, SHELEF, G. and G.J., SOEDER Eds, Elsevier/North, Holland Biomedical Press, Amsterdam.

(4) E.E.C., Project : Cost 48 (1982). Final Report of Ad Hoc working Group of Expert on research on Marine Primary Biomass. Provision and Conversion, to The Committee of Senior Official. EEC. DGXII.

(5) Report of ad hoc panel on utilization of aquatic weeds (1977). Making aquatic weeds useful : some perspectives for Developing Countries, National Academy of Sciences, Washington, D.C.
(6) C.E.M.A.G.R.E.F. (1985). L'exploitation des lagunages naturels. Guide Technique à l'usage des petites collectivités. Documentation Technique. F.N.D.A.E. Ministère de l'Agriculture. Paris n. 1.
(7) GRANELI, W. (1981). A comparison between the chemical composition of Reed shoots in summer and winter. I° International workshop on aquatic macrophytes. Illmitz, 3-10 May, Austria.
(8) WOLVERTON and Mc. DONALD (1978). Water Hyacinth (Eichhornia crassipes) Productivity and harvesting studies. U.S. National Aeronautics and Space Administration ERL Report N. 171, Bay St. Louis, Mississippi.
(9) BOUDOURESQUE, M. (1983). Les Posidonies, in Biomasse Actualites. 12.
(10) BOMBELLI, V. (1983). An approach to Posidonia exploitation in the region near Punta Alga (Sicilia). First International workshop on Posidonia Oceanica beds. Porquerolles F. 12-15 October.

Fig. I - Possibilities offered by cultivation of aquatic plants - Goldman modified (1).

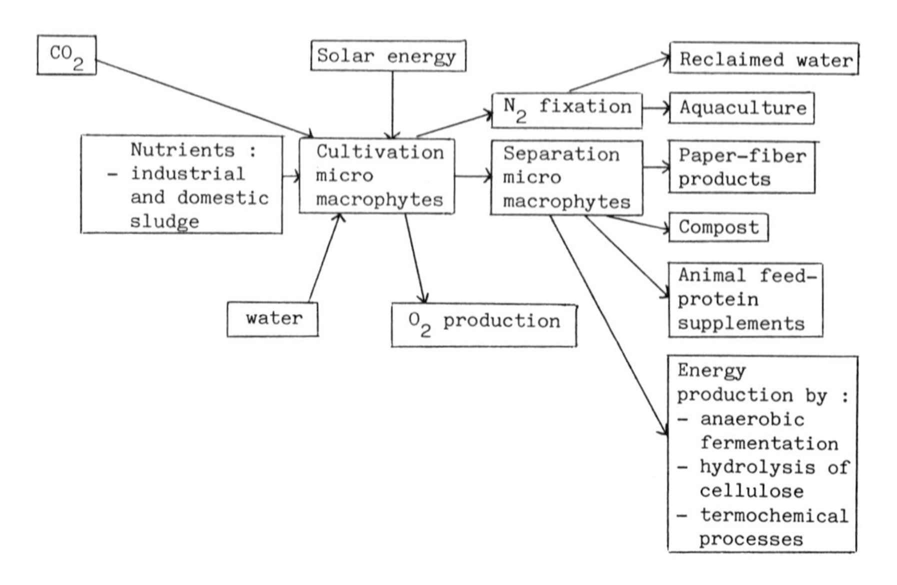

Fig. II - Spirulina large scale cultivation in Israel (Bombelli 1982).

Fig. III - Pilot culture of Rooted macrophytes in sewage water at Mulazzano-Milano (Bombelli)

Fig. IV - W. Graneli. University of Lund-Sweden energy crop cultivation of reed (Pragmites sp.) (Bombelli).

Fig. V - Marine Forest.
Posidonia on the shore - Punta Alga - Marsala-Sicilia (10) (Bombelli)

FAO/CNRE RESEARCH COOPERATION ON BIOMASS PRODUCTION AND USE FOR ENERGY

K. Kocsis
Regional Office for Europe
Food and Agriculture Organization of the United Nations

Summary
Two research networks and seven working groups have been established within the framework of the European Cooperative Networks on Rural Energy (CNRE) in order to promote the exchange of technical and scientific information and to promote the practical application of biomass production and conversion technologies in Europe and in interested developing countries. During the years 1983-86, technical consultations and workshops were organized, reports, bulletins and state-of-the-art studies were prepared and distributed among participating institutions. The bio-energy programme of the CNRE intends to play a catalytic role in the transfer of reasonable biomass energy technologies between north and south as well as west and east European regions and to provide assistance to developing countries both in related research and development activities and through the transfer of technologies and organization of professional training activities.

1. INTRODUCTION

The main objectives of the CNRE, being a continuous cooperative programme, operated by the FAO Regional Office for Europe, are to promote the more efficient use of traditional energy sources in agricultural sectors, and to increase the share of new and renewable sources of energy for and from rural sectors of the European region. The Networks cooperation comprises seven networks on Energy Conservation, Biomass Production and Biomass Conversion for Energy, Solar Energy Applications, Wind and Hydropower, Geothermal and Industrial Thermal Effluents, and Integrated Rural Energy Systems. Presently, more than 300 scientific and technical institutions, and over 700 scientists and experts, from 25 European and many non-European countries, are involved in the 21 working groups. The original aims of the Networks cooperation, namely: the exchange of scientific and technical information and the survey of related national research programmes, have recently been extended to the establishment of joint research programmes, organization of training courses and creation of demonstration sites for testing and technical-economic analysis of reasonable energy technologies among commercial conditions.

2. BIOMASS PRODUCTION FOR ENERGY

The evaluation of the energy potentials recoverable from various sorts of biomass has been estimated at 100 million t OE per year in Europe. The utilization of only 15-20% of this enormous energy potential may cover the total heat demand of European agricultural production and in a further stage, the liquid fuels of biomass origin might be a considerable substitute of fossil fuel requirements of field machinery, too. However, the fairly low level of the utilization of actually available forestry sources and agricultural residues in agro-industry of most European countries, as well as the practical experiences gained with the experimental technologies, have proved that further research should be carried out in order to develop more attractive technologies for harvesting, storing and preparation of available bio-materials and for the production of new biomass energy sources of high productivity (short rotation plantations, energy crops, aquatic plants), bearing in mind the considerable reserves of unexploited marginal lands in Europe.

The CNRE Network on Biomass Production for Energy is coordinated by Mr. M. Arnoux, INRA, Montpellier (France). At the consultations of this Network, held in 1983 and 1984 in Novi Sad (1) and Garpenberg (3), the priority topics and the establishment of various study groups were recommended and are now operated in the following two main directions:

- Forestry Biomass for Energy (short rotation species and harvesting, transportation, storing of forest products for energy)
  Coordinated by Prof. P.O. Nilsson, Garpenberg (Sweden); and
- Agricultural Biomass for Energy (identifying and growing crops for energy and evaluation of land use for energy plantations)
  Coordinated by Dr. J. Kisgeci, Novi Sad (Yugoslavia).

a. Forestry Biomass for Energy

Forestry biomass, mainly in the form of fuelwood and charcoal, is the oldest energy source from and for rural areas. However, with the rapid development of agricultural mechanization and the widening of intensive production technologies, the share of forestry biomass in the energy balance of rural areas has become small (1-4%) in most of the industrialized European countries. Since the oil crisis of the early 1970s, most European countries have examined the possible use of various alternative energy sources and forestry biomass is now again seen to be a potentially significant source of energy for rural hèat supply.

In Scandanavian countries where climatic conditions are especially favourable, considerable results have been achieved both in semi-commercial adoption of short rotation forestry technologies, and full-mechanization of related work processes. However, in order to extend the positive results to other countries, and to increase the economics of production of fuels from forestry biomass, considerable research and development still needsto be carried out in most European countries.

The main goals of the research cooperation within the CNRE Working Group on Forestry Biomass for Energy, are:

- the selection and improvment of short- and medium-term species (Poplar, Willow, Black Locust, Mediterranean species, etc.), for energy and existing forestry;
- the development of technologies for harvesting, transportation, storing and preliminary processing of wood and by-products for energy purposes; and
- the preparation of studies on economics and working efficiency of energy forestry processes.

The joint research activity in the above topics has already resulted in the preparation of state-of-the-art studies regarding the present situation and on-going research activities on short rotation cultures and fuelwood harvesting and processing technologies. These studies have been prepared by the requested study group leaders and discussed at specific workshops (5) in order to prepare recommendations for practical application and for further research and development programmes of priority importance.

One of the most important features of the research cooperation in this field, is the systematic survey of forestry energy research and development programmes carried out in various international organizations, as FAO, FAO/ECE, EEC, OECD/IEA, IUFRO, etc. by the Forestry Energy Secretariat (Garpenberg), in order to avoid duplication and overlapping of activities in Europe and to promote the transfer of reasonable technologies to developing countries.

b. Agricultural Biomass for Energy

The identifying and growing of traditional and new species of agricultural crops for producing technical energy sources (mainly liquid fuels), is a fairly new field of agricultural research. As a consequence of the energy price increase and the exhaustion of crude oil reserves, during the last decade, considerable attention has been drawn to the liquid bio-fuel producing technologies. It is obvious that agricultural lands should be used firstly for food production, however, among specific ecological and socio-economic conditions, the marginal and even traditional agricultural lands may be used for technical energy production, if profitable technologies can be offered for the farmers, without harmful effects on food supply and food prices.

The potential of growing traditional and/or new crops for energy differs from country to country. Those countries which produce sufficient amounts of food and have plenty of arable lands might grow common crops on average or poor soils where food or feed production is not efficient.

Conversely, the countries which suffer both shortages in food and energy, having limited arable land areas, are forced to choose those energy crops which grow successfully on marginal lands or on very poor soils.

To facilitate the better understanding of the subject of cooperative research in this field, according to our explanation, the production of agricultural biomass for energy covers the following groups of crops:

- traditional crops and their by-products (sugarbeet, corn, grain, oil plants, etc.);
- specific crops grown mainly for energy (Jerusalem Artichoke, Euphorbia Latiris, etc.);
- spontaneous harbaceus plants (Amarathus spp., Chenopodium spp. etc.); and
- aquatic plants.

The main objectives of the cooperative research activity in this CNRE Working Group are as follows:

- survey of the on-going research activities in Europe in order to identify the potential bio-energy sources for various regions in Europe;
- study and evaluation of experiences with small-scale techniques for growing crops for energy among various climatic conditions; and
- ecological and economical analysis of land use for production of biomass (agricultural and/or forestry biomass) for energy.

The joint research activity has already resulted in the exchange of available results of relative research programmes in cooperating countries, and hopefully, will contribute to the increase of scientific knowledge regarding the practical adoption of energy cropping and liquid biofuel production in Europe and in other regions. The collaboration will be extended to the systematic collection and dissemination of various cultivars, the exchange of seeds for testing under different climatic conditions and the establishment of comparative trials for genetic improvement of selected crops.

3. BIOMASS CONVERSION FOR ENERGY

The biomass conversion technologies for energy are basically thermo-chemical (direct combustion, pyrolosis, carbonization, gasification, etc.) and biochemical processes (anaerobic and aerobic digestion, hydrolisis, alcoholic fermentation, etc.), with the aim of producing solid, liquid or gaseous fuels for heat and electricity generation and for the operation of tractors and other mobile machines. According to the preliminary studies, the theoretically obtainable energy from the recovery and conversion of 40-50% of the forestry and vegetal by-products, and 30-40% of the animal wastes available in the European agriculture is some 45-50 million tons of oil equivalent, being more than the total direct energy requirement of rural sectors in this region.

Taking into consideration the potentials of energy forestry, energy cropping and expected surplus of certain traditional crops, the technical energy production may be developed into a profitable branch of agro-industrial activities in a number of European industrial countries during the next decades. However, the realistic achievements in this specific field of rural energy development requires further research on (a) more complete and economic recovery of various biomass sources; (b) the intensive development of practically applicable, reliable and economically feasible decentralized biomass conversion technologies; and (c) meeting the agricultural and other rural energy requirements by development of a flexible and appropriate energy supplying system including the establishment of a reasonable market for biomass energy sources.

The CNRE Network on Biomass Conversion for Energy is coordinated by Dr. A. Strehler, Technical University of Munich, Freising, (F.R. of Germany). At the technical consultation held in 1985 in Freising (6), the establishment of the following five working groups was recommended by the coordinators:

- Handling and Processing of Biomass (harvesting, storing, compaction of straw and other plant residues)
  Coordinated by Mr. J. Flieg, Gödöllö (Hungary)

- Biogas Production (identifying bio-chemical and bio-engineering factors in order to develop a second generation of more efficient biogas technologies)
  Coordinated by Prof. Dr. W. Baader, Braunschweig (F.R. of Germany)

- Heat and Power from Biomass (further development of combustion and gasification technologies to meet the conditions of wide-scale practical application)
  Coordinated by Dr. A. Strehler, Freising (F.R. of Germany)

- Liquid Fuels from Biomass (improvement of biofuel production technologies to increase the energy efficiency and decrease production costs)
  Coordinated by Dr. J.F. Molle, Antony (France)

- Charcoal Production (new technologies for improvement of the energy efficiency of charcoal making, mainly in developing countries)
  Coordinated by Dr. P. Thoresen, Kjeller (Norway)

a. Handling and Processing of Biomass

The agricultural by-products having large C/N ratio and low-moisture content (as crop residues) can be used for thermo-chemical conversion while those having low C/N ratio and high moisture content (as animal wastes or green materials) can be used for bio-chemical conversion. The related studies carried out show that on average, 60-70% of crop residues are not utilized at all and after removal from the fields are destroyed or buried. On the basis of this evaluation, it is evident that the future use of by-products for energy production could essentially contribute to the improvement of overall energy input/output ratio of agro-biological systems. The estimated amount of potentially available crop residues is about 30-35 million tons of oil equivalent which possibly could be used for direct combustion or gasification in rural areas.

The complete agronomic balance of recuperation, the possibility and the acceptable limits of removal of crop residues from cultivated land is still under discussion. No generally valid indication can be given and the ratio of unused by-products depends on the local soil and agronomic conditions. In fact, the most common plant production technologies have neglected the importance of by-product utilization and only concentrated on the main product. Therefore, new agronomical aspects should be introduced to utilize the unnecessary surplus of crop residues.

The main problem of large-scale utilization of most crop residues is the lack of suitable technologies for storing and preliminary processing of crop residues. The light specific weight, the large cubic capacity and moisture content of certain residues as well as ensuring the proper size and form of particles suitable for long-term storing and perfect combustion, all require new technologies in order to profitably utilize by-products for direct heating purposes.

The subject of scientific cooperation in this Working Group can be outlined as follows: development, testing and semi-commercial application of techologies for harvesting, storing and processing of crop residues (straw, corn stalk, sunflower stalk, etc.), available as needless surplus on agricultural lands, in order to obtain fuel-type raw material, with adequate humidity, good transport, storing and burning characteristics with low energy requirement and low production costs.

The testing and semi-commercial application of experimental technologies completed with the relevant economical and agro-energetical analysis, will certainly promote the practical adaptation of well-developed techniques among various farming, production and agro-technical conditions in Europe and developing countries. The technical, agronomical, energetical and institutional barriers and constraints concerning the utilization of crop residues for heating is being carefully examined and studies on the results of cooperative research are in progress. Further information on potentially available crop residues in cooperating countries, and the analysis of by-product cost and technical economical efficiency of by-product handling systems are also among the aims of cooperative research.

b. Biogas Production

As a result of the intensive research work carried out by micro-biologists and bio-chemists during the last decades, the level of knowledge on the anaerobic conversion of organic matters to produce biogas is relatively high. Hence, the efficiency of biogas plants could be further improved and more advanced technical solutions of digesters, and new systems, especially for digesting solid organic matter, are in an experimental stage. In view of the considerable knowledge on the various processes of anaerobic digestion, the efficiency and economics of available technologies could be, and should be further increased in order to identify more deeply the biological, chemical and engineering characteristics of digestion processes and to develop the second generation of biogas technologies for wide-scale practical application.

From the related literature and the papers presented at various national and international meetings, an overall informative survey have been made on the dynamic research and development carried out in European regions (7). However, many gaps in the knowledge relating to bio-chemistry of the whole digestion process and the connected engineering problems of reliable plants are existing. By further close collaboration between the researchers brought together in this Working Group, much more progress and success in the activities could be achieved.

The most important topics of joint research activities can be briefly listed as follows:

- identification of biological, chemical and engineering factors controlling the different types of biogas generation technologies

operated with various biomass resources;
- biogas generation from specific residues, e.g. lignocellulosic wastes and water weeds by using simple, efficient and cheap technologies suitable for developing countries; and
- development and improvement of technical devices, engines and vehicles to be run by biogas.

c. Heat and Power from Biomass

As a result of the relative national R & D programmes, the large variety of experimental technologies for the utilization of biomass for heat and power production (combustion, pyrolisis, carbonization and gasification), is available. The production technologies of different energy carriers, as heat and electricity, as well as solid, liquid and gaseous fuels for stationary and mobile use, have developed very quickly during the last ten years. However, only the direct combustion of wood and straw reached the phase of wide-scale practical application in certain European countries. Intensive national research programmes and close international cooperation is demanded for the transfer of the results of related research and development work into practice in more and more European countries and the reasonable technologies to developing countries.

In the field of biomass combustion for heat production, wood has many applications, big resources are available and many of them have already been used. Improvements are necessary concerning the technical functioning and efficiency, reduction of emission and automatic control of these processes. In the case of straw, significant applications have only been carried out in countries with high potential and consistent renewable energy policy (e.g. Denmark), where straw is available at a reasonable price, investments are subsidized and environmental regulations for straw burning in the field are anticipated.

In the development of gasification technologies for stationary electricity production, a lot of R & D work has already been carried out but technical problems with some types of fuel, overall efficiency and relatively high investment costs are being experienced. Nevertheless, efficient technologies have been demonstrated in France, F.R. of Germany and Italy. The gasification technologies for mobile machines are also under development; however fuel supply is fairly difficult. Some good technical solutions are available e.g. in Belgium and France, although in the future, further research should be carried out for more economic practical application.

The joint research and development activity is focused on the following topics of priority:
- further technical development of biomass combustion units of middle and high capacity and the promotion of their wide-scale practical application; and
- cooperative study, development and semi-commercial application of biomass gasification technologies for stationary and mobile use in order to provide heat and power co-generating units for rural areas.

d. Liquid Fuels from Biomass

Despite the current decrease of oil prices, a third oil price increase is expected in Europe before the end of this century. By this time, the cheap oil resources will be exhausted in many countries, the average production costs will be much higher than today and most probably low oil prices during the late eighties will slow down investments in order to find new oil fields. For these reasons, liquid fuel supply in Europe could be a critical point again in the following periods.

Transformation of agricultural surplus production (cereals, sugarbeet, etc.), into liquid fuels is one possible way, however, because of the low-efficiency and fairly high production costs of existing technologies, furthermore due to the world's need for all food stuffs available against starvation in developing countries, new ways should be discovered for liquid biofuel production. In the longer-term, liquid fuel production in large quantitites, from agriculture could provide up to 25% of European liquid fuel consumption without subsidies if the expected progress is achieved with lignocellulose hydrolysis.

Apart from the survey of the present state-of-the-art and the future potentials of biofuel production, the following topics were proposed for cooperative study:

- development of new production technologies and to increase the energy efficiency and to decrease the production costs of various ethanol, methanol and plant oil producing techniques;
- study of the potential utilization of liquid biofuels by current or future engines and the consequences of this utilization on the oil industry; and
- from the macro-economical point of view, the insertion of liquid biofuel production in European agricultural sectors has to be studied too.

The question of production of liquid fuels from agriculture is very controversial in Europe, and one can expect from the research collaboration, that technical and economical points of view of each country may become clearer. This research cooperation could also provide a series of technical references for developing countries regarding the production of liquid fuels on their own with a view to decrease their dependence on oil imports.

e. Charcoal Production

Charcoal has been a part of human life as long as any trace of civilization has existed, and it is still an important part of the life of a large part of the world population in the developing countries. In the industrialized countries, more convenient energy carriers like oil, gas and electricity have come into use and charcoal is used only for special purposes. In research programmes however, efforts are being made to replace oil and electricity by biomass, where charcoal is one of the possibilities.

In many developing countries, production of charcoal is forbidden in order to reduce deforestation. If the use of charcoal is to continue, highly efficient production and utilization technologies will have to be developed, and it must be inexpensive. Price competitiveness is also a demand in the industrialized countries, but at least equally important is comfort-giving. To satisfy these specifications, new technologies must be developed first of all for developing countries where charcoal is still a significant energy source but also for industrial countries where increasing markets can be expected.

Primitive production methods based on charcoal preserve only about 15% of the energy content of the wood. Improved technology, although comparatively simple, conserves easily 50-60% of the energy through optimized production of charcoal, not counting the energy in the charoils or process heat. Small, light weight and inexpensive units (2-4 t/day) and transportable units of medium sizes, (10-30 t/day) having an improved efficientcy should be developed and tested for developing countries.

4. CONCLUSIONS

The biomass resources should be considered as the most efficient and promising renewable energy sources for and from rural sectors. The potential resources in the 29 European FAO member countries for direct combustion are in the range of 100 million tons of oil equivalent. The utilization of only 15-20% of these reserves may cover the total heat energy demand of rural sectors and another 20-25% of this potential can meet all the heat demand of rural communities and local industries.

The bio-energy sources are decentralized, cheap, renewable energies suitable for continuous energy supply and for conversion of them into various more valuable secondary energy carriers. The transportation and storing of biofuels are much easier compared to solar, wind or geothermal energy. The caloric value of absolutely dry bio-materials (17-18 MJ/kg) equal to that of the medium quality coal and even the air dry biomass sources, having 14-16 MJ/kg caloric value at 15-20% of moisture content, are suitable for providing small, medium and concentrated heat supply in the range of 10 kW and 10 MW (thermal) capacity at reasonable labour cost of material handling.

Apart from the traditional utilization of about 16.5 million tOE of fuelwood in European countries (Table 1) more than 1.5 million tOE of forestry residues are used for direct combustion mostly in Scandinavian countries for heating of dwellings and in district heating systems. Among the agricultural crop residues, straw is the most important resource with a potential of some 21 million tOE in Europe (Table 2) of which nearly 0.5 million tOE is actually utilized for agricultural heat supply and in rural communities especially in Denmark (250 000 tOE), and in Austria, F.R. of Germany, Sweden and the U.K..

The more advanced biomass conversion technologies, like gasification for heat and electricity supply of remote areas, mainly in developing countries, or biogas technologies for utilization of animal waste or green crop residues for generation of heat or/and electric power are still in the development phase. The potential resources in Europe are estimated to an amount of 20 million tOE (Table 3), hundreds of experimental and semi-commercial technologies have been demonstrated and tested. Nevertheless, the practical application of these techniques into the existing agro-industrial systems is still fairly contradictory. Problems are connected with the matching of consumption and production characteristics, with the technical operation of fairly complicated equipment and with the overall economics of these technologies. For instance, plenty of biogas can be produced during summer-time when the heat requirement is limited, and on the other hand, the self-consumption of traditional systems during the winter period is too high for intensive heat production. Electricity production in decentralized biogas or gasification units needs specific knowledge and technical skills of the farmers who are normally fully engaged in their agronomical business and without having a considerable extra income do not wish to operate sophisticated techniques. The human factors influencing considerably the decision making of farmers are becoming more and more important in practical applications of renewable technologies.

The liquid bio-fuel production from traditional and specific energy crops is an even more difficult field of biomass utilization for energy purposes. The potentials are also considerable in this field: the traditional plant oil production technologies may yield 2.0-2.5 t oil/ha, ethanol production can reach 2.8-3.0 t/ha from sugarbeet, 1.8-2.0 t/ha from corn and 1.3-1.5 t/ha from wheat. Specific energy crops not grown for human food supply yield enormous amounts of green material (60-90 t/ha) and the ethanol yield may be increased up to 2.6-3.9 t/ha. The estimated potentials of liquid bio-fuel production in Europe varies between very large limits (20-100 million tOE) depending on the amount of unutilized crop residues and surplus products taking into consideration also the size of marginal and other lands which in the distant future can be utilized for ethanol or plant oil production. Since the energy and cost efficiency of these technologies were not even acceptable during the periods of highest energy prices, many European countries reduced or cancelled the related research and development activities.

The energy efficiency of biomass collection for combustion is excellent. The net caloric value of biomass residues, taking into account moderate combustion efficiency, varies between 0.18-0.22 kgOE/kg and the total energy yields between 350-550 kgOE/ha (Table 4). The fuel requirement of collection, transport and processing of biomass sources is not more than 0.3-0.6 MJ/t which compared to the net energy output (7.5-9.1 MJ/t) gives a very good energy output/input ratio (Table 5). The energy balance of anaerobic digestion is positive but less attractive. A medium size biogas plant with cogeneration of heat and power producing 120 m^3 biogas per day may provide some 200 kWh/day and 45 kgOE of heat (2600 MJ/day) but the average energy required by the plant is about 80 kWh electricity and 20 kg OE of heat (equivalent to 1000 MJ/day) i.e. at least 40% of the produced energy is used for the running of the system. The energy efficiency of alcohol production is much smaller and in some cases even negative. For instance, energy input for producing corn in the agricultural sector is 22-27 GJ/ha (3.6-4.5 GJ/t corn) 12-15 MJ/kg of alcohol. The energy input of processing technologies is nearly equal (12-14 MJ/kg alcohol having an energy content of 28-29 MJ/kg). The energy balance is more favourable if we take into consideration the energy content of the distillation by-product (3.5-3.8 MJ/kg) which can be used for animal feeding. The limited energy output/input ratio of the existing biogas and alcohol production technologies should be considered as the most important barrier of their practical application (Table 5).

The cost-benefit efficiency of biomass combustion is appropriate compared to the costs of light heating oil utilization. The payback periods of fuelwood, forestry by-products and straw combustion are acceptable for the farmers in many countries and the wide-scale distribution of these technologies now depends mainly on the national energy price policy and the introduction of subsidiary systems which are still very necessary to make these technologies more favourable for the farmers. Mainly due to the low energy efficiency, the real benefits of gasification and biogas technologies are not yet high enough to expect their considerable spread in rural sectors of European countries. Apart from the 4-5 times higher investment costs (Table 6), in most cases difficult to reach net income compared to oil burning or operating of diesel aggregates for electricity production. If there is any cost benefit the pay back periods are very long being in the range of 10-15 years. Concerning the biofuel producing, plant oils for substitution of diesel oil may be economical at farm level in many European countries with considerable subsidy, however, ethanol production from traditional resources, especially at the present energy prices, cannot be considered realistic. Of course, if the new bio-engineering development programmes result in practical technologies and the energy prices again reach the earlier or higher levels by the end of the next decade, all these technologies might again become potential sources of liquid fuel supply in Europe.

The importance and possible contribution of rural sectors to the future energy supply of the European region, should be evaluated in relation to the liquid fuel demands because electricity production will always be greatly concentrated in big power stations. It is obvious that forest raw-materials and agricultural residues, firstly straw, may cover the total heat demand of agricultural sectors and rural communities contributing with some 30-35 million tOE to the partial substitution of the 250 million tOE liquid fuel consumption used at present in Europe for heating purposes (Table 7). This potential can be increased with 4-5 million tOE of heat produced through biogas technologies in the distant future if the adoption of new, more economic techniques become available. The motor gasolene consumption in Europe is presently about 120 million t which should be compared with possible ethanol production (20-25 million tOE) if more advanced technologies could be provided. The diesel oil consumption in the same countries is in the range of 215 million t from which about 20-22 million t is used in agriculture and forestry. This amount of liquid fuel can also be provided by farmers without any harmful effect on food production and supply in Europe. To summarize, during the next decades, the energy self-sufficiency of primary food production, providing considerable extra energy sources for rural communities and other production sectors can be a realistic programme if the related technological development will be successful and the governments decide to support this programme as an organic part of their national energy policy.

REFERENCES

1. - Report of the CNRE Technical Consultation on Forestry Biomass for Energy, Garpenberg, Sweden, June 1984, FAO/CNRE Report No.5, p.39

2. NILSSON, P.O., MITCHELL, C.P. (1984): Forestry Biomass for Energy, Proceedings of the Consultation, FAO/CNRE Bulletin No.5, p.78

3. - Report of the CNRE Technical Consultation on Energy Cropping Experiments, Novi Sad, Yugoslavia, Oct. 1983, FAO/CNRE Report No.2 p57

4. KISGECI, J. (1983): Energy Croping, Proceedings of the Consultation, FAO/CNRE Bulletin No.3, p.52

5. MITCHELL, C.P., NILSSON, P.O. and ZSUFFA, L. (1986): Research in Forestry for Energy, Proceedings of the Joint IEA/FAO/IUFRO Forest Energy Conference and Workshop, Rungstedgaard, Denmark, October 1985, Uppsatser och Resultat Nr. 49/1986, p.243

6. - Report of the CNRE Technical Consultation on Biomass Conversion for Energy, Freising, F.R. of Germany, October 1985, FAO/CNRE Report No.10, p.57

7. BAADER, W. (1986): Biomass Conversion for Energy (Biochemical Conversion), Proceedings of the Consultation, FAO/CNRE Bulletin No.10a (in printing)

8. STREHLER, A. (1986): Biomass Conversion for Energy (Thermochemical Conversion), Proceedings of the Consultation, FAO/CNRE Bulletin No.10b (in printing)

9. - Report of the CNRE Workshop on Heat and Power from Biomass, Ormea, Italy, October 1986, FAO/CNRE Report No.19 (in printing)

Potential Biomass Resources for
Direct Combustion
million tOE

Table 1

	Agricultural Residues	Forestry By-Products	Fuel Wood	Total
North Europe	1.0	10.1	2.0	13.1
West Europe	11.9	11.9	4.4	28.2
South Europe	9.8	5.8	5.5	21.1
East Europe*	11.0	9.5	4.6	25.1
Total	33.7	37.3	16.5	87.5

*without GDR and USSR

Potential Biomass Resources from Major Crops
for Direct Combustion
million tOE

Table 2

	Cereal Straw	Corn Residues	Fruit Tree Cuttings	Others	Total
North Europe	1.0	-	-	-	1.0
West Europe	9.7	1.5	0.6	0.1	11.9
South Europe	5.2	1.3	1.1	2.2	9.8
East Europe*	5.6	2.6	0.1	2.7	11.0
Total	21.5	5.4	1.8	5.0	33.7

*without GDR and USSR

Potential Biomass Resources from Green
Residues and Animal Wastes for Anaerobic Digestion
million tOE

Table 3

	Green Crop Residues	Animal Wastes	Total
North Europe	0.2	0.4	0.6
West Europe	3.5	4.8	8.3
South Europe	2.4	2.2	4.6
East Europe	3.0	3.4	6.4
Total	9.1	10.8	19.9

Energy Conversion Factors for Biomass Combustion

Table 4

	Collected Residues t/ha	Moisture Content %	Caloric Value MJ/kg	Useful Energy* kgOE/kg	Energy Yields kgOE/ha
Cereal Straw	1.5-3.5	10-15	13.6-15.3	0.20-0.22	315-735
Corn Residue	3.5-5.5	35-45	8.5-10.2	0.12-0.15	470-750
Sunflower	1.9-3.5	25-35	11.0-12.7	0.16-0.18	325-600
Rice Straw	1.3-3.2	25-35	11.0-12.7	0.16-0.18	220-540
Cuttings	1.5-2.5	35-40	10.2-11.0	0.15-0.16	230-600
Forest Residue	1.5-2.0	25-30	12.5-13.5	0.18-0.19	275-350

* Efficiency: 60%

Energy Output-Input Ratios for Biomass Combustion Technologies

Table 5

	Collection Transportation GJ/t	Storing Processing GJ/t	Total Input GJ/t	Total Output GJ/t	Total Output-Input Ratio
Wood Combustion	0.2-0.4	0.1-0.2	0.3-0.6	7.5-8.1	12.5-27.0
Straw Combustion	0.3-0.4	0.0-0.1	0.3-0.5	8.2-9.1	17.0-31.0
Biogas (Heat)	0.0-0.1	8.0-9.5	8.0-8.4	8.8-9.6	1.1-1.2
Gasification (H & P)	0.3-0.4	6.1-6.8	6.4-7.2	8.2-8.9	1.1-1.4
Biogas (H & P)	0.0-0.1	8.0-9.5	8.0-8.4	12.1-13.1	1.4-1.6
Plant Oils (Rape)	8.5-9.0	7.0-8.0	15.5-17.0	35.0-37.0	2.0-2.5
Ethanol (Corn)	12.0-12.5	12.0-15.0	24.0-27.5	28.0-29.0	1.0-1.2

Estimated Energy Costs of Various Biomass Conversion Technologies

Table 6

	Fuel Price US$/t	Efficiency %	Energy Price US$/tOE	Specific Investment US$/kW	Total Energy Costs* US$/tOE
Wood Combustion	30-40	70	135-180	150-220	265-350
Straw Combustion	25--30	60	120-145	250-300	215-375
Biogas (Heat)	0-5	25	0-10	600-900	465-705
Light Heating Oil	230-270	90	260-305	80-130	320-405
Gasification (H & P)	30-40	45	215-285	1000-1400	1690-1370
Biogas (H & P)	0-5	35	0-10	1800-2200	1395-1705
Diesel Aggregate (P)	230-270	35	660-770	400-600	930-1235
Plant Oils (Rape)	210-240	20	1050-1200	400-600**	1090-1270
Ethanol (Corn)	150-250	20	750-1250	500-700**	850-1390

* Operation: 1500 hrs/year
** per 1000 litre/year capacity

Liquid Fuel and Total Energy Consumtion in Europe

Table 7

	Motor Gasolene mill.t	Gas-Diesel Oil mill.t	All Liquid Fuels mill.tOE	Total Consumption mill.tOE
North Europe	6.8	12.7	31.9	63.0
West Europe	77.4	130.3	333.8	718.7
South Europe	24.5	47.6	147.9	240.5
East Europe*	13.5	24.4	74.6	358.2
TOTAL	122.2	215.0	588.2	1380.4

* without GDR and USSR

HARVESTING PROBLEMS IN SPAIN

L. ORTIZ TORRES
Biomass Department
Renewable Energies Institute
Madrid-Spain

Summary

There are many difficult problems in developing technologies that are adapted to real problems in our country.

Rugged orography and great fractioning of land limits the use of rentable machinery, and the public and private policies make technological innovation difficult.

Nowadays the obtention of an additional energy benefit, may make certain operations and sylvicultural labours profitable, particularly in depressed areas where there is still no electricity.

Energy from biomass in Spain is directly conditioned by the current price of fuel and natural gas, thus reducing interest in this sector.

Here we discuss the methods for biomass removal and generated waste material exploitation ; and we show technology applied to waste material management, technological deficiences and stockage problems.

1. INTRODUCTION

No indigenous machinery exists in Spain for the removal and complete processing of traditional forest products and residues. This situation is due to both, structural and temporary characteristics.

Rugged orography and steep slopes are a common characteristic of a large part of the forest stands, which in many cases play a primarily protective role for hydrological basins, the productive criteria being banished to a second level of social and economic interest.

These inherent characteristics of our forests define accessibility and manoeuvring conditions which limit the use of compact and versatile machinery.

Depressed areas are defined in Spain as those standing at an altitude higher than 1000 meters above sea level, with slopes greater than 20 %. It is in these areas where energy from biomass has a great social interest.

On the other hand, of our total forest surface (26 mill. hectares) only 5 % belongs to the State, thus limiting the management capability of public offices.

Private companies and forest owners have not carried out any renewal policies, due to generalized impoverishment, depopulation and decapitalization of rural areas. Thus, the removals are still done using adapted traditional agrarian machinery ; with substituting technological deficiencies in the majority of cases.

In many cases, the great fractioning of forest properties and surfaces makes mechanization investments unpracticable.

The situation is worsened by the forest policies that were followed for decades, according to which large pine and eucalyptus stands where established, beginning in the fifties, and planted in great density without previously planning an adequate network of removal tracks or firebreaks, for maintenance and treatment tasks.

As a consequence of later procrastination, the current state of the forest stands is deplorable, and forests are deteriorating since no complete landuse planning has been carried out.

As a consequence of domestic migrations and the concentration of population in the cities during the sixties, vast agrarian and forest areas where abandoned in search of a better quality of life, and firewood and waste materials, formerly used in domestic heating, remained in the forest, favouring and promoting infections, pests and fires.

The state of the forest makes sylvicultural and removal tasks difficult, thus making all operations expensive and complicating their mechanization.

The energy exploitation of forest waste products plays an important role in this situation, since the obtention of additional energy benefits may make certain operations and labours profitable.

Today, the energy supply from forest waste products is again being considered as a suitable alternative within national energy planning, since even though quantitatively not enough resources are available to substitute an important percentage of conventional fuel, qualitatively it may shortly solve pending life quality problems in small towns to which electric energy has not arrived due to the high cost of required investments. There is a need to develop machinery capable of working efficiently in harsh conditions, and with sufficient autonomy to mechanize the different phases of the forest and integrate biomass exploitation.

2. PRESENT STATE OF THE ENERGY FROM BIOMASS SECTOR

A junctural situation of excess forest products appears today at all levels, having promoted a generalized price decrease in wood derivated products, and a high stockage in large industries, so that a part of these resources are being burnt, since better buying prices are obtained. This situation is transitory and is provoked by the great quantity of burnt wood present.

When the market stabilizes it may be the case that the burning of commercially usable wood is more profitable than its use for pulp or boards, thus producing the paradox of a wood importing country burning national wood and consuming foreign wood and elaborate products.

Energy from biomass in Spain is conditioned directly by the price of fuel and natural gas, and at the present moment the price decrease in petroleum derivatives has reduced investments and interest in this sector.

3. FOREST BIOMASS REMOVAL METHODS

Whole Tree Method

Commercially Exploited stands :

A method used when appropriate removal conditions and tracks are available. The tree is removed with a skidder or selfloader and it is debranched and debarked on track or at bolting plant, where waste separation is done.

Unexploited stands :

In cases of excess wood extraction, sylvicultural tasks, regeneration thinnings, guidance cuttings, etc, wood is removed to track, where products are chipped or lined up in streets where movable chipping equipment grinds them down.

LAND USES

	UNIPRIGATED 000 has	IRRIGABLE 000 has	TOTAL 000has
FARMING LANDS			
Grassy	8.361,1	1.980,4	10.341,5
Unproductive lands	5.145,7	142,8	5.288,5
Shrub farming	4.405,6	567,9	4.973,5
TOTAL	17.912,4	2.691,1	20.603,5
MEADOWS & PASTURES			
Natural Meadows	1.276,9	202,1	1.479,0
Pastures	5.504,8	--	5.504,8
TOTAL	6.781,7	202,1	6.983,8
FOREST LAND			
Timber-yielding forest	6.604,4	--	6.604,4
Clear forest	3.917,6	--	3.917,6
Wooded forest	4.720,7	--	4.720,7
TOTAL	15.242,7	--	15.242,7
OTHERS LANDS			
Deserted & Pastures	3.565,0	--	3.565,0
Esparto fields	436,7	--	436,7
Unproductive land	1.263,6	--	1.263,6
No agricultural surface	1.851,8	--	1.851,8
Rivers & Lakes	524,1	--	524,1
TOTAL	7.641,2	--	7.641,2
TOTAL	47.578,0	2.893,2	50.471,2

MARGINAL LANDS

AGROENERGETIC ADVANTAGE (POSSIBLE)

DISTRICT	FORESTRY LAND 000 has	PASTURES 000 has	UNPRODUCIT 000 has	TOTAL 000has
Galicia	919,9	160,8	13,1	1.123,8
Norte	516,2	104,8	242,8	863,8
Ebro	949,1	1.136,8	518,9	2.604,8
Nordeste	460,3	265,7	207,4	933,4
Duero	1.332,9	1.076,4	793,7	3.203,0
Centro	1.295,0	877,5	748,2	2.920,7
Levante	721,4	119,4	349,9	1.190,7
Extremadura	985,1	831,3	195,4	2.011,8
Andalucía Oriental	468,2	395,9	560,1	1.424,2
Andalucía Occid.	935,7	492,4	164,5	1.592,6
Canarias	24,5	43,8	207,7	276,0
TOTAL	8.638,3	5.504,8	4.001,7	18.144,8

Half Tree Methods

When the size of trunks prevents removal due to the tracks narrowness, extraction is carried out in several pieces to be processed as explained above.

Bolt Method

Wood is debranched and pieced down to commercial size (2-2.5 m) in the forest, and bolt pieces are removed to the track. Generated waste material is piled or lined up until chipped, or removed to fixed chipping equipment on the track.

Bushes, Shrubs and Pruning Wastes (Brushing)

All different waste materials generated by traditional sylvicultural tasks are processed either in the forest or on the trackways. Use of such materials should be moderate, since an excessive removal would provoke a progressive defertilization which could become unreversible.

Removal to track is done by horse draw, small tractors, forest tractors, winches, etc.

It is important to avoid incrustation of mud, sand or stones that stick to the bark and melt during combustion phase, creating silicate crusts in boiler.

Also, considerable productive time is lost in different container adjustment, and tractor drivers still add a fraction of cost to the chips.

The currently obtained yield is of 7-8 m^3 per hour with 8 workers per chipping team.

All these limitations make their use restricted to very strict management conditions.

As of forest harvestors, we may highlight the trials carried out with models in the basque country, with unfavourable results.

The introduction of a line of versatile machinery and adequate technology for our forests would be of primary importance.

4. TECHNOLOGICAL CONSIDERATIONS ON OBTAINED PRODUCTS

Spanish broad leaved species'wood have a calorific power higher than 4200 Kcal per Kg, as opposed to conifers, with 4500, due to their high resin percentage, which makes the use of conifers more profitable for energy means. Broad leave species are also more dense and compact, and have a high resistance to chipping, so that obtained yields are less. As for thermal efficiency, coniferous chips burn faster but less sustainedly than those from broad leave trees.

These characteristics determine the differential use of waste materials, since it may be interesting for industrial installations to maintain fast and effective combustion processes, while a slower sustained combustion would interest small installations.

Spain has still to solve the problem of energy use of eucalyptus bark, when fibers remain between blades, hindering effective chipping.

For wastes coming from agrarian crops ; cereal straw produces some 11 million tons of unused material, since no type of gasifier adapted to industrial energy has been developed.

5. STORAGE

Chipped products may reach the user with a high humidity percentage, so that intermediate treatment consists of natural indoor or outdoor drying.

Pile management for drying consists of stirring and airing out the products under favourable environmental conditions, or mechanically compacting to reduce hollow volume thus partially avoiding degradation phenomena.

Beheading of piles favours drying, since humidity, microbes, spores, etc, concentrate in high parts due to chimney effects.

A good method of avoiding rain penetration into chips involves the use of plastic covers on the tops of piles, since it has been proved that the 1 to 1 sloped sides drain off fastly, and water only penetrates 20 or 30 centimetres, while the inside remains dry.

As for forced drying in its different varieties, let us notice that drying prior to combustion may be profitable with values up to 30 or 40 % of dry weight.

It is very important to eliminate metallic elements and stones that may later obstruct installations and automatic transportation elements.

Different size fractions may be separated during the storage phase according to particular interest in each case, and also to eliminate coarse elements, sticks, branches, etc...

Grinding is used to obtain fine materials such as grains, powder, pellets, ...

If thes residues are submitted to compaction of pure or mixed products, different density bricks are obtained, which have an adequate use in small installations.

All phases of primary chip manufacture add value to the product and thus diversifies its use, but, on the other hand, they add costs to the process which is strongly reflected in the selling price, thus making these elaborate products less competitive than traditional fuel at todays lower prices.

REFERENCES

- I.C.O.N.A. - National forest Inventary - 1977
- I.D.A.E. - 1985 Memory Activities
- ENERGY STUDIES CENTER (ADARO Enterprises). Biomass Evaluation 1980.

CURRENT FOREST WASTE PRODUCTION (1000 Ton/year)

500-250	250-100	250-100 (continuación)	100-20	100-20 (continuación)	20
Badajoz	Avila	Pontevedra	Alava	León	Almería
	Ciudad Real	Santander	Albacete	Lérida	Las Palmas
	Córdoba	Soria	Alicante	Madrid	Sevilla
	Coruña	Valencia	Baleares	Málaga	Zamora
	Cuenca	Vizcaya	Barcelona	Orense	
	Huelva		Burgos	Salamanca	
	Huesca		Cáceres	S.C.Tenerife	
	Jaén		Cádiz	Segovia	
	Logroño		Castellón	Tarragona	
	Murcia		Gerona	Teruel	
	Navarra		Granada	Toledo	
	Oviedo		Guadalajara	Valladolid	
	Palencia		Guipuzcoa	Zaragoza	

POTENTIAL FOREST WASTE PRODUCTION (1000 ton/year)

750-500	500-250	500-250 (continuación)	250-200	250-100 (continuación)	100-20	20
Badajoz	Albacete	Lérida	Alava	Sevilla	Alicante	Las Palmas
Cáceres	Barcelona	Lugo	Avila	Tarragona	Almería	
Coruña	Burgos	Navarra	Baleares	Toledo	S.C.Tenerife	
Cuenca	Ciudad Real	Orense	Cádiz	Valencia		
Oviedo	Córdoba	Pontevedra	Castellón	Valladolid		
	Gerona	Salamanca	Guipuzcoa	Vizcaya		
	Granada	Santander	Logroño	Zamora		
	Guadalajara	Soria	Madrid	Zaragoza		
	Huelva	Teruel	Málaga			
	Huesca		Murcia			
	Jaén		Palencia			
	León		Segovia			

I

■ VIABLE INSTALLATIONS
● LARGE GRINDING AND STORAGE PLANTS

LA CORUÑA
LUGO
PONTEVEDRA
ORENSE
OVIEDO
SANTANDER
BILBAO
SAN SEBASTIAN
LEON
PALENCIA
BURGOS
VITORIA
LOGROÑO
PAMPLONA
HUESCA
LERIDA
GERONA
BARCELONA
TARRAGONA
ZAMORA
VALLADOLID
SORIA
ZARAGOZA
SALAMANCA
SEGOVIA
AVILA
GUADALAJARA
TERUEL
MADRID
CASTELLON
CUENCA
TOLEDO
CACERES
VALENCIA
PALMAS
BADAJOZ
CIUDAD REAL
ALBACETE
ALMERIA
MURCIA
CORDOBA
JAEN
HUELVA
SEVILLA
GRANADA
ALMERIA
MALAGA
CADIZ
SANTA CRUZ
LAS PALMAS

LEBEN PROJECT
THE CONTRIBUTION OF THE LEBEN PROJECT TO WORK ON THE PRODUCTION AND HARVESTING OF PLANT BIOMASS

Prof. Franco GHERI
Coordinator, Leben Project
Abruzzi Region
E.R.S.A.
(Ente Regionale di Sviluppo Agricolo)

1. AIM OF THE PROJECT

The "Agricultural development" programme of the Abruzzi region (1) calls for a series of projects to improve the technology of agricultural production and to stimulate the agro-industrial sector.

Such projects, which will involve a significant increase in energy consumption, are planned to improve production by approximately 50 %.

This situation, coupled with the shortage of energy sources in Italy and the high cost of imported energy, incited the Agricultural Department of the Abruzzi region to foster an agro-energy project - "the Leben Project" - with the aim of identifying available energy resources in the region from alternative and renewable sources and to make it easier to use these in agricultural and agro-industrial processes.

The project has been :

- drawn up with the technical cooperation of DG XIII and DG XVII of the EEC;
- included in the development programmes for the region ;
- entrusted, as far as implementation is concerned, to ERSA - Ente Regionale di Sviluppo Agricolo (Regional Association for Agricultural Development) ;
- submitted to the Ministero Interventi Speciali del Mezzogiorno to obtain funding under Italian Law No 64/1986 and EEC Directive 1786/84.

2. REGIONAL SOURCES OF ALTERNATIVE ENERGY

In the Abruzzi, the available sources of alternative energy are falling water, biomass, sun and wind.

The project has concentrated on falling water and biomass, in view of the widespread availability of these two energy sources in the region.

Nevertheless, studies are also under way on the utilization of solar and wind power.

2.1. Energy from falling water

A preliminary study under the auspices of the Agricultural Department (2) has drawn attention to the possibility of producing hydroelectric power from the river-diversions supplying the region's irrigation networks. The irrigation networks in the Abruzzi are predominantly pipe systems with high-country intakes supplying water under pressure to low-country users. The resulting pressure is almost always in excess of that required for irrigation, so that in some cases it is essential to dissipate the energy, by interrupting the descent of the water, before supplying the user.

When there is no call for irrigation (October-April inclusive), no use at all is made of the water supply system, while natural flows exceed their transport capacity.

In both cases, the excess hydrostatic pressure can be used to produce electrical energy.

The above study, carried out by ERSA in 1984 (2), enabled the installable power capacity (located in 12 power stations) to be estimated at approximately 22 130 kW, while available energy was estimated at 108 million kWh per year.

The first two power stations of the Vomano irrigation network are under construction, representing an installed power of 4 600 kW and a production of 24.5 million kWh per year. The average cost of energy produced by the entire hydroelectric system should be Lit 60/kWh.

2.2. Biomass energy

Particular attention has been paid to two aspects of the regional agro-energy system :

- the production of biofuels (biopetroleum and bio-charcoal) from agricultural and silvicultural residues ;
- the production of ethanol from the plants grown for the production of alcohol.

a) Production of biofuels

Given the particularly high concentration in the Abruzzi of agricultural and silvicultural plant biomass with a high wood content, the Leben Project turned to thermal conversion (pyrolysis) systems which produce biofuels with a high calorific value for use in thermal power plants either on their own or as a partial replacement for fossil fuels.

Studies were made of the availability as feedstock for such process of the following :

- silvicultural biomass already in existence or producible ;
- plant biomass from prunings and grubbings ;
- residues from wine and oil production.

Particular attention was paid to coppices and to marginal agricultural land which could be used for rapid-growth forest.

The regional forestry service (CRF) carried out a study, in cooperation with ERSA, to determine the current level of effective use of forests and the quantity of biomass potentially available for thermo-chemical processes (3).

The study concluded that unused wood matter totalled 3 431 400 m^3.

Subtracting those forests scheduled as conservation areas, to meet urban needs or to be managed as high forest, reduces the available biomass to an estimated 1 940 000 m^3 (approx.), spread over an area of approximately 68 000 hectares. The Forestry Service has calculated that such a density of forest would permit the following annual production figures for plant biomass :

- during the first 12 years — 186 000 tonnes/year
- after the 20th year (operational period) — 135 000 tonnes/year
- during the transitional period from years 12-20 (adjustment period) — 20 000 tonnes/year

In the Abruzzi, plans called for approximately 105 000 hectares of abandoned and marginal agricultural land to be planted in trees destined for energy generation (1).

DG XII - E1 is providing technical and economic assistance for a study currently underway to determine the species and varieties that could be cultivated in the Abruzzi region (4).

The initial draft calls for the cultivation of 20-30 000 hectares of marginal land, yielding 160 000 tonnes per year of plant biomass over a ten-year cycle.

Orchards can provide phytomass from prunings and grubbings. A study of the total extent of this resource in the Abruzzi region has been carried out using the vegetation density values calculated by the statistics service of the Department of Agriculture and the parameters determined experimentally by Professor Enrico Baldini (5).

This study has shown a regional supply of biomass, from prunings and grubbings, in the order of 165 000 tonnes/year.

The processing of agricultural products, particularly wine and oil manufacture, generates an estimated 130 000 tonnes/year of residue in the Abruzzi. Given the fragmentary nature of land holdings in the region, which greatly hinders the collection and concentration of agriculturally-derived plant biomass it was considered prudent to assume that only a quarter could be used for thermal conversion.

On the basis of these assumptions, tree-crops and the processing of related products can be expected to yield approximately 70 000 tonnes/year of plant biomass.

The total quantity of such biomass available in the Abruzzi for transformation into energy can therefore be assessed as approximately :

(in round figures)

- First 12 years			
from forests	(m^3/year)	180 000	
agricultural residue	(m^3/year)	70 000	
	(m^3/year)	250 000	250 000
- From year 12 - 20			
from forests	(m^3/year)	20 000	
agricultural residue	(m^3/year)	70 000	
energy crops	(m^3/year)	160 000	
	(m^3/year)	250 000	250 000
- After year 20			
from forests	(m^3/year)	130 000	
agricultural residue	(m^3/year)	70 000	
energy crops	(m^3/year)	160 000	
	(m^3/year)	360 000	360 000

As previously stated, it is planned to use pyrolysis-based processes for the treatment of plant biomass in the Abruzzi.

From one tonne of biomass with 20 % humidity, such processes yield biofuels (charcoal, oil) equivalent to a minimum of approximately 0.300 TOE (6).

It is planned to use the gaseous fraction, not included in the useful production figures, to power the process and to meet local energy needs.

Once the system is in operation, production of biofuels in the Abruzzi will be equivalent to 100 000 TOE/year. The average cost per kilocalorie of pyrolysis-generated biofuels has been estimated for the Abruzzi at Lit. 0.030 (7).

b) Production of ethanol

Regional agricultural development programmes (1) call for the expansion of irrigation from the present 64 000 hectares to 120 000 hectares over the next 10 years.

This change will also affect the coastal hill zone currently used for drought resistant tree-crops and cereals.

The need to avoid excess agricultural production (market gardening), which tends to emerge in southern regions wherever there is irrigation, imposes the concommitant requirement to introduce less frequently grown crops (such as high-protein, oleaginose and alcohol-generating plants) into the newly-irrigated area.

It is the last-named crop that is of interest to the Leben Project, particularly sweet sorghum, which is especially suited to the climate and soils found in the Abruzzi.

Preliminary trials have confirmed the feasibility of growing sweet sorghum on 10 000 hectares, producing approximately 60 tonnes per hectare of usable plant biomass.

With such a programme in operation, it is estimated that 120 000 tonnes/year of sugar would be available from the canes and from treatment of the bagasse, corresponding to an ethanol potential in the order of 70 000 tonnes.

It is planned to build a pilot plant in the Abruzzi to produce alcohol from sorghum, in order to determine the processing cycles and the yields (8). A preliminary cost-benefit analysis has estimated the production costs of anhydrous ethanol at approximately Lit. 500/litre.

2.3. Thermoelectric energy

Biopetroleum and biocharcoal, obtained from the pyrolysis of biomass, will be mixed to form a stabilized liquid fuel, capable of being used in this state to fire thermal systems as a replacement for liquid fuel of mineral origin (9).

The fuels produced by the Leben system will be used predominantly to fire the 27 MW thermoelectric plant in Fucino, which currently uses mineral oil and methane. The estimated annual generating capacity of this plant is 200 million kWh.

The replacement fuel, and some technological changes planned as part of the restructuring of the plant (10), will make it possible to produce electrical energy at an approximate cost of Lit 85/kWh.

It is also planned to recover the waste heat and to use it in agricultural and agro-industrial processes (such as dehydration, glasshouse-heating etc.).

3. OPTIMIZATION OF THE SYSTEM

In cooperation with DG XII - E1 of the EEC, the region is conducting a series of full-scale studies with the aim of optimizing the regional agro-energy system (reduction of production costs and an increase in the benefits obtained from using the produced energy (4 - 7).

In particular, a study is being made of production and harvesting of agricultural and silvicultural plant biomass.

This study is being carried out in two phases :

- the production of agricultural and silvicultural energy crops ;
- the harvesting of the plant biomass.

The various studies will be conducted in accordance with the following diagram :

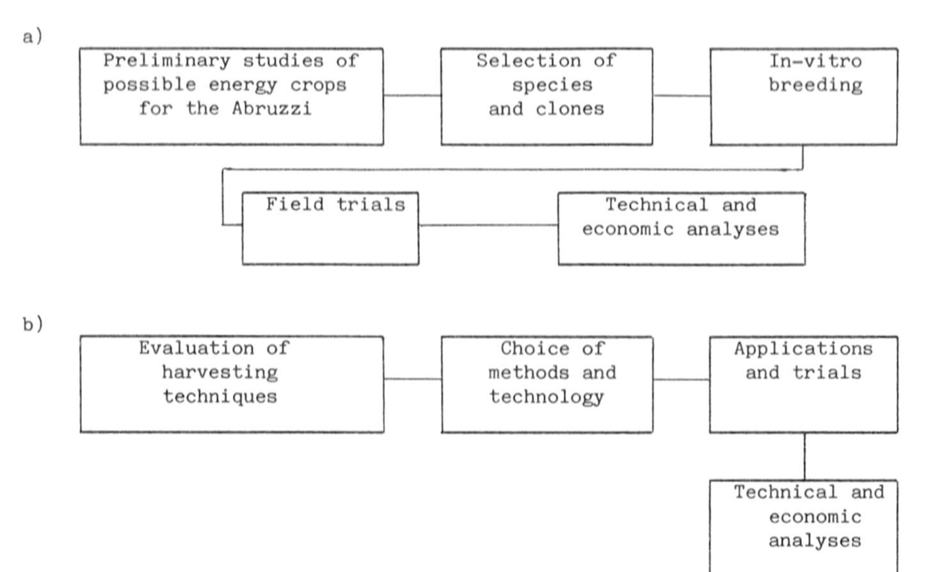

In carrying out the study, ERSA will make use of the consultancy services and cooperation of specialized institutes and Italian and Community, experts. A techno-scientific committee will be entrusted with carrying out the study.

In particular, the project plans call for the following approach to the "harvesting of plant biomass" phase :

- a systematic survey of the harvesting machinery and methods in use in Italy and abroad, in order to provide an up-to-date assessment of the "state of the art" ;
- on the basis of the study already carried out by ERSA and the CRF (4) (which identified the type, quantity and distribution of plant biomass as well as soil and climatic conditions), harvesting equipment and methods will be selected that are available and appropriate for conditions in the Abruzzi ;
- cutting and harvesting trials under varying local conditions will then be carried out in order to :
  . calculate the efficiency of the technology
  . determine the most effective harvesting methods
  . calculate the operational costs for cutting, harvesting and transport
  . supply recommendations that will aid the development of mechanical cutting and harvesting ;
- for the various types of equipment and varying conditions, the trial results will be subjected to an economic analysis in order to determine the harvesting and production costs of the plant biomass.

The entire project is expected to last approximately 30 months. The "plant biomass harvesting" phase should generate its first technical and economic results by the end of 1987.

The cutting trials, which will be carried out at test sites typical of the morphology and vegetation of the forested zones in the Abruzzi, will also serve to calibrate the satellite survey currently being carried out under the auspices of the Centre de Télédétection et d'Analyse des Milieux Naturels (11).

This study is significant in that it enhances the economic value of the Leben Project.

The preliminary feasibility study has in fact shown that harvesting and preparation represent 77 % of the cost of the plant biomass (Lit 45 - 50 000/t).

Running concurrently with this project, the professional training centre at Sulmona, under the auspices of the regional administration, is providing a course for trainers and operators associated with the Leben Project. This course trains technicians and specialized operators in the production, harvesting and preparation of the biomass.

The first group of eight technicians responsible for training the operators will complete the course by the end of this month.

Next year, training courses will be organized for the first thirty operators to be employed in the production phase of the Leben Project.

As defined by the regional programmes, the training course is intended to provide a stable structure open also to those from other regions or countries (particularly developing countries) interested in adopting the Leben system.

4. CONCLUSIONS

The Leben Project is of a special economic and social significance to all zones where :

- unused plant biomass exists ;
- fertile but uncultivated land can be used for rapid-growth forestry.

Implementation of the project makes it possible to :

- exploit agricultural and/or forestry potential as yet unused ;
- improve local conditions as a result of the crop-growing activities ;
- increase employment and the income of the population in the area directly involved.

A preliminary study of the last-named aspect, has made it possible to predict the creation of approximately 400 additional jobs, in the production and processing of the plant biomass, in the Abruzzi region.

The reasonable cost of the energy made available will also generate spin-off industrial activity. At the present time, such employment can only be estimated on the basis of the energy that can be produced and on the average quantity of energy required per job, this figure being 12 kW for the Abruzzi region (12).

On the basis of these parameters, 2 000 jobs can be expected to be created in the industrial sector.

REFERENCES

(1) Linee Programmatiche di Sviluppo Agricolo (Dipartimento Agricoltura 1982) Piani di Sviluppo Agricolo Zonale (Dipartimento Agricoltura 1986).

(2) A 1984 study, carried out by ERSA in partnership with FIME and TECNO-SCIS, on the technical and economic feasibility of energy generation using the Abruzzi irrigation networks.

(3) Disponsibilità di fitomasse forestali detinabili alla termoconversione (CRF-ERSA June 1986).

(4) Research Contract EN 3B/AA/298 I.

(5) Prof. Enrico BALDINI - Analisi preliminari delle destinazioni energetiche alternative dei prodotti e sottoprodotti agricoli : Il

possibile contributo dell'arboricoltura da frutto italiana - Informatore Agrario 1984.

(6) Trial results from the Raiano pyrolysis plant (CRITA 1983).

(7) EC Research Contract - DG XII - E1 No concluded with the Universities of Louvain, Pisa and

(8) Pilot Project (EEC - DG XII - E1 No).

(9) Scheda tecnico-economica di un impianto di pirolisi (ERSA 1986).

(10) EC research contract - DG XII - E1 No

(11) DG XII E1 contract No.

(12) Figures drawn from the 1985 census of Abruzzi agriculture, carried out by COTEI to provide the basis of zonal planning.

SESSION REPORT

Session II was divided into three parts, covering three different aspects, viz. :

- production and technology of energy crops (Lyons, Kocsis, others part of the time, Belletti and Bombelli, although not present for the discussion);
- harvesting machines (Ridacker, Van Landeghen, others part of the time) ;
- economic aspects (Abeels, others part of the time).

In addition there were two general papers on the effects at national and/or regional level of the development of energy crops (Gheri, Ortiz Torres).

Interest was being shown in research institutes and other bodies in the production and technology of energy crops. The crops which hold out the most favourable prospects for production were short-rotation varieties, (willow, poplar and alder, a number of shrub species such as false acacia and broom, species which have been tried more recently) and sweet sorghum and Jerusalem artichokes. Some interesting research is also being done on aquatic plants. The objective is to achieve an annual production level of the order of 10 to 15 tons/ha of dry matter (7-8 % moisture content). On the technical side, research is being done to avoid harming the environment by restricting the use of fertilizers, avoiding excessive monoculture etc. The difficulties of extending and developing these crops are, however, linked to the scarcity of farmers - since the prospects of making a profit are too long-term - and above all the scarcity of machines for automated harvesting and the subsequent processing.

With regard to the problem of harvesting machines, the paper on the "Scorpion" developed by CIMAF of Paris was of particular interest.

The papers on the economic aspects highlighted not only the direct benefits of developing energy crops, but also the indirect or induced benefits, such as environmental protection, the revival of agriculture on marginal land, production of ecologically acceptable fuels, use of local labour, etc.

In the course of the discussions which followed the papers a number of questions were put to the speakers. In particular, more information was requested on the use of the Scorpion on coppice crops in relation to the possibility of the formation and growth of stool shoots, on the productivity of a number of tree species, and on the exploitation processes mentioned by the speakers. Although the speakers' replies were exhaustive, it became clear that, in general, more studies and research are required for the full-scale cultivation of these energy crops.

With regard to the general papers, much interest was aroused by Professor GHERI's paper on the "Leben" project in the Abruzzi region of Italy, which involves a number of bodies and projects at an estimated cost of the order of more than Lit 400 000 000. In the course of the subsequent discussion Professor GHERI gave details of the timetable and the phases of the projects planned and the possibilities of financing the whole scheme.

SESSION III - CROP BY-PRODUCTS, STRAW

Chairman: *K. KOCSIS, FAO*
Rapporteur: *F. DI PAOLA, SVIM Service*

STRAW FOR ENERGY PURPOSE

P. KELLER
Statens jordbrugstekniske Forsøg
National Agricultural Engineering Institute
Bygholm, DK 8700 Horsens

Summary

The only agricultural by-products available in Denmark are straw from cereals.

There are on average 2.1 million tons of straw in surplus each year in Denmark, and great efforts are being made to utilize this surplus instead of burning it in the fields.

Traditional straw has been used on farms for bedding and feeding, but during recent years greater attention has been paid for energy purposes. Straw is used in about 15.000 straw burners on farms and in about 15 district heating plants. The latter utilization is expected to rise considerably in coming years.

During recent years new types of straw balers have been brought on the market, and they differ from earlier types by making cylindric round bales or quadrangular big bales weighing 200 to 500 kg each.

The small bales may be used everywhere with or without the use of advanced handling technology. However, much manpower is required, and the capacity is moderate in connection with transport of small bales to the farm. Also the manual work may be strenuous, especially manual loading and unloading and stacking at the loft.

The use of round bales or big bales increases the capacity considerably in connection with the transport, but front loader or a similar technical device is required for loading and unloading. The pressing capacity of the round baler is the same as that of the high-density baler, while the capacity of the big baler is more than twice as high. When the bales are stored indoors, much space is required in the barn or stack shed for handling the bales.

Fifteen systems of harvesting and collecting the straw are compared. The methods are very different as regards capacity and manpower requirements. This difference is most obvious in connection with the handling of round and big bales, where capacity is high, labour requirements small, and the manual work for gathering in the straw has been eliminated. This may also explain, why these methods have become very popular during recent years.

1. INTRODUCTION

The only by-products in agriculture available in Denmark is straw from cereals.

Table 1 : Straw surplus in 1000 t

Production		
Winter cereal	2650	
Spring cereal	2300	
Rape and seedgrass	550	
	----	5500
Consumption		
Bedding	1100	
Feeding	1280	
Covering of beets	270	
Fuel (heating)	670	
Industry	80	
Surplus	----	3400
		2100

There is on average 2.1 million tons of straw in surplus each year in Denmark, and great efforts are being made to utilize this surplus instead of burning it in the fields.

Traditional straw has been used on farms for bedding and feeding, but in recent years greater attention has been paid to its use for energy purposes. Straw is used in about 15.000 straw burners on farms and in about 15 district heating plants. The latter utilization is expected to rise considerably in the coming years.

2. MECHANIZATION

During recent years new types of straw balers have been brought on the market, and they differ from earlier types by making cylindric' round bales or quadrangular big bales.

2.1. Round bales

Round bales have been known for at least 15 years.

The net-capacity for baling straw in round bales is about 10 tons per hour, while the gross-capacity is only about 5.5 tons per hour, as the baling itself occupies only 55 per cent of the total time in the field. The remaining time is used for turnings, stops and twining the yarn around the bale.

The loading of the bales in the field on to a wagon takes place with a front-loader mounted on a tractor. Normally the capacity of the wagon is 11 round bales in two rows at the bottom, and one row at the top. With a weight of the bales of 250 kg the total load will be 2750 kg.

When unloading, which is carried out with a front-loader, the bales are placed in a barn or in a stack outside.

Figure 1 shows the labour requirements when handling straw in round bales, for some of the most interesting methods. The distance between field and barn : 500 m.

The figure shows that the labour requirement differs from 12 to 14 man-min. per ha. This depends on the size of the load of the wagon.

2.2. Big bales

Straw baled in big bales has special interest where the straw has to be transported over a long distance, for instance when the straw is used for energy purpose in large heating plants.

The net-capacity for baling straw in big bales is measured to be 18 tons per hour on average, while the gross-capacity is 11 tons per hour on average.

Loading of the bales in the field is done with a front-loader mounted on a tractor and can be carried out in two ways :

- One man system, where one man takes care of both the loading and the removal of the wagon in the field.
- Two-man system, where one man takes care of the loading, while another man follows with the wagon. Normally the wagon contains 8-12 big bales.

2.3. Labour requirement and gross-capacity

A comparison of the labour requirement and the gross-capacity is shown in table 2 (1).

Table 2 : Labour requirements and gross capacity for some common methods for handling small and big bales of straw. Yield 3 tons per hectare, distance of transport 500 metres

	Number of hands	* Manpower requirement		Gross capacity tons per hour	
		man-hours per hectare	man-minutes per ton	press-ing	trans-port
High-density small bales					
Common farmwagon, manual loading	1+2	3.4	67	5.3	2.4
Tipping trailer, bale thrower	3	2.2	43	4.6	4.6
Bale boggie, slide	2	2.0	40	3.3	3.3
Stack-liner	1+1	1.4	26	5.3	5.0
Round bales					
8 bales/load, front loader	1+1	1.3	26	5.2	4.9
11 bales/load, front loader	1+1	1.3	25	5.2	5.5
14 bales load, front loader	1+1	1.2	23	5.2	6.2
14 x 2 bales/move, front loader	1+2	1.1	22	5.2	12.8
Big bales					
12 bales/load, front loader	1+1	0.7	14	12.4	8.2
12 x 2 bales/move, front loader	1+1	0.7	14	13.8	9.5
8 bales/load, front loader	1+1	0.8	15	11.2	6.9

* Required manpower :
If pressing and transport are carried out simultaneously, the two figures shall be added up, otherwise not. The second figure refers to the manpower for transportation of the straw.

3. ECONOMY

In order to be able to utilize straw for energy purposes calculations of the cost for harvesting, storing and transportation to heating plants have been carried out (2). All prices below are expressed in øre per kg of straw.

By evaluating the results, the following should be noted :

- The calculation comprises all costs, for instance the variable costs for fuel, maintenance etc., as well as wages and capital costs.

- Some of the tasks might be executed by contractors. In that case the calculated costs should be increased by 30-50 per cent to include administration, transportation of machines etc.

Figure 4 shows the costs in øre per kg of straw for baling, handling (loading, transportation and unloading on the farm), storing and transportation to the heating plant.

The figure shows that the costs for baling straw in small bales vary from 10.4 to 11.8 øre per kg of straw, while the costs for baling round bales are 9.3 øre and big bales 12.0 øre.

The costs for handling the straw on the farm (loading, transportation and unloading) vary for small bales from 7.5 to 14 øre per kg of straw, for round bales 4.1 øre and for big bales 3.2 øre per kg of straw.

The costs for storing the bales in a pole barn, which is rather common, are for small bales 14.5 øre per kg of straw, for round bales 10.8 øre, and for big bales 7.2 øre per kg of straw. If the bales are stacked outside and covered with a plastic film, the costs will be reduced by about 5-6 øre per kg of straw. However there is a great risk of a reduction in the quality of straw in bad weather conditions during the winter period. The transportation costs are calculated for a distance of 10 km to the heating plant. When the straw is transported by tractor with two wagons, the costs for small bales are 23.5 øre per kg of straw, for round bales 13.0 øre, and for big bales 7.3 øre per kg of straw.

4. REFERENCES

(1) Villy NIELSEN : Håndtering af halm. Statens jordbrugstekniske Forsøg, Beretning No. 25.
(2) Søren FOGEDGÅRD : Økonomien ved bjaergning of lagring af halm. Statens Jordbrugsokonomiske Institut. Rapport No. 25.

Fig. 1: Labour requirement by handling Roundbale

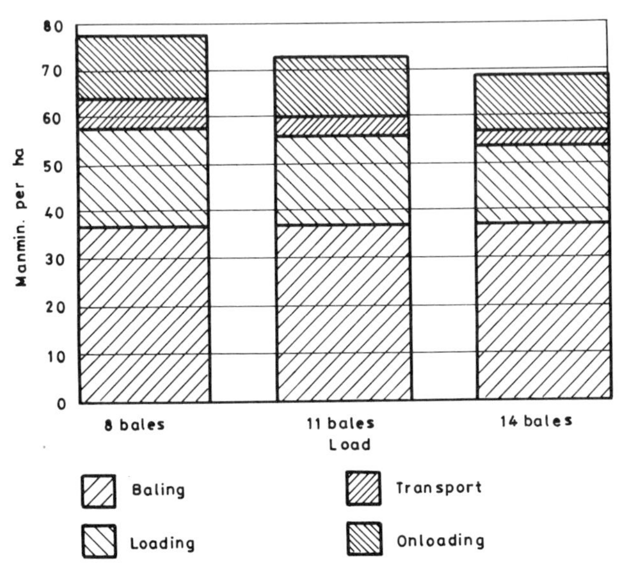

Fig. 2: Labour requirement by handling Bigbales

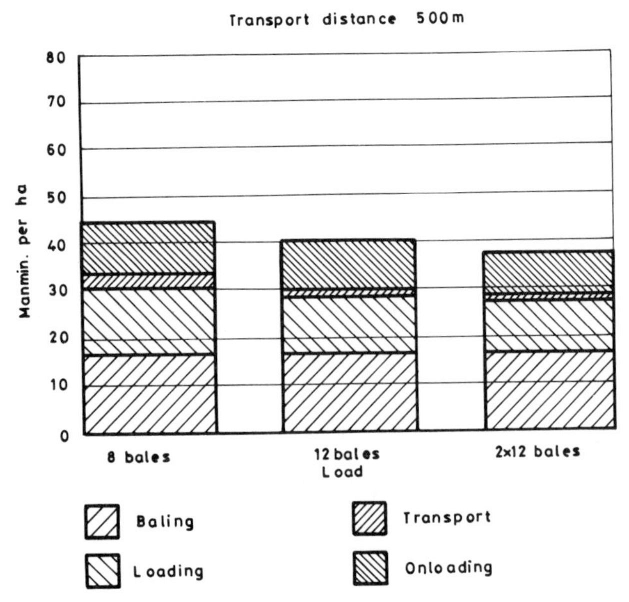

Fig.3: Labour requirement by hand. Smallbales

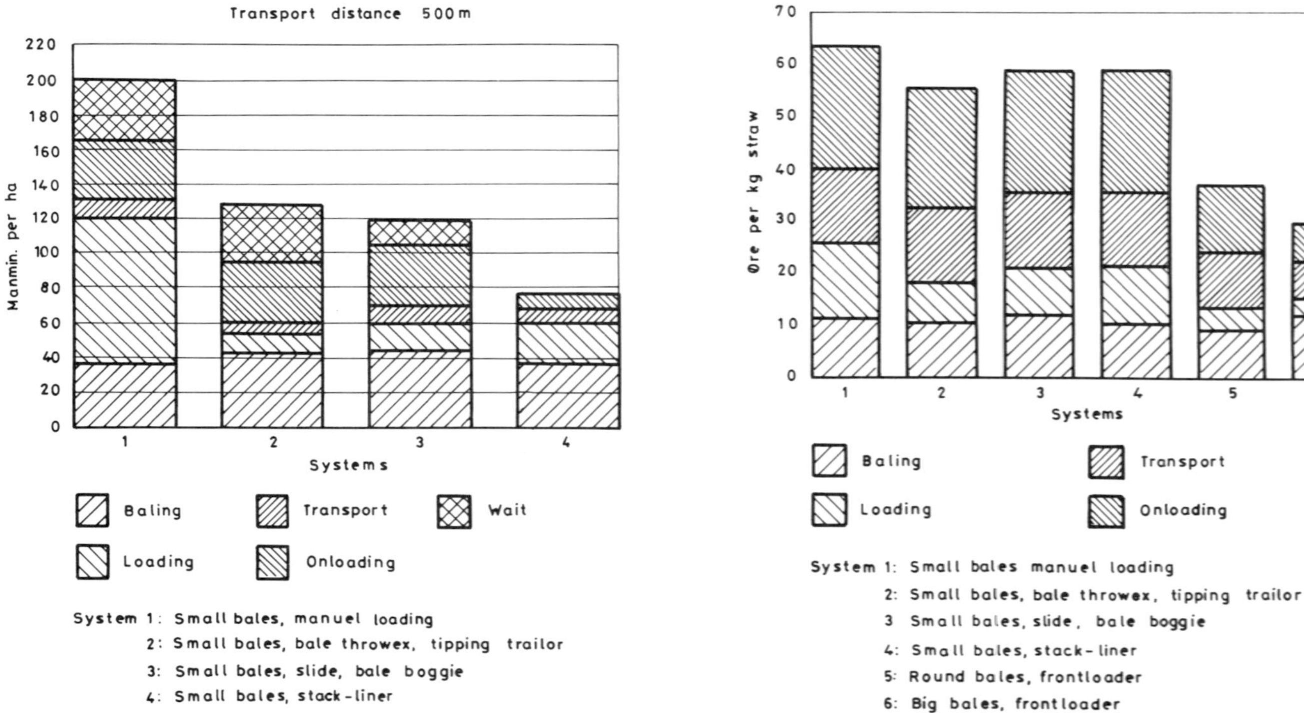

RICE STRAW HARVESTING MECHANISATION CHAINS

A. FINASSI
Istituto per la Meccanizzazione agricola del CNR - Torino

SUMMARY

The world produced about 461 million tonnes of paddy in 1985, together with 500 to 550 million tonnes of straw with an energy equivalent of 8.10^{12} MJ (i.e. 190 million TPE). Combine harvesting enables about 60-65% of the straw produced to be recovered. It has been estimated that the amount of rice straw that could be collected in the United States is 3.6 million tonnes (as dry matter). The figures for Europe and Italy are 1.2 million and 0.65 million tonnes.

Since rice straw is a low-value product with a low weight density, harvesting and transport costs considerably influence the cost of its utilisation.

In this study, the operating efficiency and running costs of three types of working chains for the gathering of rice straw are assessed. Two ricegrowing areas are examined: one in California, the other in the Po Valley, Italy.

The three chains are composed of:

- rectangular baler (standard dimensions) + handling with loader + transport with trailer
- round baler + loading with front loader or mobile pick-up
- handling with self-loading truck + round baler + handling with self-loading-unloading truck.

No substantial difference in baling capacity was observed. The figures ranged from 4 to 6 tonnes per hour for every working chain considered. More marked differences were found with respect to handling and bale transportation, both in the field and on the road. These differences may reach a peak of 80%.

The factors with the greatest incidence are the mass and shape of the bales and their individual mass.

Harvesting costs are essentially influenced by the cost of the chain and the mass of the product handled. Choice of the chain is often influenced by the distance between the points where the straw is harvested and where it will be used, and above all by the purposes for which the straw will be utilized.

An account is also given of the problems associated with the conservation and storage of baled straw.

1. INTRODUCTION

1.1. Availability

In 1985 rice production was about 461 million t, as a by-product, 500-550 million t of straw was produced, possessing a thermal energy potential equivalent to 8×10^{12} MJ (190 million TEP).

Straw makes up about 50% of the dry weight of rice plants, with a significant variation from 40 to 60% according to the cultivar and growing method.

The proportion of straw recoverable depends on the technique of reaping and harvesting (manual-mechanical and on the condition of the field or crop (flooded or lodged).

In this report we consider combine harvesting techniques, as mechanization of straw harvesting is an operation that, logically, should form part of an advanced production system.

Available data (1) show that in California the amount of rice straw recoverable is about 4 t/ha (DM) while in Italy and European rice-growing countries, the figure reaches 3.5 t/ha.

In consequence, the total amount of rice straw recoverable is estimated to be 3.6 million t (DM) in the USA, 1.2 million t in Europe, and 0.65 million t in Italy.

These potential figures must be reduced in practice due to losses incurred because of bad weather, which may cause damage before harvest (lodging) or post-harvest, by hindering natural drying or the use of machinery in the field.

Unfortunately all estimates of the actual availability of the straw turn out to be strongly variable, with a range so wide as to totally upset the use.

1.2. Present day utilization

In the USA the straw is, in most cases, burned in the field, to avoid propagating fungal diseases or to assist cultural operations; in some cases it is chopped and ploughed in, but only rarely or occasionally is it harvested.

In Europe, in some areas in Spain (Andalusia, where 2/3 of the rice is grown) it is buried by puddling, since harvesting is done on submerged fields. The situation is similar in France (Camargue). In Italy, at present only 10% of the straw is harvested, for use as bedding for cattle.

For the most part, 70-80% is burned at the side of the fields, and not more than 10-15% is buried after being chopped by a chopper mounted on the combine.

1.3. Main characteristics of the straw

Rice straw is fibrous, with a high ash content, varying according to the state of conservation, from 13 to 20%. The ash contains 75% S_iO_2, 10% K_2O, 3% P_2O_5, 3% Fe_2O_3, 1.3% CaO, and smaller amounts of Mg, S and Na. The melting point for the ash is about 1250°C, and the warpage point is about 1060 °C (10).

The thermal upper calorific value is about 16 MJ/kg (DM).

2. HARVESTING

2.1. Constraints

The farmers do not tolerate any interference with harvesting operations, and in any case are not disposed to change the method of working to aid in harvesting straw unless they are assured of a guaranteed market that is economically favourable for a sufficiently long time. Thus straw harvesting must be put off until the end of the rice harvest proper, or limited to those hours in the day when rice harvesting has been suspended. Such considerations are valid for all areas where rice growing is mechanized.

2.2. Reduction of water content

At harvest, the straw has a water content of 60-70% (w.b.); however one must wait until this falls below 25% before harvesting and storage can begin. This drying must be left to occur naturally, in view of the low intrinsic value of the product.

Where the field is not lodged, the straw windrowed by the combine reaches a water content of 18-20% in three days, without further operation, under the best weather conditions, in California (1) (Fig. 1) while in Italy the water content can fall to 25% after 4-8 days (Fig. 2-3). This difference is due, essentially, to differences in climate, and in particular to different daily changes in temperature and air humidity.

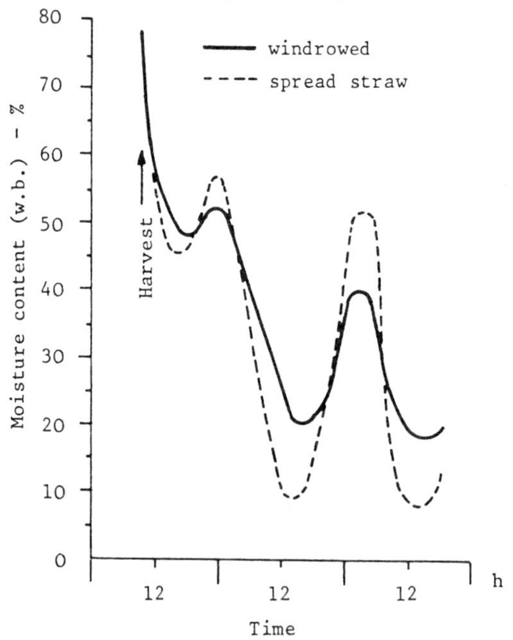

Fig. 1 - Typical drying curves for spread and windrowed straw in good curing weather conditions (Dobie, Miller, Parsons (1))

We should note that if the water content of the baled straw is above 25%, fermentation begins, with loss of dry matter and consequent worsening of quality. Dobie (8) has noted a loss of 5-6% in calorific value and 20% in total proteins.

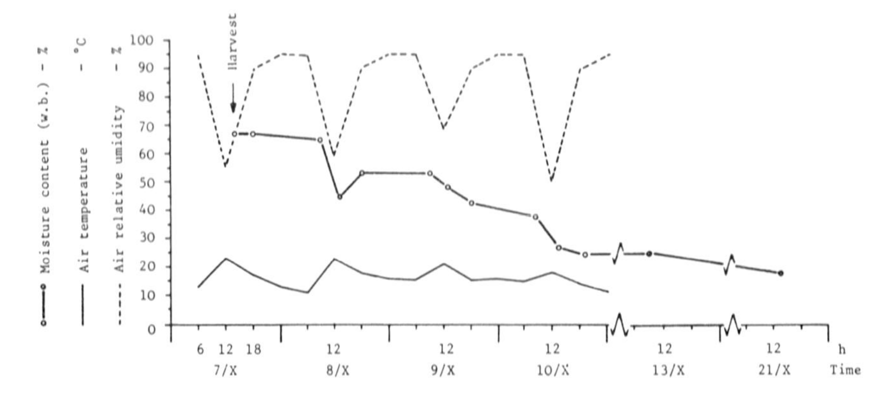

Fig. 2 - Field drying curves for rice windrowed straw in good weather conditions for Po valley

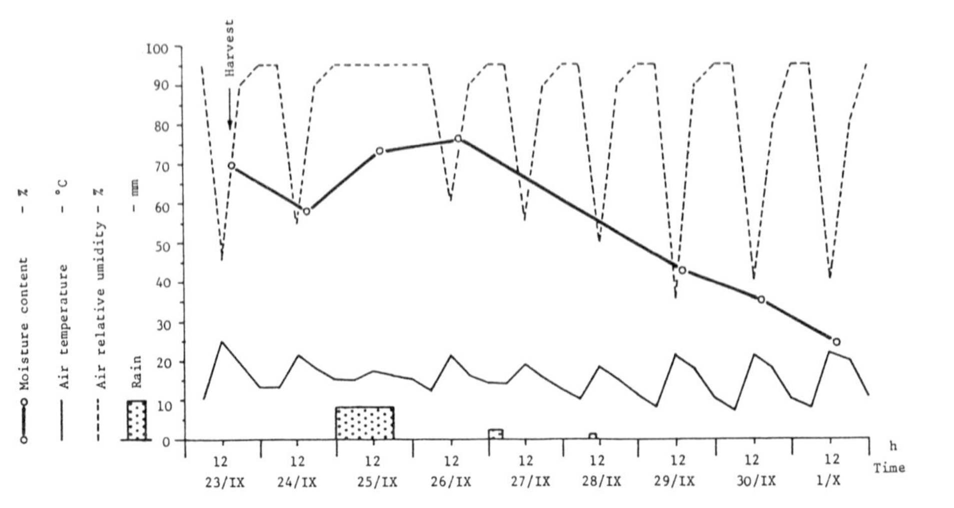

Fig. 3 - Field drying curves for rice windrowed straw in normal weather conditions in Po valley

2.3. Usability of machinery

The feasibility of mechanical harvesting depends on the carrying capacity of the soil; often, due to long and heavy rain, the soil turns to mud and this limits the use of machinery. The problem is particularly important in those areas (in the USA and Europe) where rice is harvested in autumn, when evaporation is poor and sunshine limited.

In California the carrying capacity of the soil has been defined by correlating different soil types with the machinery, and the specific pressures exerted by the vehicles used in the straw harvest (6).

The results underline the importance of substantially reducing specific pressures by using half-tracks, double wheels or special tyres in order to get pressures down in the range of 3.5-4.3 kPa/cm^2. Thus the vehicles should be equipped with large-section tyres, more axles or half-tracks. In fact the technical solutions exist, but they may measurably increase costs and maintenance expenses of the machinery.

3.3. Criteria for the choice of harvesting chain

As straw is a product of low value, limited thermal energy and low density, harvesting methods must add as little as possible to the costs added before final utilization. In fact the destined utilization (combustion, anaerobic digestion, litter or animal feed) conditions the choice of harvesting method.

Technically, the possibilities can be summarized as follows:

- harvest and transport with self-loading trailer;
- with chopper and trailer;
- by rectangular baler (stack hand) + trailer;
- by big roll baler, then field storage;
- by big roll baler, stored at road side;
- by big roll baler, stored at road side, then trailer; + site storage;
- by standard rectangular baler + self-loading trailer + road side storage;
- by standard rectangular baler + self-loading trailer + site storage.

The possible practical combinations are summarized by Jenkins (3) in Fig. 4.

In fact, some chains are not practical, due to particular circumstances in the field, which raise costs too high.

In particular, those chains are not economic where chopping and cubing is done in the field. In the other hand, harvesting with a rake, or with a self-loader on loose straw can be feasible in a few cases when the straw is to be used as litter on the same farm, and for small quantities (about 30 t/year). As the density of dry compressed straw is only 28 kg/m^3 for straw collected with a self-loader, and 52 kg/m^3 for a stacking wagon, any transportation outside the limits of the farm will not be economic.

In strictly practical terms, straw harvesting is reduced to pressing operations on the windrows left by the combine. From analyses done by Ferrero (5), the mean time taken for harvest is 1.08-2.10 h/t.

Today, the harvesting systems giving the most reliable practical results are composed of unspecialized machinery of limited weight involving low inputs in costs and energy, which can work in sub-optimal field conditions without damage to the levelling of the paddy.

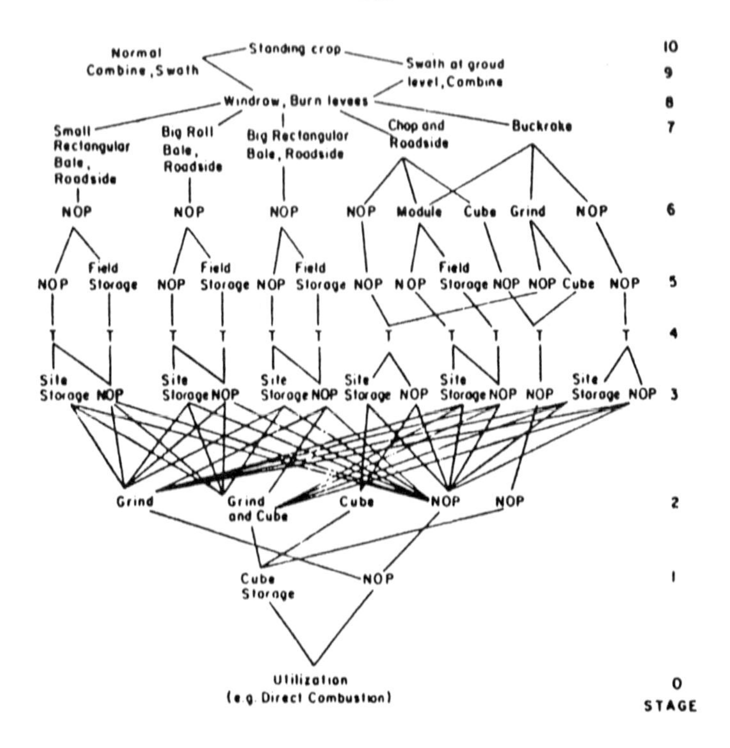

Fig. 4 - Network of alternatives for handling and utilizing rice straw. T = Transport, NOP = No operation (Jenkins, Arthur (3))

4. OPERATIONS

4.1. Baling

This is carried out when the straw has attained a water content of less than 25%.

Irrespective of the type of bales used, the bales are left in the field.

In California some tests were made with a special rotobaler, half-tracked and self-propelled, capable of transporting to the edge of the field two rotobales in addition to that held in the baler (1).

The prototype was operated in 1975, with a work capacity of 3.5-4.5 t/h in muddy fields, and proved to be somewhat difficult to work with when conditions became critical.

In Italy, working with the pick up baler, the mean working time for pressing only was 0.8-1.2 h/ha (Table I), (Fig. 5) falling to 0.6-1.0 h/ha using the rotobaler, with a working capacity equivalent, respectively, to 2.8-3.8 t/ha (DM) for the pick up baler (standard dimensions: 0.36x0.40x0.90 m) and 3-5 t/h (DM) for the rotobaler (bales 1.5x1.2 m).

In both cases, tractors of 60-70 kW were used, with fuel consumptions of 6-8 l/h, without great differences between the two working systems.

4.2. Transport to the road-side

This requires the same amount of time as the baling. With standard rectangular bales weighing an average of 18-20 kg (DM) the loading with bale pick up and two men required 1 h/ha; using a self-loading trailer the operation can be done by one man only, but the time saved is negligible (Tab. I, Fig. 5).

Tab. I - Working chains and operating capacity in Italy (Po valley)

Operating chain		Baling		Road siding or loading		Unloading		Average capacity	Chain operating capacity	Investment cost
	workers no.	bales/h	t/h (1)	bales/h	t/h (1)	bales/h	t/h (1)	t/h (1)	t/h (1)	Lit.10^3
Tractor 4 WD - 70 kW +										45,000
a) strandard pick up baler	1	150÷200	2.8÷3.8					3.3		12,000
1)- trailer + pick up bale loader	2			160÷180	3÷3.5	200	3.3	1.8	1.17 (a1)	17,500
2)- self loading-emptying trailer	1			160÷180	3÷3.5	400	7.6	2.2	1.32 (a2)	27,000
b) round baler	1	12÷20	3÷5					4		17,000
1)- front or rear mounted fork loader + trailer	1			25÷30	6÷7.2	25÷30	6÷7.2	3.3	1.80 (b1)	32,000
2)- self loading-emptying trailer	1			30	7.2	48	11.5	45	2.12 (b2)	33,000

(1) Dry Matter

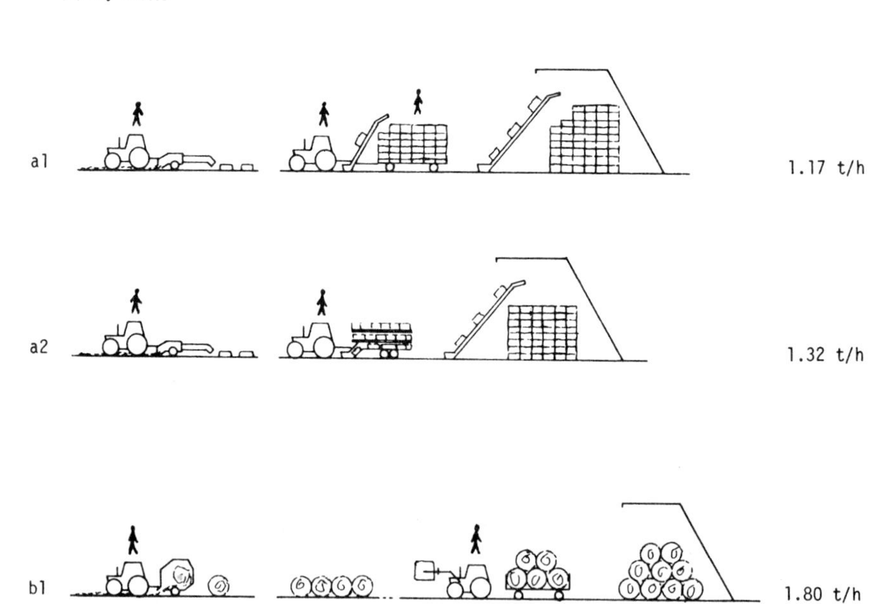

2.12 t/h

Fig. 5 - Operating chain scheme

In fact the self-loading wagons can save an appreciable amount of time only during unloading, whereas loading depends on the operating conditions in the field.

Using rotobales, transport times per ha can be reduced by 50% due to the greater unit weight, and are about 6-7 t/h, using forks loader at the front or back. In italian working conditions, it did not prove convenient to use double forks (at the front and the back) because the length of the fields is on average less than 500 m, and so the advantage gained in transport was not enough to offset time lost in manoeuvring and the increased wheel tracking due to the weight.

4.3. Loading on a trailer and transfer

With the traditional rectangular bales, loading is done in the field, and the straw is transferred using the same trailer. Only when transport distances are longer than 20 km is a truck used, and in this case the load is brought to the farm road or, more often, to the farm.

Experience shows that transfer of straw collected from one ha (3.5-4 t) from the trailer to a truck needs 35 min, using two men. If a self-loader-unloader is used, this time can be reduced by 50% (Tab. I).

Working with rotobales each weighing about 300 kg, transport on a edge of field with fork-loader needs 0.4-0.5 h/ha, and likewise the subsequent loading on the trailer, in this case however, as the operations, take place at different times, the work can be done by a single person.

As regards transport, it was found that rectangular bales of standard size can be loaded to a height of 7 units, with a specific load of 0.4-0.5 t/m^2, while with round bales, a standard trailer can carry 8 bales each 1.5x1.2 m, with a total weight of 2.5-3 t, corresponding to a specific load of 0.2-0.3 t/m^2.

4.4. Storage of harvested straw

Only the rotobales can be stored in the open for a long time without protection, and then it is necessary to place them on a well-drained base, not in contact with each other and tied with string or netting made of plastic. If the bales are stacked, they must be protected form rain.

There are at present available tractor-mounted front fork-loaders that can stack as many as 6 bales; in this case the specific load reaches 1 t/m^2.

However, the large bales produced by the stack hand are almost impossible to stack, and storage in the open causes appreciable losses.

5. ECONOMICS

5.1. The cost of harvesting straw depends greatly on whether the farm machinery is underused.

Under italian conditions, it does not seem that, on average, it is possible to use machinery for more than 200 h per annum, working on little more than 250 ha, and harvesting about 800 t (DM) of straw. These values are about half those encountered in California (1).

Specifically referring to real conditions in Italy, and assuming the current custom rates, it appears that the cost of harvesting and transport at a distance of 60 km is about 50-55,000 Lit/t (DM) (Table II), to which one must add 15,000 Lit/t paid to the farmer, giving a total cost of 65-70,000 Lit.

In California, Dobie and Miller (1) have calculated that costs are only 30-40% of those in Italy, the difference being mainly due to economics of scale derived from the size of the farms and the intensity of use of machinery.

Tab. II - Average rice straw harvesting and transport cost in Italy for 1986

Average cost for custom operation		
	Lit/ha(*)	Lit/t (DM)
Baling		
- standard rect. bales	85,000	25,700
- round bales	75,000	22,700
Road siding		
- standard rect. bales	30,000	9,100
- round bales	15,000	4,500
Transport (**)		
- standard rect. bales	60,000	16,600
- round bales	80,000	21,200
Total cost		
- standard rect. bales	175,000	53,000
- round bales	170,000	51,500

(*) assuming 3.3 t/ha DM straw

(**) assuming 50 km transport distance

6. CONCLUSIONS

Harvesting of rice straw is strongly influenced by the weather, which may impede drying or reduce access of machinery to the fields. The harvesting, and hence the availability of the straw in Italy is therefore quite variable.

Similar problems are met with in California (2). The low value of the product at the moment dictates that available machinery must be used, without the possibility of constructing or adapting machinery, especially for working on wet muddy ground.

The work is separated into different phases: baling and then transport of the bales from the field.

The operating capacity of the different harvesting systems using baling is strictly related to the quantity of straw in the field, to the suitability of the ground for machinery, and to the field size, more than to the technical characteristics of the machinery.

In fact, conditions in the rice field impose notable limitations (e.g. velocity) so this exerts a certain levelling effect on performance.

Regarding the type of machinery, most interest has been raised by round balers, though there is little to choose in either work capacity or reliability between types with a fixed or variable-dimension compression chamber.

Recently, round balers making bales 1.2 m long have been preferred because they allow two bales to be carried side by side on the trailer without overhang.

The baling technique with rotobales allows operation by one person only.

In the best field conditions and weather, in Italy, the yearly working capacity of the better performing harvest chain (round baler + self loading- unloading trailer trailer and fork front loader) range from 700 to 800 t (DM) rice-straw corresponding to 250 ha.

In Italy, the cost of harvest and transport over 50 km is around 60-70,000 Lit/t. The final destination of the straw may also influence the choice of baling technique.

The use of cuber, stack hand, and all machinery above a certain mass is excluded due to the limited load-bearing capacity of the rice fields, and the absolute need to leave the levelling of the ground undisturbed.

REFERENCES

(1) DOBIE J.B., MILLER G.E., PARSONS P.S. - Management of rice straw for utilisation. Transactions of the ASAE (1977) 20 (6), 1022-1028.

(2) JENKINS B.M., SUMMER H.R. - Research needs and priorities for harvest and handling of agricultural residues. ASAE Summer Meeting 1985, paper No. 85-3066.

(3) JENKINS B.M., ARTHUR J.F. - Assessing biomass utilisation options through network analysis. Transactions of the ASAE (1983) 26 (5), 1557-1559.

(4) JENKINS B.M., HORSFIELD B.C., DOBIE J.B., MILLER G. - Agricultural residues: renewable energy for utility power companies. Transactions of the ASAE (1981) 24 (1), 197-207.

(5) FERRERO A. - Raccolta delle paglie. Tecniche attuali e nuove tendenze. L'Italia Agricola, (1981) 118 (2), 100-136.

(6) KAMP J., DONKERS J., CHANCELLOR W., DOBIE J. - Mobility of rice straw collection equipment. Transactions of the ASAE (1983) 26 (2), 372-377.

(7) JENKINS B.M., TOENJES D.A., DOBIE J.B., ARTHUR J.F. - Performance of large balers for collecting rice straw. Transactions of the ASAE (1985) 28 (2), 360-363.

(8) DOBIE J.B., HAQ A. - Outside storage of baled rice straw. Transactions of the ASAE (1980) 23 (4), 990-993.

(9) BENGTSSON N. - Hay-handling - Swedish Institute of Agricultural Engineering, Meddelande n. 381, Uppsala 1979.

(10) OSMAN E.A., GOSS J.R. - Ash chemical composition, deformation and fusion temperatures for wood and agricultural residues. ASAE Winter Meeting 1983, paper No. 83-3549.

(11) JENKINS B.M., EBELING J.M. - Thermochemical properties of biomass fuel. California Agriculture (1985) 39 (5, 6), 14-16.

(12) JENKINS B.M., ARTUR J.F., MILLER G.E., PARSONS P.S. - Logistics and economics of biomass utilisation. transactions of the ASAE (1984) 27 (6), 1898-1904.

HANDLING AND STORAGE OF STRAW AND WOODCHIPS

Dr. A. STREHLER
Technical University Munich
Bayerische Landesanstalt für Landtechnik
Weihenstephan
D-8050 Freising - FRG

Summary

Combine harvesters leave straw in swathes in the field. Depending on the moisture content, baling or chopping is done 1 to 4 days later to ensure that the straw is dry enough, i.e. a moisture content below 18 % (low losses, no self-heating and high quality combustion). Chopping in the field requires a high labour demand. High density small bales are easy to handle manually, but lead to a high labour demand at harvest time. With big bales labour demand is much lower. For high density big cubic bales, increased transport weights are possible and the storage volume is less than with roto bales. The production of straw briquettes and pellets is the basis for heat generation in small furnaces for use outside of agriculture.

Weak wood is utilized as woodchips with particle size of 1 to 10 cm. Chipping usually takes place in the forest using pre-dried trees from thinning. Chips or chunk wood has to be dried additionally in storage with ambient air, usually pre-heated.

1. INTRODUCTION

Energy from biomass is an important way to improve security in energy supply for the future in the European Community. There are two groups of resources. A : waste biomass like surplus straw in agriculture and weak wood in forests, B : biomass produced as energy carrier in order to reduce problems of financing the agricultural market of the European Communities. There are some different types of energy carriers like oilplants (rape, sunflowers, euphorbia), reed, fast growing cereals. Harvesting and handling technology is similar to straw technology. The basical problem is low energy content of wood and straw, causing high technical effort in handling and processing.

2. STRAW HARVEST

Straw is harvested mainly from cereals, seldom from rape, beans and maize because of moisture content, storability and value for combustion. Table 1 shows medium moisture contents of different types of cereals, rape, beans, maize in harvest time and the days after under South German conditions.

Table 1

crop	harvest time	moisture content day of harvest	days after harvest 1	2	3	4
barly	July-August	16-20	14-18	13-16	13-15	12-14
rye	August	16-22	16-19	14-16	13-15	12-14
wheat	August	18-24	16-21	14-18	13-16	12-15
oats	August	20-26	18-23	16-20	14-17	13-16
rape*	July	40-60	22-30	18-22	16-18	15-17
beans	September	35-60	28-45	22-30	20-25	19-23
maize	October	55-70	50-60	45-50	40-47	37-45

*) Apfelbeck (1)

For proper combustion and storage without losses the moisture content should be below 20 % (w.b.). The reduction of moisture content is strongly correlated to weather conditions. Beans and maize are harvested late, therefore the reduction of moisture content is very slow. When drying in swathes, reduction of moisture is relatively slow, as shown in table 1. Additional tedding is improving drying velocity. Spreading few cut straw directly with the combine harvester has the same goal, this will be investigated in the frame of an EC research contract. Figure 1 shows the decided equipment for a combine harvester.

Background of research work decided :

It has been observed, primarily by big bale combustion, that operational problems with the de-baler arise due to moisture in the straw and additionally, the combustion quality is no longer satisfactory, producing a negative effect on the environment (high emissions) as well as a drop in performance and tar accumulation in the system. The reason is the high and very differing moisture content of the straw in the field, mainly caused by the thick swathes thrown out by large combines which dry very slowly especially when re-wetted by rain. Spreading the straw for drying in an additional operational step is too time consuming in the harvest season, so that a loosening and spreading directly from the combine is needed. A prototype is to be constructed, which will be mounted on a large combine. The big bale combustion system at this place will be used for measuring the effect. The chopper mounted on the combine harvester is to be equipped with only around 20 % of the usual cutting blades, whereby the straw is pre-cut to the point that it is easily spread but can still be picked up well with the machine. This pre-cutting should also facilitate de-baling and straw feeding to the combustion chamber. The straw spreader will be attached under the chopper ; initial tests will be made with a receiving plate having radially mounted prongs as demonstrated in figure 1.

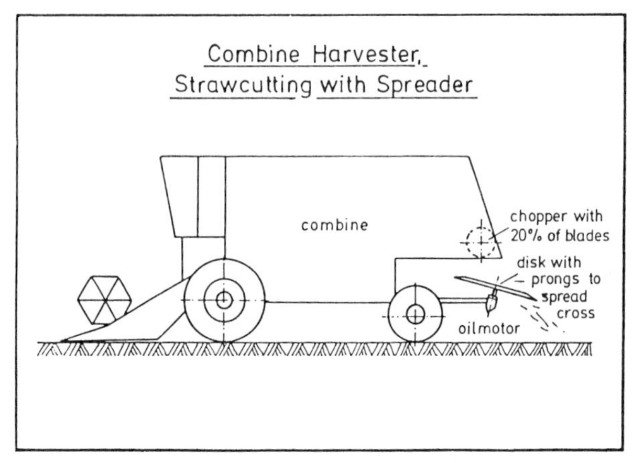

Figure 1 : Mounting a straw wide spreader under a modified chopper.

After drying, broad spread straw has to be windrowed. For harvesting many systems had been developed, as shown in figure 2.

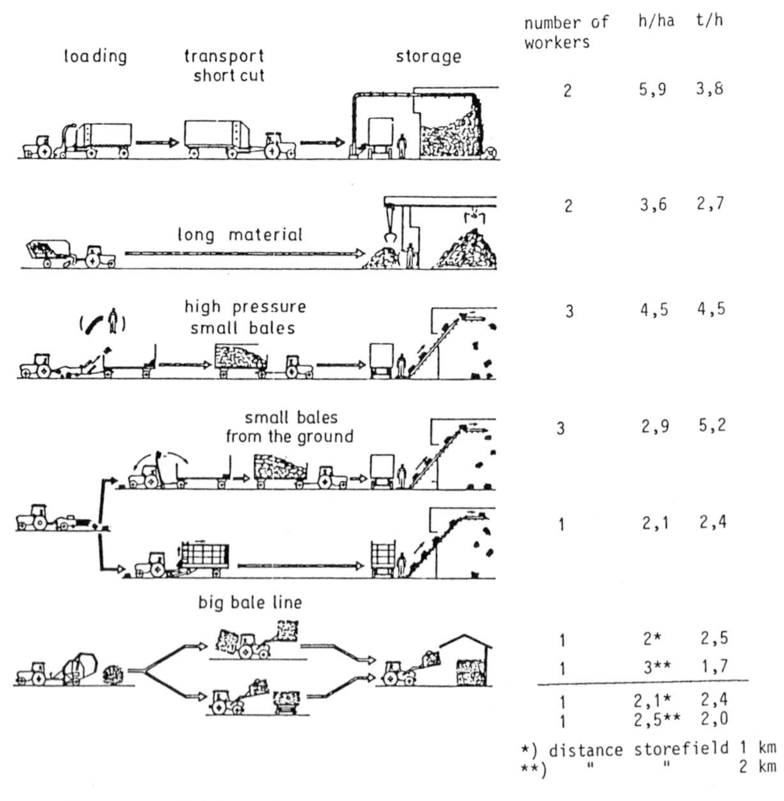

Figure 2 : Systems of Straw Harvest

Two main systems are engaged as shown in figure 3 :

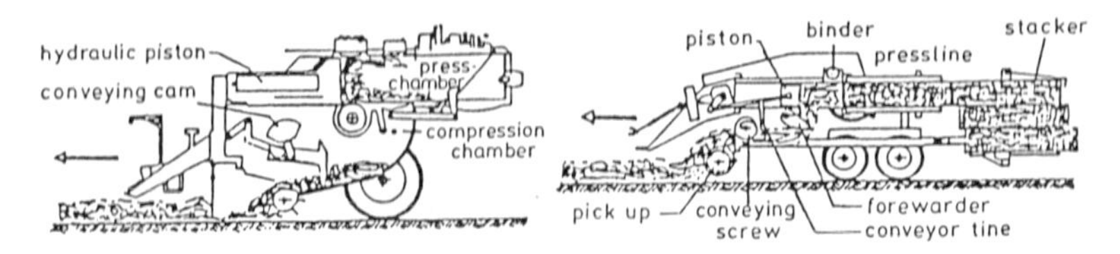

Figure 3 : Cubic big bales

Big bales are becoming increasingly important, because there is less labor demand. The trend in big bales is towards cubic high density big bales, when long distance transport is necessary.

In the first case one big bale is set on the ground, in the second case 3 bales are put down together. The smaller bales are more easy to handle on farms.

In cases of short distance to application of straw or energy carriers the chopping system is engaged with a self-propelled high power chopper, using different attachments, as shown in figure 4.

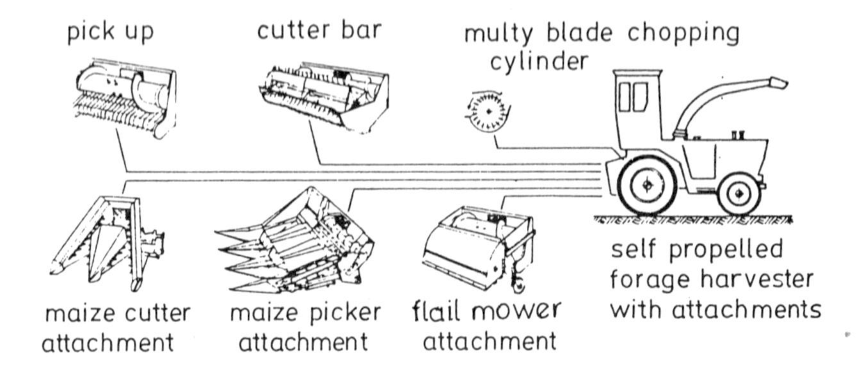

Figure 4 : Forage harvester

Some research work had been done in briquetting cereals completely (corn + straw). Engaging a screw press the moisture content could be in the range of 20 to 25 %. The briquettes became extremely hard while pressing the starch out of the kernel. Also combustion results had been very promising. Therefore it seems to be sensible to create a self-propelled briquetting machine harvesting the whole crop as energy carrier. For this machine 1000 horsepowers are necessary to harvest 1 ha total plants (15 t/ha) in 2 hours. Straw briquettes are a prerequisite for making excess straw available to consumers outside of agriculture. Test results have also shown, that rape straw can be compacted with less effort than cereal straw. To use rape as an energy plant for relieving farm market problems, a series of further tests are necessary to determine exactly the work load from the harvest to the compaction of the straw. Previous attempts at straw briquetting have indicated one great disadvantage, this being the excessive need of equipment and energy for straw chopping. By employing an exact chopper (used for chopping maize) the machinery and energy requirement should be reduced. Up to now, more energy is required for pre-chopping (50 - 70 kWh/t) than for pressing (40 - 65 kWh/t) using a piston press. The energy extensive grinding and beating is to be replaced by an energy saving exact cut. An essential point in improving the product quality lies in pre-heating the raw material which reduces the resilient forces after pressing. Initial tests are very promising. Further research work in this field will be done at the Technical University of Munich.

In the production of energy carriers, the moisture content of the complete plant is of urgent interest. When producing rape,

- oil will be used as liquid fuel,
- the presscake as animal feed,
- the straw as solid fuel.

The disadvantage is the high moisture content of straw, when harvesting. 2 to 3 days drying on the ground diminishes the moisture content from near 50 % (w.b.) at harvest time to 16 - 20 %, as shown in figure 5.

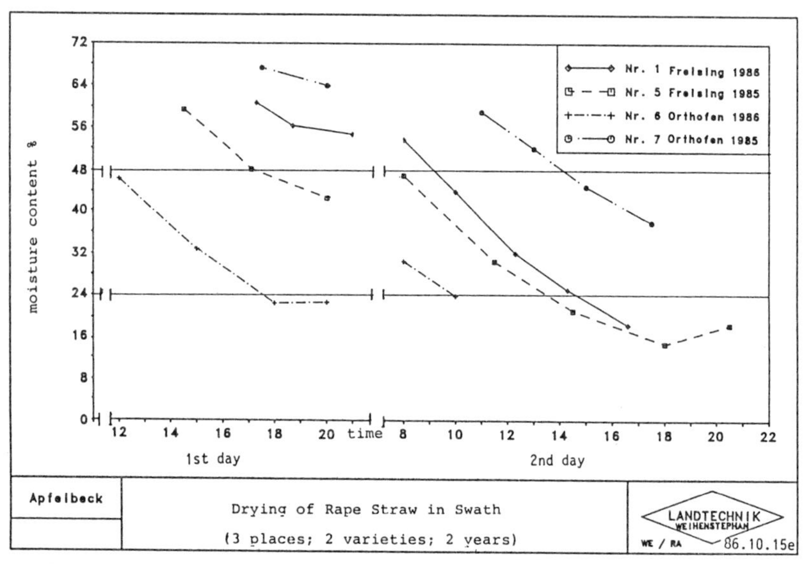

Figure 5 : Result of rape straw drying on the ground.

When using the whole plant of annual crops as solid fuel, different systems for harvest, storage, grinding and calibration are possible, as shown before in the case of straw. New research activities on TU Munich showed best results in briquetting annual crops including kernels and straw with a total moisture content of 18 to 26 %, when engaging a screw press. The briquettes had a density of 1,1 t/m^3, the combustion quality was much better than with pure straw. Using material with higher moisture content in the case of screw presses, will require a relatively long harvesting time. Therefore a briquetting machine added to a self-propelled chopper causes lower fixed annual costs when only used some days. Therefore a self-propelled chopper combined with a briquetting device seems to make sense. Figure 6 shows the moisture content of total wheat plants in the last growing period to the harvest time.

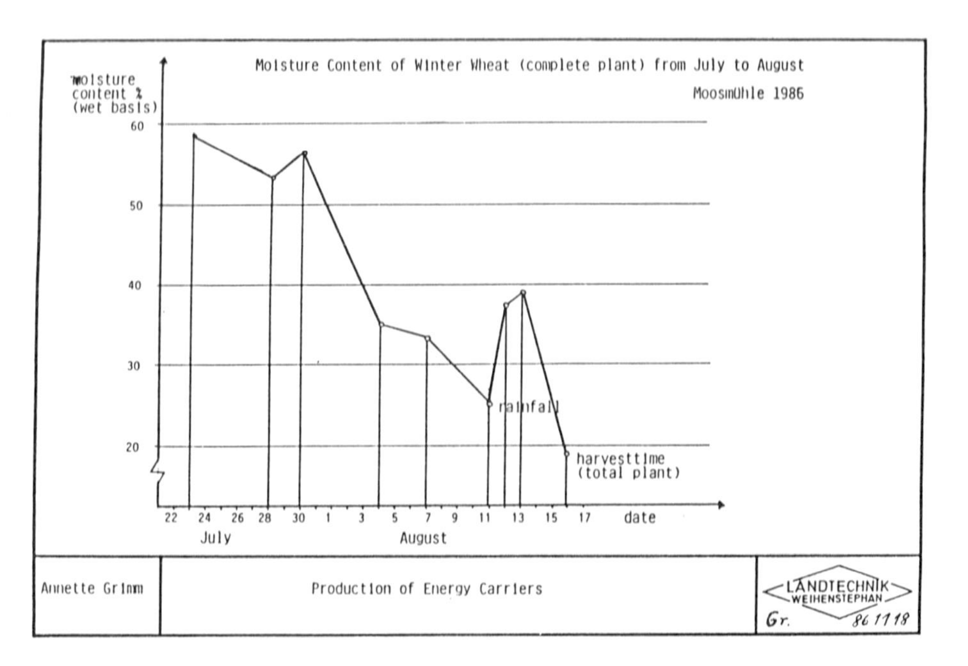

Figure 6

Harvest time could start on August 10th up to September. Only one variety of cereals could have 30 harvest days. Other annual crops could have harvest times up to 60 days a year. In this case a self-propelled briquetting machine, costing 500 000 DM would cause annual fixed costs of about 70 000 DM or 116 DM/h respectively 29 DM/t (4 tons/h). Energy and labour costs would be roughly 120 DM/h respectively 30 DM/t. These main cost factors are relatively low, therefore it is worthwhile to construct a self-propelled chopper-auger-press. The basis machine could have a multi blades cylinder and an attachment for cereals (No. 2, figure 7) or an attachment for maize (No. 3, figure 7).

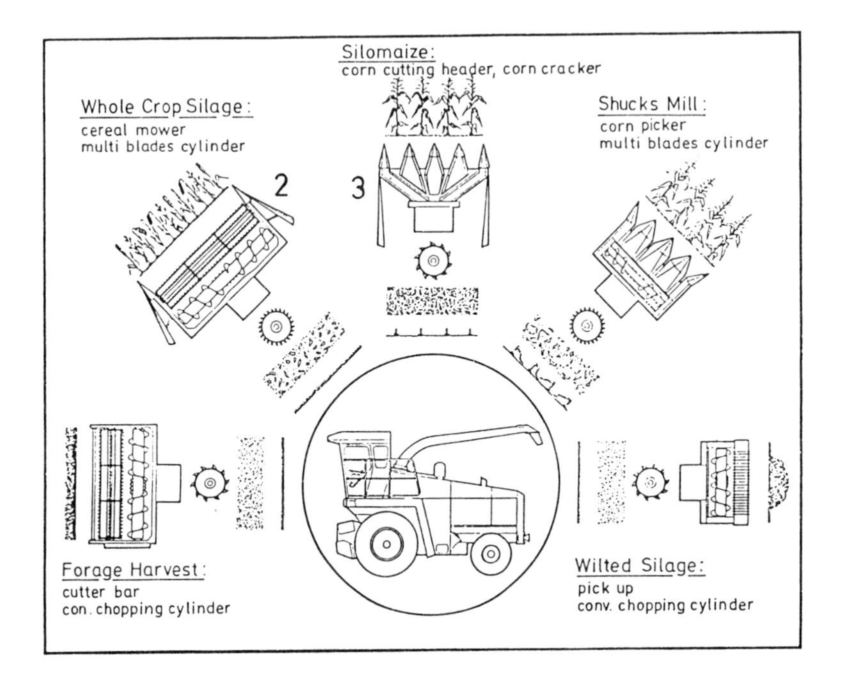

Figure 7 : Basic self-propelled chopper with several attachments, different cylinders.

3. WOODCHIPS PRODUCTION : HARVEST, TRANSPORT, CONVEYING, DRYING

Many systems of producing wood chips are well known, and can be divided into cylinder-, disk- and screw-chippers. They might be mounted on tractors or pulled on their own wheels. Charging by hand requires 3 manpower/h when chipping 1 t/h, charging by crane requires 2 man/h when chipping up to 5 t/h. In case of energy plantations via short rotation forestry self-propelled choppers can be engaged as shown in figures 8 and 9, cutting 1 to 3 rows at once.

When precut and predried (about 1 year) a pick-up can be used instead of handcharging (fig. 9) as already used in Denmark and Sweden, when thinning complete rows of forest for the first time.

Power demand for chipping is about 5 - 10 kWh/t for wood-chipping. Pre-dried wood needs very sharp blades, taking more power than wet wood. Special high performance self-propelled wood-chippers are to be constructed for South European conditions. The Scandinavian types will be improved.

Transport of chips is done best in tipping trailers.

For discharging woodchips from store, rear- and frontloaders are very reliable. Automatic mechanical discharging equipment for special containers is very expensive because woodchips become dense when stored longer than 2 months.

Conveying is done best with belt conveyors, cheaper but less reliable with screw conveyors.

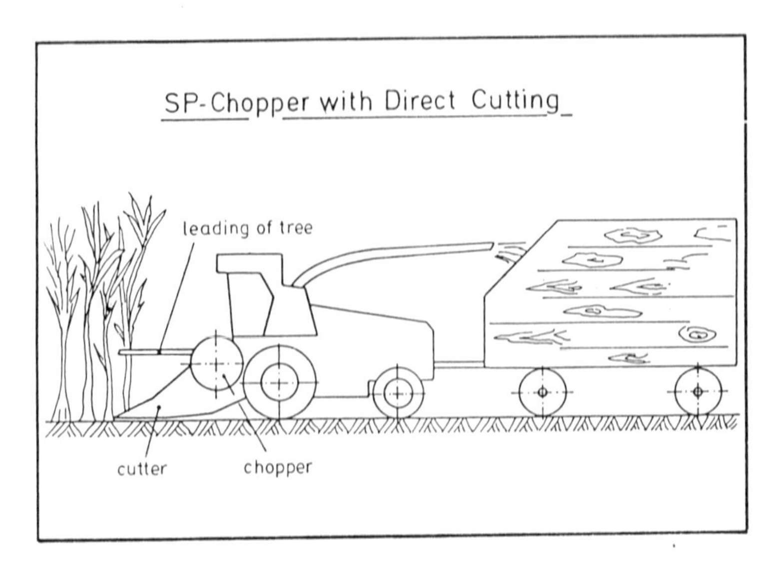

Figure 8

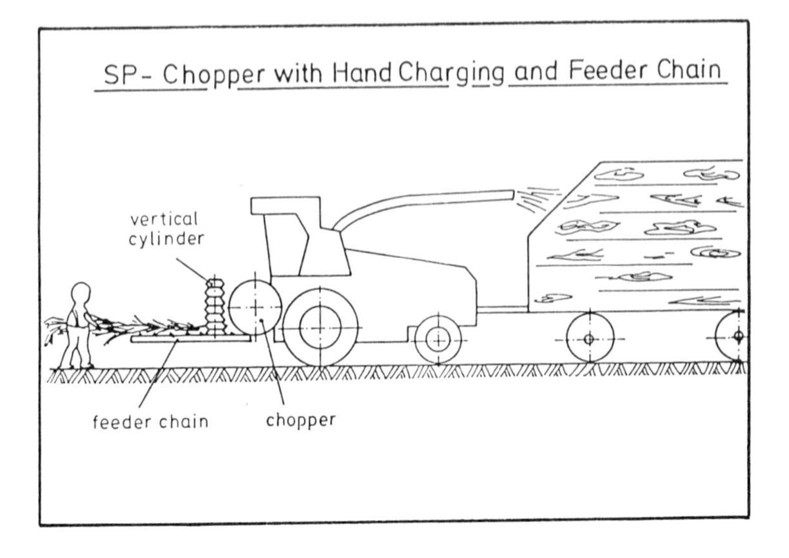

Figure 9

Woodchip drying :

To operate combustion systems at high efficiency without polluting the environment, the moisture content must be reduced to under 20 % (wet basis). For weak wood, the possibility of pre-drying on site exists, thus leaving the needles which contain minerals in the forest. Various experiments have already been conducted to determine the optimal cutting time, which is a compromise between drying effect and prevention of the bark beetles. However, unfavourable weather in a given year can force the immediate collecting of moist weak wood if a pest epidemic is foreseen. Then it is necessary to dry the collected wood chips mechanically. One alternative also exists in de-limbing the weak wood and transporting the poles out of the forest to a prescribed distance (protection from bark beetles) and allowing them to dry in the ambient air. Later, the wood is chipped at this location and stored for final drying. Site-dependent tests of pre-drying are still continuing.

Woodchips are made in many different sizes. Fine chips, which can be dosed very exactly, are relatively difficult to dry, as they resist air circulation. Warm air driers are suited such as those used for grain drying. Coarse wood chips, in contrast, have a very small air flow resistance ; appropriate fans can be driven with little power demand. If simple wooden blocks are split (hydraulic splitting device) e.g. in lengths of ca. 150 mm, only a relatively small flow resistance must be overcome by the drying fan. Here, one could even dry with pure convection. The wood would have to be piled on a grate, so that a vertical airflow can be established. This last method, however, is not applicable to very thin weak wood, as it has too many fines, which hinder natural convection. In most applications, coarse and fine wood chips will be present.

Basic experiments on the drying rate and determining the energy requirements had been done at Weihenstephan. Larger units have been installed on different farm operations. A combination of warm air solar collector and wood fired ovens is an energy saving solution. Wood chips are especially well suited for solar collectors, as the available drying time is much longer than for grain or hay. Figure 10 shows an example for measuring drying time with wood chips independent of the original moisture content, air velocity and drying air temperature.

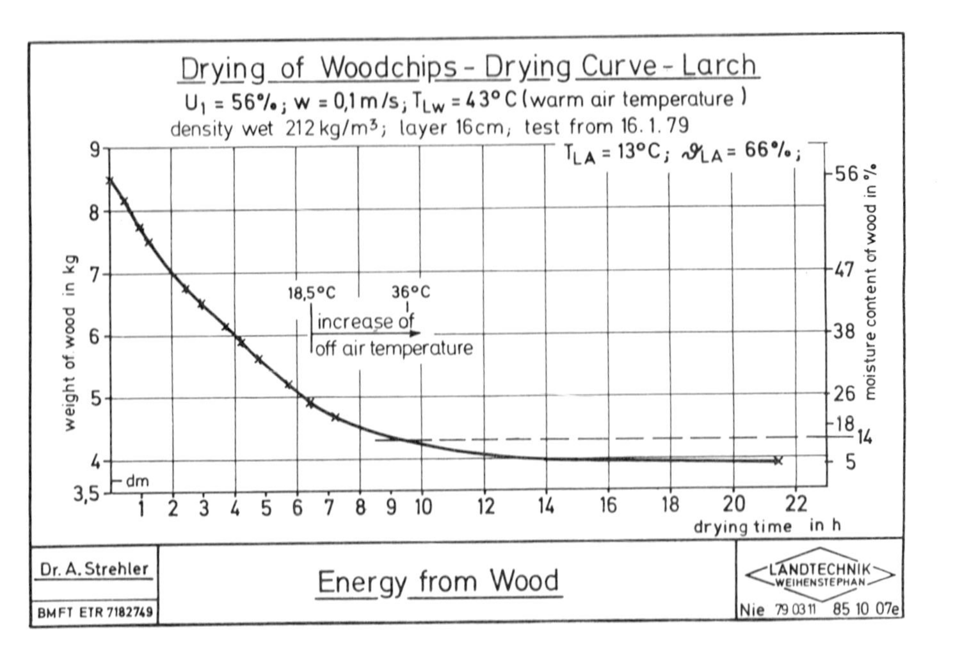

Figure 10

4. REFERENCES

(1) APFELBECK, R. : Unpublished results of research work carried out in Landtechnik Weihenstephan.

(2) STREHLER, A. : Die Trocknung von Holzhackschnitzeln ; Schriftenreihe Landtechnik Weihenstephan, März 1984.

AVAILABILITY OF BY-PRODUCTS : FEASIBILITY STUDY ON PRODUCING ALCOHOL FROM BIOMASS

E. CRESTA, A. GIOLITTI
Department of Chemistry, Siena University

Summary

This paper examines a number of basic aspects of utilising ligno-cellulosic agricultural by-products for producing fuel alcohol in the Province of Siena. Since we thought that a study based on statistics was not enough to get a real picture of the use of biomasses, we decided to make a survey in a limited area.

Data collected from direct sources (stores and local firms) showed that current supplies were of the order of 9 660 t dry weight per 18 000 ha. In an effort to determine future supply figures for the by-products a survey was carried out on the basis of a land use plan which shows the amount of irrigable land within the area concerned. This survey shows that the increased production of biomass would be such that a limited area (approximately 3 000 ha) would yield greater quantities than those produced at present in the total area covered. This paper confines itself to quantifying the by-products available, as a basis for evaluating their energy potential.

At the same time we determined the successive planning phases required for the appropriate use of these resources in terms of the environment, economics and development. These are :

- rational organization of harvesting and storage ;
- detailed study of the effects of extraction of organic material on soil fertility ;
- integration of new industrial activity and animal husbandry.

INTRODUCTION

Among the raw materials which can be used for producing fuel alcohol the ligno-cellulosic waste from various agricultural crops is of particular interest in that it is produced in large quantities, but is rarely used and, under the present farming system, involves additional transport costs.

It should also be stressed that the non-utilization of the by-products means a considerable reduction in useful photosynthetic efficiency. For example, the photosynthetic efficiency of grain maize and other cereals is 0.2 and 0.1 respectively, compared with 1.1 for sugar beet which is used completely.

This paper is concerned with quantifying agricultural by-products of herbaceous crops in a limited area in the Province of Siena.

The data we give are taken partly from a research project carried out on behalf of the ENEA by the Team di Roma company, in which the authors were involved, and partly from studies carried out by the Chemistry Department of the University of Siena, which form part of a project involving entrepreneurial, agricultural, institutional and scientific aspects. This was sponsored by the Province of Siena, and concerned the installation of a plant for producing ethanol from residual agricultural biomass.

Since we thought that it was not enough to make a quantifying study based on statistics, as this would be too general to pinpoint the by-products and to define the problems relating to their utilization, we restricted our study to smaller areas, in which it was possible to obtain more reliable and detailed data through direct observation and contact with the local farmers.

Only by studying the actual farmland in this way is it possible to identify the factors which determine the effective availability of a given by-product, i.e. its location, concentration, distribution and any alternative outlets, and at the same time to assess the factors favouring the recovery of such material, i.e. the lie of the land, the number and type of harvesting machines, the structure of the farms and farm association, and the road system.

Some aspects of the utilization of by-products for energy, such as preserving the fertility of the soil which produces them and integrating the new industrial activity with existing agriculture and animal husbandry are so complex that a specific phase of the study must be devoted to them, since they form the basis for proper energy planning, which will promote the development of the rural areas and the exploitation of resources while taking due account of the environment and the farming community.

METHODOLOGY

Firstly we determined the total quantities of ligno-cellulosic farm by-products in the Province of Siena, using ISTAT data. This study showed that there would be 303 475.8 t dry weight of by-products of herbaceous crops. Further details were obtained by concentrating our activities on specific areas where :

a) the ligno-cellulosic material was present in considerable quantities and the yield per hectare was good ;
b) the difficulties of the recovery operations (harvesting, transport and storage) were at a minimum ;
c) the use of advanced cultivation techniques and high levels of entrepreneurial and associational activity in both the agricultural and animal husbandry sectors could be used as points of departure for the field study, organizing the harvesting of the by-products, and programming integration with the new industrial activity (production of bio-ethanol).

An area was selected on the basis of the following criteria in order to ensure that these conditions were met :

1) Irrigation : Irrigation, which is not very widespread in the Province, has a beneficial effect on yield per hectare and tends to go together with a reasonable level of mechanization, generally reflecting greater predisposition to innovation.
2) Production : i.e. prevalence or expected development of the crops under consideration.
3) Lie of the land : In a predominantly hilly district such as Southern Tuscany we selected areas with slopes with gradients of less than 6 %, both because yield per hectare is greater and also to avoid at this initial stage of the study the problem of hill farming, such as higher costs, soil erosion etc.
4) Mechanization : The suitability of machines already available in the area for recovering the by-products was considered a favourable factor in that it helped to reduce the initial investment costs and at the same time increased efficiency thus reducing running costs.

FIELD STUDY IN THE DISTRICT OF SOVICILLE (Province of Siena)

The area of Sovicille has a number of significant features, the importance of which was confirmed in subsequent field studies, i.e. :

- extensive maize-growing, irrigated for the larger part and concentrated in the Rosia Plain (approximately 3 600 hectares) which straddles the Merse river and where increasing areas are being used for a number of industrial crops such as sunflower ;
- widespread and expanding irrigation (on the Plain) ;
- agricultural mechanization in terms of above the Associazione Intercomunale average hp/ha of arable area ;
- a highly developed system of agricultural and animal husbandry associations : two communal storage plants (Coop. maidicola and Cons. agrario Provinciale) with a total capacity of 12 000 t, a communal stable of average dimensions and a services cooperative.

The area studied comprises approximately 18 000 hectares and is bounded to the north by the outskirts of Siena and to the west and south by average to high hills, large areas of which are covered with vegetation. To the east, where the main crops are wheat and barley, grown in an area of low hills and gentle slopes, the line was drawn so as to include the farms which send their cereals to the storage facilities mentioned above.

AVAILABILITY OF LIGNO-CELLULOSIC BY-PRODUCTS. Fertility of the soil

The base-line values for the quantities of crops (maize, wheat, barley, sunflower and rice) produced in the area under study were obtained from the storage plants (current estimates) (Table I). The figures for the corresponding by-products were calculated by applying the coefficients for the ratio between by-product and main product, as given in the literature (Table II). It was found that the area of Sovicille produces 12 200 t of by-products dry weight. The sunflower stalks and barley and rice straw are not at present used, being generally burned in the field, and therefore 100 % of this is available. The same applies to the wheat straw, of which only a small fraction (3 %) is used in animal husbandry locally.

This does not apply to maize stalks and cobs, which in this area are usually chopped up and ploughed in. This practice increases the organic content of the soil. In this connection it should be noted that the quantity of humus formed in each cycle depends to a large extent on the quality and type of the organic material ploughed in, its distribution in the soil and the biological and chemical/physical properties of the soil itself.

We observed at Sovicille that, although the vegetation was chopped, it was not ploughed in evenly. There were noticeable points at which it accumulated so that besides obstructing the farming work it reduced the efficiency of the rotting-down process into humus and encouraged agronomically harmful processes, such as the spread of pathogenic agents and eremacausis.

Thus, if the remainder is distributed more uniformly in the soil, removal of some of the by-products need not reduce fertility.

The quantity of maize stalks and cobs available for use is estimated at 65 %, the mechanical efficiency of the harvesting being 50-80 % and the remaining 35 % being ploughed in. The organic content of the soil should be checked regularly so that appropriate organic fertilizers may be used where necessary.

This means that the quantity of by-products effectively available is approximately 9 600 t dry weight (Table III).

Table I - QUANTITIES OF MAIZE, WHEAT/BARLEY AND SUNFLOWER STALKS IN EXISTING PLANTS IN THE PIANO DI ROSIA

	MAIZE (t)	WHEAT/BARLEY (t)	SUNFLOWER (t)
Maidicola Cooperative (Malignano district)	5 200	600	300
Consorzio Agrario Provinciale (Rosia district)	5 500	6 000	750
Azienda Agricola "Correto a Merse"	1 500	-	-
T O T A L S	12 200	6 600	1 050

Table II - TOTAL QUANTITIES OF BY-PRODUCTS FROM THE MAIN HERBACEOUS CROPS

CROP	TOTAL grain (t)	TYPE OF BY-PRODUCT	RATIO BY-PRODUCT (%)	QUANTITY OF BY-PRODUCT fresh (t)	dry (%) (t)
Maize	12 200	stalks and cobs	1.5	18 300	7 300
Wheat/barley	6 600	straw	0.6	3 960	3 360
Sunflower	1 050	stalks	2.0	2 100	1 250
Rice	400	straw	1.0	400	300
T O T A L S	20 250			24 760	12 210

Table III - AVAILABLE QUANTITIES OF BY-PRODUCTS SHOWN

BY-PRODUCT	TOTAL DRY WEIGHT (t)	PRESENT USE	% USED	AVAILABLE DRY WEIGHT (t)
Maize	7 300	ploughed in	35	4 750
Wheat/barley	3 360	burned in field/ animal husbandry	3	3 260
Sunflower	1 250	burned in field	0	1 250
Rice	300	burned in field	0	300
T O T A L	12 210			9 560

(%) see text

QUANTITIES OF ETHANOL OBTAINABLE FROM AGRICULTURAL BY-PRODUCTS IN THE AREA OF SOVICILLE

From the quantities of by-products available it is possible to obtain 2 140 t/annum of ethanol (Table IV).

The types and quantities of associated products obtainable from this process will depend on the technology used but it is realized that proper use must be made of them if the overall operation is to be economically viable.

A number of associated products, such as lignin and furfurol, may be supplied to the chemical industry, outside the area, whereas yeasts and pulp, correctly processed, can be used as protein additives in fodder for locally reared livestock.

QUANTIFICATION OF BY-PRODUCTS ON THE BASIS OF LAND-USE DISTRIBUTION IN THE PIANO DI ROSIA AND THE PIANI DELLA RANCIA

In addition to quantifying by-products by surveying the local storage facilities and farms land-use in the area of Piano de Rosia and the Piani della Rancia was plotted.

This technique permits forecasts and technical calculations of the availability of by-products.

In fact present production does not reflect the increase which would be possible by extending irrigable surface area (at present out of 2 300 hectares approximately 1 000 are irrigated) and the consequent rise in unit production.

This plotting technique was applied to the land on both sides of the River Merse.

The land-use map for the Tuscany region was taken as a basis, and the areas plotted in terms of their gradients.

The reason for this was that most of the fairly flat areas in the valleys are irrigable and have a high yield of biomass, since they do not suffer from seasonal water shortages.

In addition, the very lie of the land is such that removing the by-products has no ill-effect as regards soil erosion, and makes their protective function superfluous, although this may be vital on the hill slopes.

The boundaries were determined by reconstructing the contour lines, within the limits of the topographical survey, on an ordnance survey map and excluding slopes steeper than 6 %.

The flat areas measure 15 km north to south, whereas east to west they cover a maximum of 5 km and taper off towards the extremities.

A number of surveys were carried out which showed that the map was accurate and reliable.

The total area was approximately 3 600 ha - or 35 km^2 in terms the units used for transport distances.

As may be observed, even if the plant is located near the edge of the plain, transport is less than 15 km. 67 % of the area (Table V) is ordinary irrigable arable land, approximately 50 % of which is in fact irrigated.

The whole area is given over to either cereals, industrial crops or intensive animal husbandry.

On the irrigated land maize is the major crop and only in the past few years has cultivation of the sunflower become widespread and more intensive. In the Piani della Rancia a large area is used for the intensive cultivation of rice. In the non-irrigated areas wheat and barley are grown and sometimes also maize and sunflowers. The potential yields of alcohol per hectare from by-products of irrigated crops are 1 582, 1 345 and 949 respectively for maize, sunflower and rice (Table VI).

Table IV - QUANTITIES OF ETHANOL RECOVERABLE FROM BY-PRODUCTS AVAILABLE

BY-PRODUCT OBTAINED FROM	CELLULOSE I	HEMICELLULOSE I	CELLULOSE (t)	HEMICELLULOSE (t)	RECOVERABLE ETHANOL (1 000 litres) from cellulose and hemicellulose		TOTAL ETHANOL (1 000 litres)
Maize	40	25	1 900	1 190	893	500	1 393
Wheat/barley	45	23	1 460	750	686	315	1 001
Sunflower	45	19	560	240	262	101	363
Rice	40	20	120	60	48	12	60
T O T A L							2 817

Table V - LAND USE IN THE PIANO DI ROSIA AND PIANI DI RANCIA (1978)

	ha	%
Ordinary arable irrigated	2 389	67.0
Ordinary arable not irrigated	734	20.6
Special vines, Special olives, Vines, Fruit	81	2.3
Forestry	63	1.8
Woodland coppice or full height trees	240	6.7
Poplar	43	1.2
Grassland	12	0.4
	3 562	100

Table VI - BY-PRODUCTS AVAILABLE PER HECTARE ON IRRIGATED LAND AND RECOVERABLE ALCOHOL

PRODUCT	BY-PRODUCTS	T DRY/HA	HARVESTED [1]	CELLULOSE %	HEMICELLULOSE	ALCOHOL 1	PRODUCED 2	(TONNE/HA) 2 + 1
MAIZE	STALKS AND COBS	7.2	4.68	4.0	2.5	8.79	4.91	13.70
SUNFLOWER	STRAW	5.0	5.0	4.5	1.9	9.45	4	13.45
RICE	STRAW	3.5	3.5	4.0	2.0	6.5	2.94	6.5

[1] For maize harvesting a loss of 35 % is assumed.

If, as planned, irrigation were extended to cover 2 300 ha, this area, minus, the area under rice (currently 131 ha) - i.e. the remaining 2 169 hectares - divided between production of cereals and industrial crops (sunflowers) could yield by-products in dry form from which it would be possible to obtain approximately 3 174 331 kg alcohol, from an area of 36 km^2.

CONCLUSIONS

A direct survey indicated a quantity of approximately 9 600 t dry weight by-products for Sovicille, which is equivalent to 17 200 t of total by-product, over an area of 18 000 ha. In this case a supply of 45 t per day of raw material 365 days per year could be guaranteed for a plant taking material over a radius of 7 - 8 km, if it was located in the centre of the area where this crop was grown.

Estimates based on land-use indicate that, if irrigation was extended to all the land classified as simple irrigated arable land, the quantities of by-products would increase steadily and a similar quantity to that estimated for the entire area could be produced just from the area devoted to more specialized agriculture.

Other aspects of the recovery of agricultural by-products should also be examined, i.e. :

- the removal of organic substances from the earth and the question of conserving its fertility ;
- different industrial technologies may be adopted and consequently the by-products derived from the processing may be consumed by an outlet outside the area (lignin and furfurol) or may be used in livestock farming locally as protein additives ;
- it is of prime importance to establish a degree of integration between plant and territory to ensure the economic viability of the process and to guarantee the development of an agro-industrial system based on the preservation and exploitation of the natural resources.

TECHNICAL AND ECONOMIC COMPARISON OF DIFFERENT METHODS OF HARVESTING MAIZE RECOVERING THE COBS FOR USE IN GRAIN DRYING

G. VAING (Long)
Engineer for the "Energy" programme of the
Research Centre for Tropical Agricultural Mechanization
(= Centre d'Etudes et d'Expérimentation du Machinisme Agricole Tropical)
(CEEMAT)
Antony, France

With the consent of CEMAGREF (Centre d'Etudes du Machinisme Agricole, du Génie Rural, des Eaux et des Forêts = Research Centre for Agricultural Mechanization, Engineering and Forestry Construction) the author has reproduced in this document part of his original report on the same subject previously published in CEMAGREF Information Bulletin No 295-296 of August 1982.

I. INTRODUCTION

The present state of French maize production is the result of a combination of many factors, both favourable and detrimental to its development, which have raised the annual production level from less than 20 million tonnes from 350 000 hectares between 1950 and 1955, to between 8 and 10 million tonnes of dry grain maize from a near 2 million hectares since 1972.

Rural depopulation, leading to a scarcity of manpower, in itself promotes the development of mechanization in this field, which in turn facilitates a continual expansion of the area of maize cultivation, every stage of which can now be mechanized.

However, of all France's main farming activities, maize production is the most costly in terms of the amount of heat required to dry the crop.

If the price of oil continues to rise at its present rate, the resulting economic problems could threaten further development of maize cultivation, or even the maintenance of its present level of production.

Although maize production is high in energy consumption, it has the compensatory advantage of considerable quantities of a waste product, the cob, which can be used as a substitute fuel for heating grain dryers.

In an attempt to find even a temporary solution to this energy problem, we have carried out a technical and economic comparison of different maize harvesting methods involving the recovery of the cobs for use in drying grain, with an assessment of their respective economic advantages.

The first problem tackled was that posed by the substitution of maize cobs for conventional fuel, considered in the context of drying only.

The following questions also had to be considered :

1/ How best to recover and collect the maize cobs and transport them to the work site for the lowest possible cost.
2/ What would be the main effects of substituting maize cobs for conventional fuel in existing drying installations ?
3/ Would the exercise be technically and economically worthwhile ?

II. TECHNOLOGICAL ASPECTS OF MAIZE HARVESTING AND DRYING

1. Physical condition of the crop at the time of harvesting

Established by M. BERAUT of the ITCF (Institut Technique des Céréales et des Fourrages = Technical Institute for Cereals and Fodder).

1.1. Ratio of the initial grain moisture content to that of the cob

at the time of harvesting is represented by the following empirical values :

Initial grain moisture content (in kg of water/kg of moist grain)	0.35	0.40
Initial cob moisture content (in kg of water/kg of moist cobs)	0.57	0.70

1.2. Ratio of the weight of grain to the weight of cobs

The RAGT in Rodez provides the following values :

Per 100 kg of moist ears :
- 21.250 kg of cobs with a standard water content of 48 %
- 78.750 kg of grain with a standard water content of 25 %

A sample of whole ears was taken from a crop harvested on 15 October 1980 in the region of l'Eure and Loir, and analyzed in the CNEEMA chemical laboratory, where the following results were obtained :

	kg	% of total
- total weight of sample	1.839	100
- weight of moist grain	1.187	64.54
- weight of moist cobs	0.407	22.13
- weight of moist husks	0.245	13.33
Therefore weight of moist ears without husks :	1.594	86.67

For the same sample, the water content of the moist crop (degree of humidity) at the time of harvesting was :

34.07 % of the grain
60.19 % of the cobs
55.51 % of the husks

1.3. Net calorific value (= pouvoir combustible inférieur PCI) of the maize cobs

According to the calculation of the AGPM (Association Générale des Producteurs de Maïs = Association of maize producers) the PCI of the cobs at 15 % humidity = 4 400 kcal/kg.

According to the calculations of the RAGT, the PCI of the cobs varies between 3 500 and 4 500 kcal/kg according to the degree of humidity of the cobs.

2. Evaluation of the heat balance of the drying of maize in cob-fuelled dryers

Dryers converted to accept cobs as fuel are already available. We used two types of dryer for our assessment : dryers for maize ears and grain dryers.

2.1. Heat balance of the drying of maize ears

We based our evaluation on a stationary vertical compartment dryer with a rotation cycle of 3-4 days according to the initial humidity of the ears.

The degree of humidity in the grain was reduced in the drying process from H1 35 % to H2 13 % (where H1 is the initial humidity and H2 the final humidity).

The results obtained relate to an installation for drying maize ears tested by the SIMONS Laboratory in Toulouse for the RAGT. The estimated output of the installation was around 200 t/day of dry product or 8 t/hour.

The installation used dry cobs recovered directly during shelling of the dried ears.

The master data were as follows :

- humidity level of grain 35 %
- humidity level of cobs 60 %
- 1000 kg of fresh ears comprised :
  650 kg grain at 35 % humidity
  350 kg cobs at 60 % humidity

In order to reduce the humidity of the grain to H2 = 13 % (commercial standard for seed) and the humidity of the cobs to H2 = 15 % (usual standard for fuel) the quantity of water which must be evaporated to obtain a tonne of dry grain would be :

720 kg broken down as follows :
338.5 kg from the grain
and 381.5 kg from the cobs

For every tonne of dry grain thus produced, 339 kg of dry cobs at 15 % humidity would be obtained.

Taking the PCI as 3 800 kcal/kg, this would correspond to around 1 288 thermal units.

Given that the dryer burns 1 600 kcal for every kg of water evaporated, the energy required to evaporate 720 kg of water is around 1 154 thermal units.

The balance is therefore appreciably positive.

2.2. Heat balance of the drying of maize grain

Here we based our assessment on a new concept in the drying of grain maize involving burning cobs recovered after threshing ears which are still moist. The cobs are therefore also moist (H1 ≃ 60 %).

As shown by tests carried out by certain manufacturers, it is in fact possible to burn maize cobs, collected after threshing, in a straw burner, creating hot air which can then be supplied to a conventional continuous drying installation for maize grain by means of a heat exchange generator.

In order to do this, the originally oil or gas-fired heating system must be modified.

The equipment on which these modifications were made were middle range dryers with an evaporating power of 430 to 860 kg water per hour for maize grain, given an average consumption of 1 000 kcal per kilogram of water evaporated.

For the sample of maize ears considered previously, but where the target final humidity of the grain is 15 %, the following quantities of water must be evaporated :

	per tonne of moist ears	per tonne of dry grain at 15 %
for the grain	156.0 kg	311.0 kg
for the cobs	185.3 kg	381.1 kg
T O T A L	341.3 kg	692.1 kg

Thus 1 t of moist ears gives :

501.3 kg dry grain at 15 %
164.7 kg cobs dried at 15 %

and for 1 t of dry grain at 15 %, 328.5 kg of dry cobs at 15 % are produced.

The energy balance for 1 to of dry grain at 15 % can also be calculated as follows :

- total calorific value of the cobs :
  3.8 thermal units x 328.5 kg = 1 248.5 thermal units

- heat required for the cobs themselves :
  $\frac{2 \text{ thermal units x } 185.29 \text{ kg x } 1000}{501.3}$ = 739 thermal units

- remainder available for drying grain :
  . . . = 509.5 thermal units

- quantity of heat required for drying grain :
  $\frac{1 \text{ thermal unit x } 156 \text{ x } 1\ 000}{501.3}$ = 311 thermal units

- remainder available for other uses = 198.5 thermal units

The harvest of 9 334 000 tonnes of dry grain in 1978 was dried in the following proportions :

Method of drying	Tonnage expressed in tonnes of dry grain
A. Drying of whole ear	1,520,280
1. Natural drying in cribs	1,362,400
2. Artificial drying at low temperature	158,400
B. Drying of grain	7,813,200
3. Artificial drying at high temperature	7,813,200
Total production in 1978	9,334,000

According to the heat balance data already established, the quantity of water to be evaporated for each cycle is as follows :

Method of drying	Quantity of water to be evaporated (tonnes)		
	cobs cycle	grain cycle	total
1. Natural in cribs (ears)	519,074	423,706	935,968
2. Artificial at low temperature (ears)	60,350	53,618	113,968
3. Artificial at high temperature	(1)	2,429,905	2,429,905
Total to be evaporated for each cycle	529,424	2,907,229	3,479,571

(1) Normally, when drying grain at high temperature using oil or gas, the cobs are not dried, so they have not been taken into account. However, where the cobs are being recovered for use as fuel, their water content must be reduced from H1 ≃ 60 % to H2 ≃ 15 % (the equivalent of around 2,976,829 tonnes).

Distribution of the heat requirements for the 1978 harvest cycles are then as follows :

Method of drying	Heat requirements (in kilotherms)
1. drying in cribs	nil
2. drying of whole ears at low temperature 1.6 kth x 113,968 (2)	182,348 kth
3. drying of grain at high temperature 1 kth x 2,429,905 (3)	2,429,905 kth
T O T A L	2,612,253 kth

(2) 1.6 kth/t of water is the average energy consumption of a low temperature batch dryer.

(3) 1 kth/t of water is the average energy consumption of a high temperature continuous dryer.

In order to use the cobs for energy purposes, they must be dried to H2 = 15 %. The additional quantity of water to be evaporated is then around 2,976,800 tonnes. Assuming a very poor efficiency of 2 kth per tonne of water evaporated, the additional energy required to achieve this would be 5,953,600 kilotherms, making a total of 8,565,853 kilotherms which would be totally recouped from the cobs thus recovered.

The potential quantity of dry cobs which could be recovered per harvest, on the basis of 0.328 tonnes per tonne of dry grain at 15 %, is in the region of 3,061,550 tonnes. Given that the PCI of the cobs at 15 % water content is 3,800 kcal/kg, the calorific potential available for drying purposes obtained by burning cobs would be 11,633,890 kilotherms, compared with an actual requirement of 8,565,853 kilotherms.

If this new drying procedure were applied 100 % across the whole area of cultivated land, the residual energy produced would theoretically be : 11,633,890 kth - 8,565,853 kth = 3,068,037 kth, or around 300,000 TEP per year (TEP = tonne équivalent pétrole - tonne oil equivalent. TEP is equivalent to 10 kilotherms).

III. SOME POSSIBLE HARVESTING METHODS RECOVERING THE COBS FOR USE AS FUEL FOR DRYING MAIZE GRAIN

Utilisation of cobs as fuel for drying maize grain is possible in four specific ways according to the way in which the cobs are recovered. These are :

- Utilisation of dried cobs recovered at the shelling unit following drying of the seed ears in batch dryers ;
- Utilisation of dried cobs recovered at the shelling unit from ears dried in cribs or ventilated platform dryers ;
- Utilisation of moist cobs recovered at stationary threshing units from ears freshly harvested by corn-picker and transported to the farm or delivered direct to a collecting firm which shells out and dries the grain in one process ;
- Utilisation of moist cobs recovered behind grain harvesters.

Each method will now be considered in turn.

1. Utilisation of the dry cobs recovered at the shelling unit downstream of the ear-drying installations :

This method relates particularly to a quantity of 160,000 tonnes of seed produced annually.

The RAGT in Rodez commissioned a study of the problem by the SIMONS laboratory in Toulouse, working with seed-drying units each with a capacity of around 18,000 t/year, which in principle operate for 90 days of the year continuously.

The technical characteristics of these units are as follows :

- vertical type dryer with eighteen drying compartments ;
- slow drying in batches at low temperature, with a rotation cycle of 3 to 4 days according to the humidity of the ear ;
- energy consumption including exchanger : 1 600 thermal units/tonne of water evaporated ;
- drying normally from 35 % to 13 % humidity ;
- daily output rate of dried product : 200 tonnes or 8 t/hour ;
- evaporating capacity : 5.440 tonnes water/hour ;
- heat rating of oven : 8,700 thermal units/hour.

The SIMONS laboratory based their calculations on an assumed composition of the moist maize ears by weight at the time of harvest of :

65 % grain at 35 % humidity ;
35 % cobs at 60 % humidity.

The cost of installing a drying unit of this capacity (output rate of dry grain : 8 t/h) has been assessed as follows (in FF) :

- dryer with exchanger 700 000 F
- compartments 300 000 F
- cob-fuelled oven with rated at 9 000 th/h 700 000 F
- handling equipment for cob-feeder 600 000 F
- cob storage units 500 000 F

TOTAL 2 800 000 F

The proportion of this cost incurred by the adaptation of the installation to accept cobs as fuel is around FF 1 800 000 for 9 000 thermal units/hour.

No modification of the harvesting equipment or the means of transporting the ears for drying is necessary. The cost of converting the oil or gas-fired drying system to a cob-fuelled drying system represents about 65 % of the total cost of the batch drying installations.

As an installation with an effective heat rating of 8.7 kilotherms/hour would require per crop cycle around 1 880 tonnes of oil at FF 2 100 a tonne, or an annual amount of FF 3 948 000 for heat (PCI of the oil = 10 000 kcal/kg ; 90 24-hour days per drying cycle), and since the use of cobs recovered in this way, which are practically free, means a saving of FF 3 948 000 a year, the cost of cob recovery equipment (around FF 1 000 000) could be recouped in less than a year.

In addition, the total quantity of cobs recovered for 18 000 tonnes of dry grain processed is around 6 100 tonnes at 15 % humidity with a PCI of 3.800 kcal per kilogram. This quantity of recovered cobs produces 23 180 kilotherms ; more than the actual requirements of the installation itself, which are estimated at : 8.7 kth x 24 h x 90 days = 18 792 kth. The recent and necessary addition of an exchanger means that all the cobs are now used for the drying process.

It should also be noted that the cobs are recovered during shelling of the dried ears from the dryer and can therefore be used directly.

The cost of conversion to the above process is estimated as follows :

Equipment required for each stage of the process	Cost of conversion expressed as a percentage of the cost of :
1. Harvesting machinery	nil
2. Collecting and transporting equipment	nil
3. Special batch drying installation	65 % of total cost of installations
4. Shelling machinery	nil

It can be seen that for the entire process, conversion to the use of recovered cobs would only require modification of the heating equipment, in particular the substitution of the oil or gas-fired oven with a cob-fuelled oven with attached cob-feeding and storage equipment.

Since all seed producers already use batch dryers for maize ears, it would appear that the conversion described above, the cost of which would very quickly be recovered, would offer the advantage of a free and readily available fuel on the same scale as the product to be dried, with no attendant disadvantages whatsoever.

If we consider the hypothetical situation of using all available batch driers used exclusively for drying maize ears for the entire current maize ear harvest, which represents something like 1 665 000 tonnes (of dry grain), the quantities to be dried would far outweigh the actual capacity of the drying installations.

Economically considered, maize-producing equipment comprising a double-row corn-picker, a 70 hp tractor, a 50 hp tractor, a 30 hp tractor, a stationary sheller, and two trailers of 10 m^3 capacity, the cost of harvesting and threshing including labour in 1979 was as follows :

Site	Basic cost (F)		
	per hour	per hectare	per tonne
1. Harvesting	192.91	481.90	74.14
2. Transport	72.30	180.70	27.80
of grain	21.69	54.21	8.34
of cobs	50.61	126.49	19.46
3. Shelling	88.31	220.78	33.97
Entire process	353.52	883.80	135.97
Cost of grain	302.91	757.31	116.51
Cost of cobs	50.61	126.49	19.46

This summary table shows that the cost of recovering the cobs represents around 14.30 % of the total cost of the maize ear harvesting process, including threshing in stationary units.

On examining all the operations forming part of the maize ear harvesting process, it is evident that whichever means of recovering the cobs is used, the cost of transporting them must be included in that of transporting the moist ears from the harvesting site to the threshing station, as they must be transported whether or not they are to be recovered.

It follows logically that the cobs are free from the moment they leave the threshing. However, they may if desired be taken into account as a separate item in the transport costs, allowing the overall cost of harvesting maize ears, which is always higher than that of harvesting grain, to be reduced.

This being the case, we shall here assume the cost of recovering the cobs in the maize ear harvesting process to be the proportion of the transport costs they incur, i.e. 122.82 FF per hectare or 18.89 FF per tonne of harvested dry grain.

Given that 1 hectare of maize yields 6.5 tonnes of dry grain, and one tonne of dry grain corresponds to 0.328 tonnes of dry cobs at 15 % humidity with a PCI of 3 800 kcal/kg, the cost of producing a tonne of dry cobs in 1979 was :

$$\frac{\text{FF } 19.46 \times 1 \text{ t}}{0.328} = \underline{\text{FF } 59.33}$$

and the cost per TEP (10 kth) was :

$$\frac{\text{FF } 59.33 \times 10 \text{ kth}}{3.8 \text{ kth}} = \underline{\text{FF } 156.10}$$

2. Utilisation of moist cobs recovered at stationary threshing units from ears harvested by corn-picker and delivered to a collecting firm

This process is of potential relevance to all maize production involving the harvesting of the ears and artificial drying in cribs.

The first point to note is that no changes to the harvesting or transport methods would be required. Instead of being stored in cribs or transferred to batch dryers as described above, the harvested ears are shelled immediately by collecting firms equipped with grain drying installations which can accept the recovered moist cobs as substitute fuel.

The process suggested above is very feasible for several reasons :

- the maize ear harvest usually starts seven to ten days before the grain harvest, because harvesting and husking of ears which are still moist help avoid loss of grain due to over-aggressiveness of the cutting implements ;
- the seven to ten days gained allow use to be made of drying installations belonging to organizations which operate exclusively with grain harvesting equipment ;
- maize ear producers incur no additional capital expenditure ;
- the additional working time required of existing installations used for this stage of maize ear production is ten to twelve days, which is well within their availability ;
- it only remains to be seen whether the cost of drying can be offset by the loss sustained in crib storage, which is around 4 % of the value of the stored grain, and financial losses (around 10 %) representing interest on tied-up capital, incurred during the storage period ;
- organizations owning grain-drying installations stand to gain from participating in this additional work because it enables them to reduce their costs without disturbing their normal work programme.

3. Utilisation of cobs recovered behind grain harvesters

At present grain is harvested over 1 485 000 hectares, according to the working capacity of the national stock of harvesters. For an average yield of 5.2 t/ha, this area corresponds to a production figure of 7.7 million tonnes of dry grain, which would guarantee 2 548 000 tonnes of recoverable cobs (0.331 tonne dry cobs per tonne of dry grain).

The basic agricultural equipment used for this level of maize production can be summarized as follows, allowing for the fact that since 1977 some outdated harvesting equipment has been replaced by new models :

- for grain harvesting (1)
  2 500 corn shellers equipped with 10 000 headers
  14 722 combine harvesters with 56 000 headers
- for grain drying :
  2 413 collective drying installations with a total capacity of 4 155 000 points/hour
  8 073 individual drying installations with a total capacity of 1 369 000 points/hour.

(1) Gathering capacity : 20 to 25 hectares per header per harvest cycle.

Using the existing grain harvesting equipment, working methods could be modified as follows :

- the cobs, which are normally ejected onto the ground, could be gathered up behind the harvesters and brought to the collection point at the end of the field in the same way as the grain ;

- grain and cobs could be transported to the drying stations in the same proportions ;
- the cobs thus recovered could be used to heat the air for drying the grain from the same harvest.

There is therefore a case for suggesting material modifications in four areas : the harvesting site, harvesting machinery, transport and the drying process.

3.1. Modification of the harvesting system

Observation of the refuse ejection shows that it is expelled in two crossing trajectories :

- the cobs projected backwards brush against the upper surface of the sieve extension of the concave with bars, following a horizontal trajectory before dropping suddenly as soon as they are beyond the end of the sieve;
- the remaining debris of fine cobs, husks, leaves and even stems, follow an upward trajectory carried by an air current caused by the fan below ; this air current crosses the path of the cobs ejected beyond the upper sieve.

Thus the finest, lightest debris falls furthest away and the cobs fall nearer the machine, at the discharge point.

The cobs or all the waste products can therefore be collected according to the following diagram :

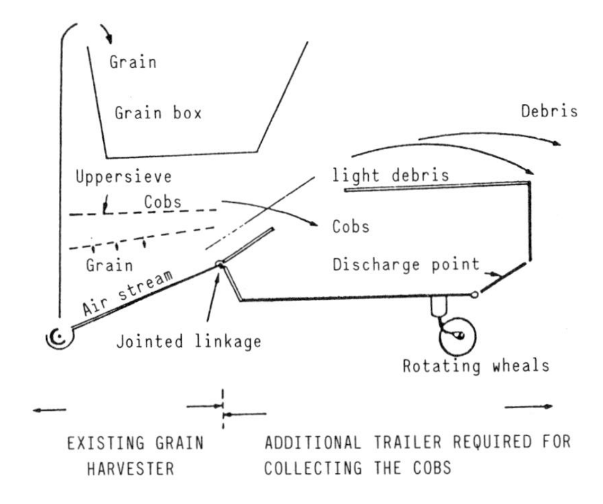

The trailer for collecting the cobs would need a capacity of 12 to 15 m^3 as they take up twice the volume of the grain, and the grain harvesters (corn shellers and combined harvesters) generally have a receiving box for 3 to 3.5 m^3 of grain per 2 or 3 headers on the machine. With a box this size it would be possible to cover the whole breadth of a field of around 1 km.

As a guide, a trailer of 12 to 15 m^3 with a discharge facility on both sides or at the back and a jointed linkage system cost around FF 30 000 ex-works without VAT in 1979-80.

The additional time lost discharging the cobs would be around 5 to 6 minutes for each discharge.

There would be no additional labour costs other than a loss of 5 to 6 minutes working time.

3.2. Modification of the harvester only

It is possible to harvest a mixture of grain and cobs simultaneously by increasing the number of sieves on the harvester. This practice is already employed by pig breeders to obtain a product suitable for ensilage (corn-cob mix). Using this system, however, only 50 to 70 % of the cobs are recovered.

It would appear to be possible to recover all the cobs by converting the harvester at little cost (by fitting hoppers and altering the gauge of the sorting sieve). This solution would not affect the transport chain.

On the other hand, it would be necessary to install additional sorting equipment at the storage depot to separate the grain from the cobs. The cost for an output of 50 t/hour would be around FF 120 000 without tax, at a rate of FF 24/tonne.

At a rate of 12 000 t quintals/year the machine would pay for itself in five years.

3.3. Modification of the transport system

Cobs occupy twice the volume of grain, while the difference in weight is negligible.

In order to ensure that the grain and cobs are transported in the same proportions, the capacity of the transporting equipment must be doubled.

A standard harvesting system has on site for a 3 to 4-row corn sheller, two high-sided trailers of 10 m^3/7 tonnes with a 50 hp tractor for transporting the grain. To transport the recovered cobs an additional 50 hp tractor would be necessary along with two more trailers each with a capacity of 12 to 15 m^3.

3.4. Modification of the drying station

It would be necessary to replace the oil or gas-fired heating system with a maize cob-fuelled stove with attached continuous feeding system.

The cost of this conversion would amount to approximately 17 % of the total cost of the existing system (estimated by the R. Roulin firm from their own sources).

3.5. Economic study of the proposed system

Let us consider a typical grain harvesting process selected on the basis of the sales statistics given earlier : a 125 hp combine harvester with a four-row picker. This machine forms the basic equipment of a farm with 80 hectares of maize and 80 hectares of cereal crops, and its use is divided equally 50-50 between the two. This distribution has the advantage of rapid recoupment of costs.

For the farm's maize production, two particular systems come under consideration :

Nature of operation	Method employed	
	With no recovery of the cobs	With recovery of the cobs
Harvesting of grain (80 hectares of maize)	Combine harvester with four-row picker + 1 combine harvester driver	Combine harvester with four-row picker + trailer + combine harvester driver
Transport of grain	1 50 hp two-wheel drive tractor + 2 trailers of 10 m^3 /7 t + 1 tractor driver	1 50 hp two-wheel drive tractor + 2 trailers of 10 m^3/7 t + 1 tractor driver
Transport of cobs		1 50 hp two-wheel drive tractor + 2 trailers of 10 m^3/7 t + 1 tractor driver
Drying of grain (520 t dry grain/crop)	Installation with output of 430 thermal units/hour (based on 1 000 kcal/kg water evaporated)	Installation with output of 430 thermal units/hour (based on 1 000 kcal/kg water evaporated) cob-fuelled oven cob feeding appliance

- 1st system : harvesting without recovering the cobs and using conventional fuel for drying the grain.
- 2nd system : harvesting recovering the cobs and using them as fuel for drying the grain.

In short, the cost of harvesting the maize is as follows according or whether or not the cobs are recovered :

Harvesting with no recovery of the cobs

Nature of cost	Cost of harvesting one hectare according to the price of a kg of domestic fuel oil (FF)	
	2.4 F/kg	2.5 F/kg
1. Fixed costs (combined)	455.95	455.95
2. Running costs :		
. Combine harvesters	81.36	83.54
. 50 hp tractor	38.60	39.44
T O T A L	575.91	578.93

Harvesting with recovery of the cobs

Nature of cost	Cost of harvesting one hectare according to the price of a kg of domestic fuel oil (FF)	
	2.4 F/kg	2.5 F/kg
1. Fixed costs (combined)	658.66	658.66
2. Running costs :		
Combined harvester	81.36	83.54
50 hp tractor (two tractors)	77.20	78.88
T O T A L	817.22	821.08
Cost per hectare of harvesting the cobs	241.31	242.15
This cost represents : 0.328 x 6.5 t = 2.13 t harvested cobs. The price of a tonne of cobs is :	113.29	113.69
Production cost of 1 TEP of cobs	296.88	297.91

* TEP = tonne équivalent pétrole
tonne oil equivalent

It is also a profitable exercise to establish the heat balance for a hectare of maize on the basis of the table below.

If the results are applied to the global production of 7.7 million tonnes of harvested and dried grain and assuming the process suggested above is applied 100 %, the overall energy balance would be as follows :

Basic values : 3,800 kcal/kg of dry cobs at 15 %
0.328 t of cobs per tonne of dry grain

1. Total calorific value of the recovered cobs	9 600 000 kth
2. Energy requirements of the cobs themselves	5 690 000 kth
3. Available for drying grain	3 910 000 kth
4. Remainder available for other heating requirements	1 524 000 kth

It can be estimated from the above that implementing a system of grain harvesting recovering the cobs would produce a heat surplus of at least 1,500,000 kth or 150,000 TEP in the form of cobs per season, after deducting the energy required for drying the grain.

Looking again at our typical example producing 80 hectares of maize, the cost of converting existing grain harvesting methods with the addition of equipment for recovering the cobs and an appliance which will accept them as fuel would be around 170,000 FF made up as follows :

- harvesting equipment :		FF 30 000
- transporting equipment :		FF 116 413
- drying equipment		FF 22 770
	TOTAL	FF 169 183

By extrapolation and assuming a drying process fuelled completely by cobs, the conversion cost for the whole 1 485 000 hectares under consideration would amount to :

$$\frac{\text{FF } 169\ 183 \times 1\ 485\ 000 \text{ ha}}{80 \text{ ha}} = \text{FF } 3\ 140\ 400\ 000$$

or around FF 2 115/ha

Heat balance for one hectare of maize

Item	Per tonne of dry maize grain	Per ha of maize	
		Yield 5.2 t/ha	Yield 6.5 t/ha
Units	Therms	Therms	Therms
1. Total heat energy value of recovered cobs	1,248	6,490	8,112
2. Heat required to dry the cobs from 0.60 to 0.15	739	3,842	4,803
3. Available for drying grain	509	2,631	3,308
4. Required for drying grain	311	1,617	2,021
5. Usable surplus	198	1,014	1,287

These figures suggest that for an additional investment of FF 3 000 000 000, 320,000 TEP in domestic fuel oil could be saved, corresponding to 832 million francs per season at an oil price of FF 2 600/t (April 1982) with a PCI of 10 000 kcal/kg.

Such a fuel saving would allow the above investment costs to be paid off in less than four years. This period would be reduced still further as oil prices increased.

From the table below the selling price of a tonne of dried cobs collected on site can be calculated, the cobs being regarded as a fuel in the same way as wood or straw :

- cobs coming from cribs or ventilated platform dryers following threshing are regarded as waste and can either be given away or sold for FF 10 or 20 a tonne at most, as is sometimes the case ;
- where the whole ear is harvested and threshed before drying of the grain, the cost of transporting the cobs may be included, in which case the production cost would be around FF 60/t, which would make the TEP around FF 160 ;
- in the case of grain harvesting using a specially developed trailer for recovering the cobs, a tonne would work out at around FF 115 and a TEP at FF 300 (applicable to 260 tonnes dry maize/year) ;
- in the case of mixed cob and grain harvesting, the production cost of a tonne of cobs would be at the most FF 60.25 and a TEP FF 161 (for a storage firm with a capacity of 12 000 t dried product/year).

Comparative table of the two types of maize harvesting

Compared aspects	Ear harvesting system	Grain harvesting system
A. Harvesting cost per hectare (FF)		
1. Where cobs are recovered	883.80	821.08
2. Where cobs are not recovered	757.31	578.13
3. Cost of recovering cobs per hectare	126.49	242.15
B. Cobs produced/hectare (t)	2.13	2.13
C. Production cost of a tonne of cobs (FF)	59.38	113.69
D. Production cost of a TEP of cobs (FF)	156.10	297.91

NB : All values are estimated on the basis of 100 % implementation of the use of recovered cobs as a substitute fuel.

IV. CONCLUSION

The above study is purely theoretical and its estimations are based on a maize production infrastructure which is no doubt already outdated.

The present conditions affecting the growing and drying of this important agricultural product once more allow reliance on fossil fuels.

However, the trends in the rate of increase in oil prices and the increasing difficulty of obtaining oil products makes research into a substitute fuel which is cheap and readily available nationally, in this case maize cobs, ever more justified.

It should not be forgotten that French grain maize production fluctuates between 9 and 10 million tonnes, which corresponds to 2.9/3.2 million tonnes of dry cobs, the equivalent of 1 100 000 to 1 200 000 tonnes of oil.

Since the drying of the maize consumes 320 000 TEP of fuel, and drying the cobs would require the equivalent of 700 000 TEP, it can be seen that the balance is positive.

The above calculations show that the operation is technically and economically viable, since a TEP of cobs for fuelling the oven would be at the most FF 300, while the current cost of a TEP of domestic fuel oil is FF 2 600 (April 1982). The outlay for converting the heating system would therefore quickly be recouped. The most cost-effective systems are, according to our calculations those involving harvesting and transporting of maize ears or harvesting and transporting of a mixture of maize grain and cobs, setting a TEP of cobs at FF 161.

In 1981, three prototypes were successfully tested in the Atlantic Pyrenees and the "Landes". These were small installations with an output of 0.8 t to 1 000 t/year. The time required for gross recovery of the capital expediture on them will vary between 3 and 7 years according to the type of installation. As from 1983, the cost of drying using cobs alone will be able to compete with oil, given the rate of increase in energy prices.

The scale of the above figures and relationship between them leads us to believe that it would be possible, as oil prices rise, to consider substituting all fossil fuels so far used specifically for cereal drying. The necessary investment and equipment would be relatively inexpensive, and the equipment could be installed very quickly.

It should be possible to recoup the outlay on new equipment within 3 to 7 years without difficulty, and possibly within an even shorter time depending on the rate of increase in energy prices.

BIBLIOGRAPHY

1. "Perspectives agricoles", Nos 6 and 7, 1977
   - special issue on maize harvesting
   - special issue on grain drying
2. "Perspectives agricoles", No 24, 1979
   - special issue on grain drying equipment
3. "Techniques modernes de séchage des grains", FL/CN, June 1979
4. "Séchage du maïs-grain par combustion des rafles", D. SENGELEN and Y. BRENDEL, ISAB (Institut Supérieur Agricole de Beauvais).
5. First thesis by Bernadette MILLET, June 1977, INAPG (Institut National Agronomique Paris-Grignon).
6. Provisional year book 1979 : Vegetable production - departmental and regional results, No 18, February 1979.
7. "La récolte du maïs", AGPM-ITCF document (AGPM = Association Générale des Producteurs de Maïs. ITCF = Institut Technique des Céréales et des Fourrages (Technical Institute for Cereals and Fodder).
8. "Le maïs-grain : récolte, réception-séchage-conservation-qualité", report of the proceedings of the Symposium held on 26 and 27 January 1972, ITCF document.
9. CNEEMA (Centre National d'études et d'expérimentation du machinisme agricole) information bulletin No. 266-267, March-April 1980 : summary of a report by the "Conseil Supérieur de la Méchanisation et de la Motorisation de l'Agriculture", page 15.
10. "Le sechage du maïs aux rafles", AGPM document, 1981.
11. French contribution to an OECD-AGPM study, Montardon 1982, X. GAUTIER.
12. CEMAGREF information bulletin No. 295-296 August-September 1982.

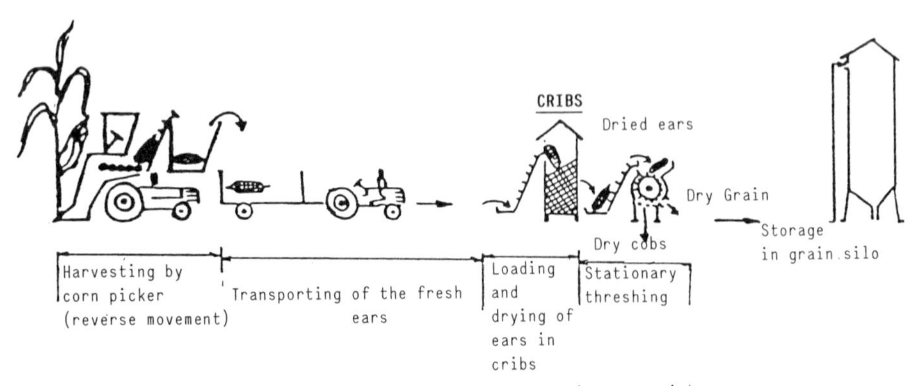

Diagram of normal ear harvesting process using corn picker, crib drying + stationary threshing of dried ears

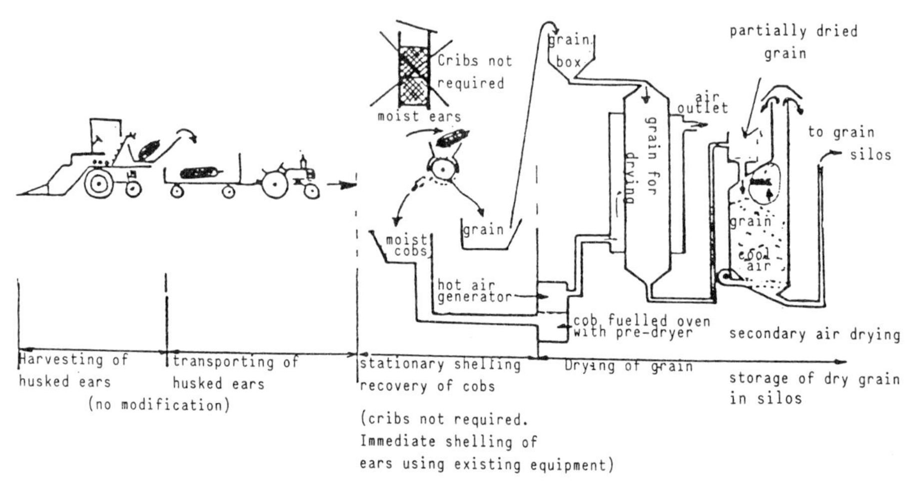

Recovery of cobs at stationary shelling unit for moist ears and drying of grain in continuous dryer fuelled by moist cobs in specially adapted oven with pre-drying attachment)

Recovery of cobs behind maize grain harvester and utilisation of cobs for fuel for drying the grain.

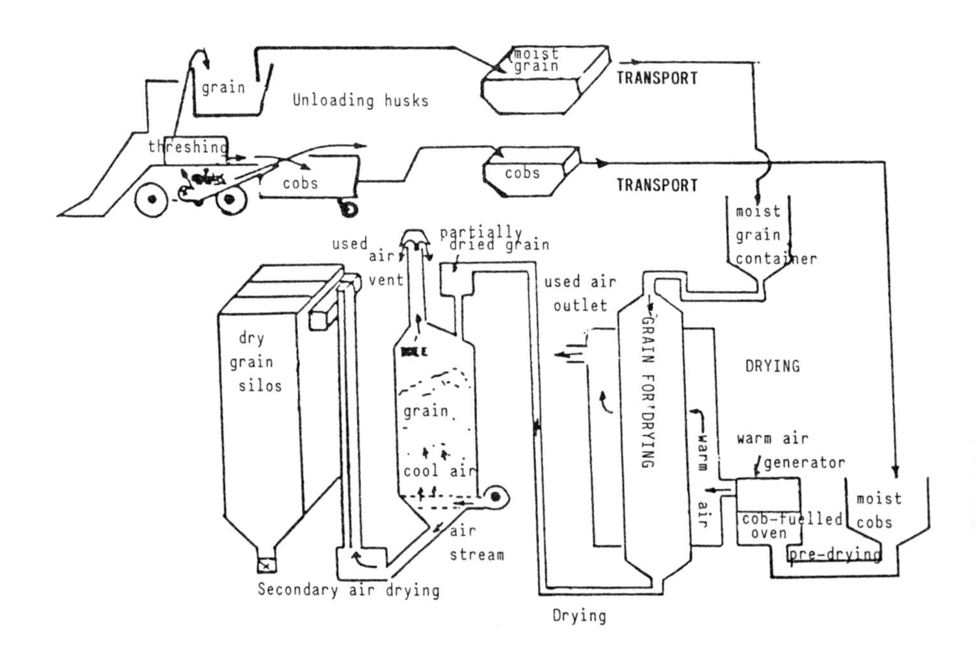

MSW AND VEGETABLE BY-PRODUCTS, GUIDELINES FOR RESEARCH : FROM THE HARVESTING PROCEDURE TO THE ANAEROBIC PROCESS

F. CECCHI*, P.G. TRAVERSO*, J. CLANCY**, J. MATA-ALVAREZ*** and C. ZAROR****

* Department of Environmental Science, University of Venice, Dorso Duro, 2137, 30123 - Venezia, Italy
** Department of Engineering, University of Reading, P.O. Box 215, Reading, U.K., RG6 ZAY
*** Department of Chemical Engineering, University of Barcelona, Diagonal 647, 6, 08028 - Barcelona, Spain
**** Department of Food Technology, University of Reading, Reading, U.K., RG6 ZAY

Summary

A review of state of the art R&D in the field of anaerobic digestion of municipal solid waste (MSW) is presented for Europe, paying particular attention to the influence of the harvesting procedures on the quality of the refuse and then on the performance of the anaerobic process. A comparison is made of the quality and quantity of refuse produced in different European countries using data available in recent literature. Good data compatibility is demonstrated amongst organic fractions of MSW (OFMSW) produced in Europe and between these and vegetable by-products.

The conclusions to be drawn from the analysis are that in programmes involving demonstration and full scale plants it is necessary to carry out detailed studies of the process and its control as well as further work on the microbiological aspects of anaerobic digestion of MSW since these areas are, as yet, not fully understood. Moreover it is indicated that the joint digestion of vegetable by-products and OFMSW seems useful since there are no negative indications.

1. INTRODUCTION

Anaerobic technology, in the field of the disposal of solid refuse (SW), can be considered one priority to be developed. This is demonstrated by the enormous efforts that, over the last ten years, in applied research have been put into the topic, not only at the USA level (1-19), but also at the European level (20-35).

Despite some perplexities, arising from extrapolations due to commercial pressures rather than scientific results, it seems useful to sustain the opportunity to plan actions, which join the contractoring efforts with those of research. Thus, the objective is to fill the existing gaps in the knowledge of anaerobic digestion of complex and concentrated substrates and also to fully optimize, from the economical point of view, the treatment plants, from the harvesting procedures to the final disposal of by-products.

The aim of this paper, limited to the evaluation of the anaerobic digestion process, is to demonstrate, through a review of the European experiences, the importance of the quality of the biomass and thus of the harvesting procedure, on the performance of the process.

Only an analysis of the European situation is presented because there is more interest in it, and because this limited analysis allows the evaluation of a more homogeneous situation. The characteristics of the MSW produced in Europe, compared with the USA refuse, seems to allow anaerobic technology to be more easily applied.

2. RESULTS AND DISCUSSION

Quality of substrates. The situation, coming from recent papers, of the quality and quantity of refuse produced in Europe is given in tables 1, 2, 3.

Table 1 - Quality and quantity of MSW produced in Europe

Country Ref	Quantity	Wt % Dry Matter		Wt % MSW ww			Note
	MSW Produced Kg/capita day	Organic	Total	Digestable Organic Matter	Other Organic Matter	Inert	
Europe (35)	0.7-0.8	60-70	55	-	-	-	(a)
Italy (36)	0.657	67*	44*	59.7	24.7	15.6	(b)
Italy (37)	0.759	79*	43*	58.2	31.7	10.1	(c)
Italy (21)	--	70*	46*	48	32	15	(d)
France (31)**	0.877	50	62.5	-	-	-	-
Spain (38)	0.9-1.2	80	60.9	49	41.5	9.5	(e)
Spain (28)	--	-	55	-	-	-	(e)

(a) European mean values ; (b) Italian domestic refuse mean value ; (c) Italian mean value of MSW collected by the Public Service ; (d) Sample of 300 families in Treviso, Italy ; (e) MSW of Barcelona city.

* Value calculated according to 20 % TS in the digestable organic material (21), 71 % TS in the other organic material and 91 % in the inert material (weighted mean value according to (7)) ;

** Values calculated according to the data in (33).

Table 2 - Main chemical characteristics of the OFMSW

Country	Ref	TS %	TVS % TS	TCOD/ TVS	C % TS	N % TS	PPO4 % TS	C/N
Belgium	(32)*	23-30	82-87	1.5	-	1.9-3.8	0.9	-
France	(33)**	35	59.6	-	-	-	-	-
Belgium	(29)**	30-35	-	-	33	1.15	-	28.7
France	(31)**	25-40	47	-	-	-	-	-
Italy	(21,23)*	20	88	1.2-1.6	48	3.2	0.4	15
Italy	(39)***	16.3	91	1.34	45	2.2	0.3	20.5
Spain	(38)****	21.7	84.8	-	49	1.61	0.44	30
Spain	(28,40)	22.5	87.5	-	-	1.5	0.26	-

* OFMSW without paper and hand separated

** OFMSW poorly separated by machine

*** OFMSW simulated with 50-50 TS of miscellaneous food and green refuse

**** Calculated from the analysis of the components of MSW.

In table 1, it is shown that at a European level, the specific municipal solid waste (MSW) production (Kg MSW/capita, day) is constant and that, when it is divided into the following three main classes of products; easily digestable organic matter, other organic matter (not digestable or digestable with difficulty in the retention time generally adopted in anaerobic digesters) and inert matter, the agreement of the data reported in the literature is good especially for the organic fraction.

Table 2 demonstrates, using parameters generally adopted for the chemical characterization of substrates to be anaerobically digested, that the quality of the biomass is strictly dependent on the selecting systems and then on the harvesting procedures adopted. In fact it is possible to find two families of data in table 2 : the first, refs. (21, 23, 28, 32, 38-40), refers to results arising from organic fractions of MSW (OFMSW) carefully selected at pilot plants for treatment, or preselected (by the families) (23), or calculated taking into account the components of the MSW; the second one, refs. (29, 31, 33) refers to the OFMSW mechanically selected and to MSW coming from unselected harvesting procedure. As well as the evident agreement between the data of the first family, from table 2 it is possible to notice, using the C/N ratio parameter as a comparison, that the source selected substrate has better characteristics than the mechanically selected one (compare data of ref. 23 to those of ref. 29, for instance). This possibly depends on the large amount of materials which are rich in carbon and poor in nitrogen (paper, plastic, etc...).

If the comparison is extended from table 2 to table 3, where some characteristics of vegetable by-products are reported, it can be seen that there is no prejudice towards the anaerobic digestion of mixed substrates : OFMSW and vegetable by-products.

Table 3 - Main chemical characteristics of some vegetable by-products

Vegetable by-products	Ref.	TS, %	TVS, % TS	C, % TS	N, % TS	C/N
Mixture (a)	(41)	18.41	88.6	-	2.3	-
Mixture (b)	(39)	10.3	-	44.2	2.0	22.1
Mixture (c)	(2)	-	77.8	54.7	3.04	18.0
Water Hyacinth	(42)	4.7	77.7	41.0	1.96	20.9
Bermuda Grass	(42)	5.1	95.0	47.1	1.96	24.0
Tomato Residues	(43)	5.60	94.40	54.75	2.25	22.0
Green Bean Residues	(43)	19.65	83.35	48.34	2.20	24.3
Geranium Flour (d)	(44)	92.41	80.42	40.61	1.35	30.08
Akalona (e)	(44)	91.85	96.20	51.59	1.03	50.09
Watermelon Residues (f)	(44)	6.88	88.95	49.42	0.36	137.28

(a) Peels and waste of potatoes, apples, oranges and salad.
(b) Supermarket vegetable wastes.
(c) Supermarket vegetable wastes consisted predominantely of lettuce leaves, a sizable amount of cabbage and smaller quantities of onion and other vegetables.
(d) Air dried residue remaining from geranium plants after extraction of the essential oil.
(e) Outer portion of the wheat grains which results from the scouring of the wheat grain during the milling process.
(f) Watermelon fruit residues after separating the seeds (i.e. rind, pulp and juice).

The anaerobic digestion process. A detailed comparison between the data in the literature is impossible due to the non-uniformity in presentation and, sometimes, to the incomplete information supplied. Nevertheless, it is possible to make, from figures 1, 2 and 3, some important considerations. These are regarding the influence of the feeds' quality on the process performance and the possibility of evaluating the advantages of operating under mesophilic or termophilic conditions. Moreover some conclusions can be drawn about the dry digestion technique when the feed of the digester is mechanically selected OFMSW.

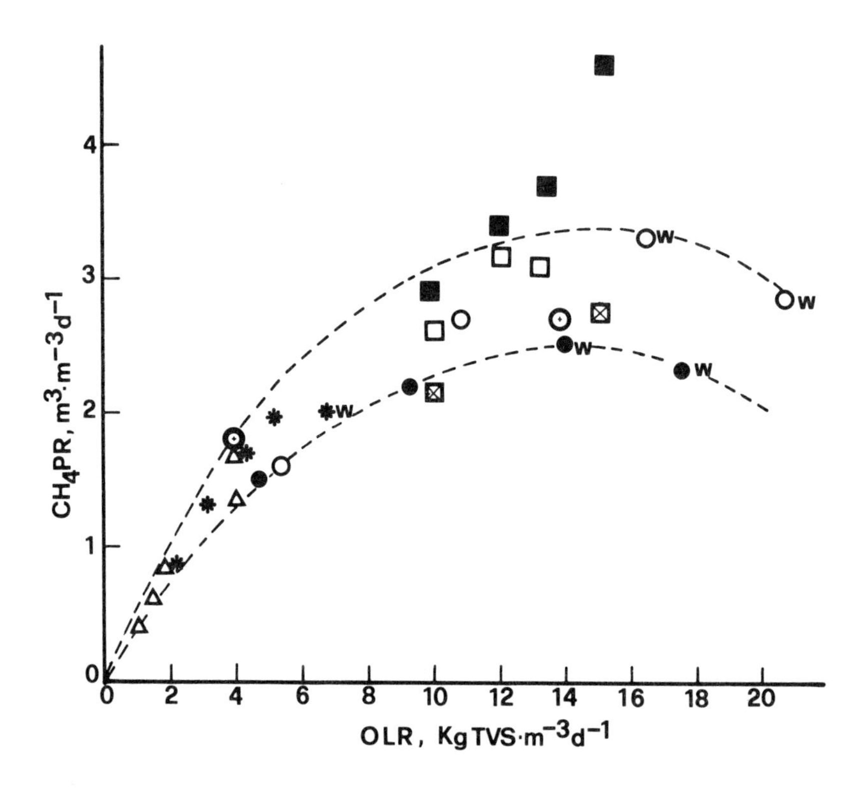

Fig. 1. Comparison of the European experiments based on the CH_4PR vs. ORL. ✱ (23) Meso; ■ (29) Thermo; ⊠ (30) Meso; O (31) Thermo, TSA=35%; ●(31) Thermo, TSA=30%; Δ (32) Meso; ⊙ (39) Meso; □ (29) Meso; ⊙ (33) Meso.

Figure 1 shows that apart from some inconsistencies in the data of ref. (29), due possibly to the fact that the digester works with large differences between HRT and SRT values (so the term retention time is improperly used by the Authors and is only referred to as the hydraulic retention time), the other ones are inside a narrow range and have a pattern that may indicate that :

- there are no substantial differences between the data obtained in mesophilic and thermophilic conditions. It does not appear, therefore, that the last conditions give actual advantages ;

- when OLR and HRT are higher or equal to 5 Kg TVS/(m^3, d) and 14-15 days respectively, the methane production rate, $m^3/(m^3$, d), depends on OLR independently of the total solids concentration in the substrate fed in the wide range reported (5/6-30/35 % TS). From the biological point of view this can mean that the digester is operating with an excess of active biomass inside it ;
- when HRT/SRT is less than 14/15 days ("w" points in the figures) the digester's behaviour changes at any OLR applied. One observes in fact, going from HRT = 14/15 days to HRT = 9/10 days, a flexion of the specific methane production rate dependence on OLR. This is independent of the total solid concentration in the feed (compare data from ref. (23) to those of ref. (31)). This possibly depends on a partial wash-out of methanogenic bacteria. The last opinion is supported by the relevant increasing of VFA concentration inside the reactor underlined in ref. (23) ;
- at OLR higher than 10 Kg TVS/(m^3, d), the specific methane production rate, in general, remains substantially constant. This means that the active biomass inside the reactor is the limiting factor.

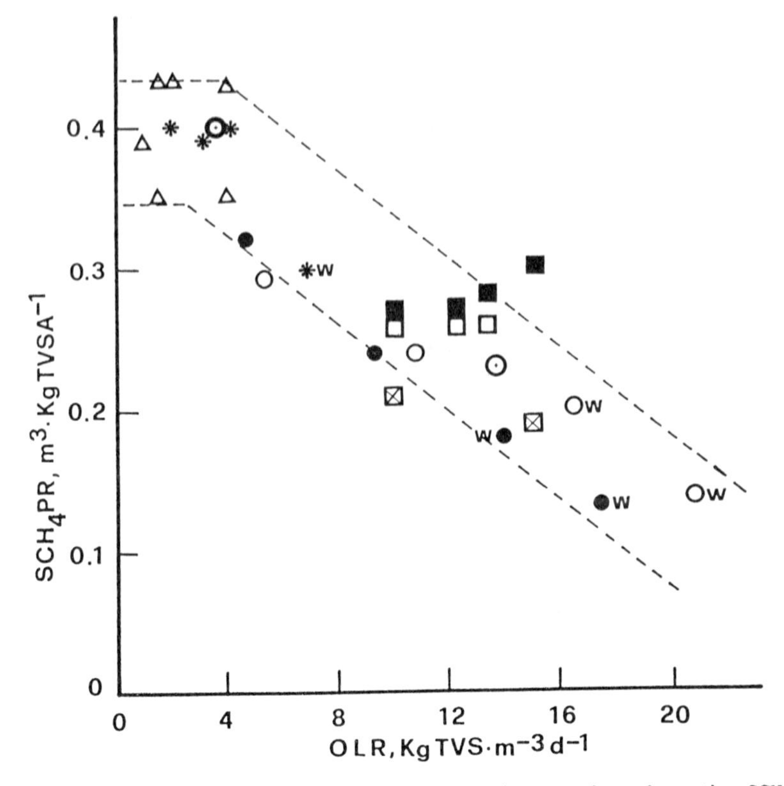

Fig. 2. Comparison of the European experiments based on the SCH_4PR vs. ORL.. ✱ (23) Meso; ■ (29) Thermo; ⊠ (30) Meso; O (31) Thermo, TSA=35%; ●(31) Thermo, TSA=30%; △ (32) Meso; ⊙ (33) Meso; ⊙ (39) Meso; □ (29) Meso.

An analysis of figure 2 demonstrates that the specific methane production rate (SCH_4PR, m^3/Kg TVA) declines sharply when the OLR increases. This occurs not only for the "w" points, but also for those obtained with an apparent HRT/SRT higher than 15 days. This later observation suggests that only a portion of TVS of the feed is utilised in the reactor. From the evidence available (PH, VFA values) it would appear that the digester's methanogenic/acidogenic bacteria equilibrium is not affected by such high OLR values. This may be due to the fact that a fraction of the organic matter fed is not biodegradable in the solid retention time used, as clearly shown in figure 3. The OLR applied is then very different from the effective biodegradable OLR in the reactor.

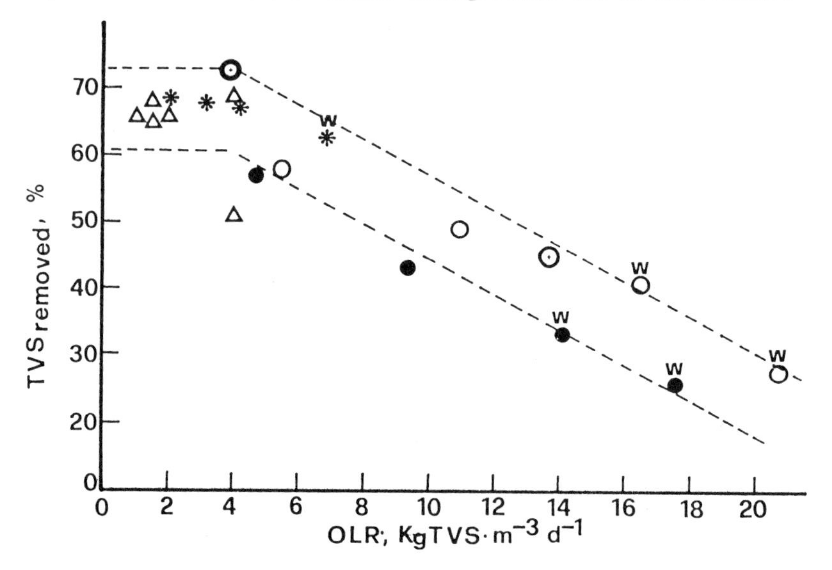

Fig. 3. Comparison of the European experiments based on the % TVS removed vs. OLR. ✱ (23) Meso; ○ (31) Thermo; TSA=35%; ● (31) Thermo, TSA=30%; △ (32) Meso; ⊙ (33) Meso; **O** (39) Meso.

The previous discussion shows that we are far from speaking in terms of a second generation of digester in the field of anaerobic digestion of solid refuse.

According to the results of ref. (23), in particular to figure 4, it is possible to make some reasonable and reliable predictions of the process performance that might be obtained using different systems and approaches. The gas production rate could be maintained, with appropriate preconditioning of the substrate fed, to the maximum level of the instantaneous gas production rate (IGPR = 0.6 m^3/h ; or 5 m^3/ (m^3, d), working with 6 % TS feed or even three times more with undiluted feed (20 % TS = actual OFMSW). The latter extrapolation seems possible if it is considered that the biomass in the reactor is directly proportional to the TVSR (45), that the TVS removal is 68-70 % (see fig. 3) and that the specific gas production is 0.08 m^3/(Kg TVSR, d) (23).

The values, coming from the later prediction, could seem surprisingly good, but they are less high than those obtained with advanced reactors : i.e 32.2 m^3 / (m^3, d) (46).

The possibility of digesting OFMSW with other organic substrates, vegetables by-products for instance, does not seem to present a lot of problems, according to their characteristics (see table 3). Nevertheless the experience in the field is limited, so it is necessary to express opinions with prudence, but data in table 4 and our results seem to confirm the absence of any contra-indications.

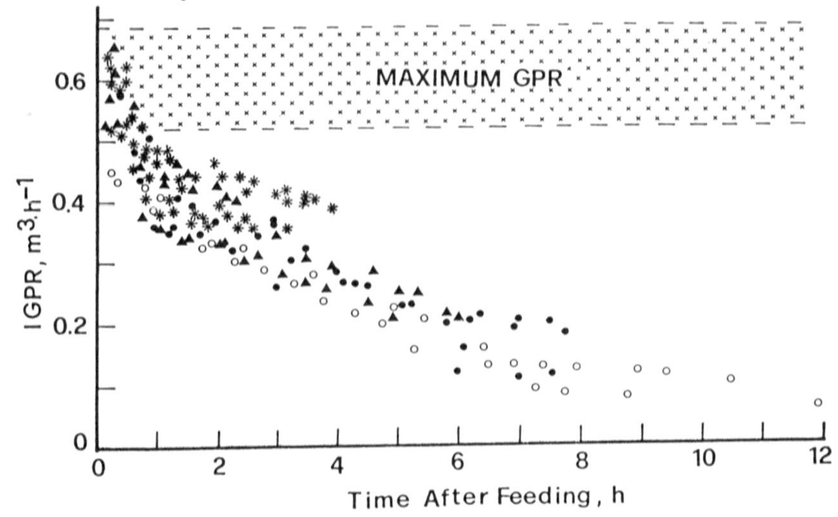

Fig. 4. Instantaneous gas production rate pattern between one feed and subsequent one obtained feeding a 3 m^3 reactor with 6% TVS diluted OFMSW (23). o : OLR = 2.1 KgTVS/(m^3d); • : ORL = 3.2; ▲ : OLR = 4.2; ✱ : OLR = 6.9.

Table 4 - Process performance of anaerobic digestion of joint OFMSW and vegetable by-products.

Vegetable By-products	Ref.	HRT/ SRT d	OLR KgTVS/m^3d	GPR m^3/m^3d	SGPR m^3 /kgTVSA	OR %	T (C
Mixture (a)	(41)	-	-	-	0.62-0.75	82	55
Water Hyacinth + Bermuda Grass + MSW	(42)	12	1.6	0.55	0.34	34	35
Green Garbage + Primary Sludge (b)	(2)	30	1.3	0.53-0.66	0.41-0.51	75	37
Mixture (c) : Vr = 130 m^3		21.7	1.4	1.035	0.937	-	33
Vr = 150 m^3	(43)	25	1.6	1.692	1.044	-	33
Mixture (d)	(39)	14	3.9	2.6	0.66	72.5	35

(a) Peels and waste of potatoes, apples, oranges and salad.
(b) 62 % supermarket vegetable wastes consisted predominantly of lettuce leaves, a sizable amount of cabbage and smaller quantities of onion and other vegetables and 28 % primary sludge.
(c) 85.5 % dry bread, 9.8 % vegetables and fruit residues and 4.7 % cooked food.
(d) 50 % food wastes and 50 % supermarket vegetable waste.

3. CONCLUSIONS

a) Experimental data indicate that the maximum specific methane production ($0.4\ m^3$/Kg TVSA) and the maximum TVS reduction (60-70 %) are achieved at OLR = 5 Kg TVS/ (m^3, d). These values are not yet obtained with the dry digestion process.
b) The few experiments carried out under thermophilic conditions and the inconsistencies amongst the available data suggest that more research work is necessary on this topic before doing any economical comparison with the mesophilic approach.
c) From the exposition made it is possible to see the difficulty in evaluating characteristics of plants and processes through the use of terms like : gas production rate (m^3/ (m^3, d), specific gas production rate (m^3/Kg TVSA) and TVS removal. These lose their meaning as process' parameters when used incorrectly and/or in a misleading way (when the organic matter is not quantified, for instance).
d) According to the homogenity of refuse produced in Europe the collaboration amongst contractors, municipal workers and researchers in a wider context than the national one seems useful. This could facilitate the R&D activities at a European level and the transfer of reliable and really significant data to full scale and demonstrative plants.

4. ABBREVIATIONS

C = carbon, %TS
CH4PR = methane production rate, $m^3/(m^3d)$
C/N = carbon, nitrogen ratio
GPR = biogas production rate, $m^3/(m^3d)$
HRT = hydraulic retention time, days
IGPR = instantaneous biogas production rate, m^3/h
Meso = mesophilic conditions
MSW = municipal solid waste
MSWww = MSW wet weight
N = nitrogen, %TS
OFMSW = organic fraction of MSW
OLR = organic loading rate, Kg TVS/m^3d
SCH4PR = specific methane production rate, m^3/Kg MSWA; m^3/Kg TVSA
SGPR = specific biogas production rate, m^3/Kg TVSA
SRT = solid retention time, days
SW = solid wastes
TCOD = total chemical oxygen demand
Thermo = thermophilic conditions
TS = total solids
TSA = total solids added
TVS = total volatile solids
TVSA = total volatile solid added to the digester
TVSR = total volatile solid in the digester
VFA = volatile fatty acids, mg CH_3COOH/l
Vr = reactor volume.

5. REFERENCES

(1) WISE, D.L. (1983). "Fuel Gas Developments", CRC Press Inc. Ed., Boca Raton Florida, 75.
(2) GOLUEKE, G.C. and Mc GADHEY, P.H. (1970). "Comprehensive Studies of Solid Waste Management" 1-2nd Annual Report SERL, University of California, Berkeley.

(3) GOLUEKE, G.C. and Mc GADHEY, P.H. (1971). "Comprehensive Studies of solid Waste Management" 3d Annual Report SERL, University of California, Berkeley.
(4) PFEFFER, J.T. (1974). "Reclamation of energy from organic refuse", EPA 670/2-74-016, National Environmental Research Center, Office of Research and Development, U.S. E.P.A., Cincinnati, Ohio.
(5) PFEFFER, J.T. and LIEBMANN, J.C. (1974). "Biological conversion of organic refuse to methane". Annual Report NSF/RANN Grant No. GI-39191, Dept. Civil Engineering, University of Illinois, Urbana, Ill. Report UIUL-ENG-74-2019.
(6) PFEFFER, J.T. and LIEBMANN, J.C. (1976). "Energy From Refuse by Bioconversion, fermentation and residue Disposal Processes", Resources Recovery and Conservation, 1, 295-313.
(7) PFEFFER, J.T. (1974). "Temperature effects on Anaerobic Fermentation of Domestic Refuse", Biotech. Bioengineering, 15, 771-787.
(8) DIAZ, L.F., KORZ, F. and TREZEK, G.J. (1974). "Methane Gas Production as Part of a Refuse Recycling System", Compost Science, 15, Summer, 713.
(9) DIAZ, L.F. and TREZEK, G.J. (1977). "Biogasification of a Selected Fraction of Municipal Solid Wastes", Compost Science, 18, March-April, 8-13.
(10) KLASS, D.L. and GHOSH, S. (1973). "Fuel gas from Organic Wastes", Chemtech, 3, November, 689-698.
(11) KLASS, D.L. (1978). "Energy Production from Municipal Wastes" Third Annual Meeting SOIL Conservation Society of America, July 30-August 2, Denver, Colorado.
(12) GHOSH, S., HENRY, M.P. and KLASS, D.L. (1980). "Bioconversion of Water Hyacinth - Coastal Bermuda Grass - MSW - Sludge Belds to Methane", Biotechnology and Bioengineering Symp., 10, 163-187.
(13) GHOSH, S., HENRY, M.P. and CHRISTOPHER, R.W. (1985). "Hemicellulose Conversion by Anaerobic Digestion", Biomass, 6, 257-269.
(14) COONEY, C.L. and WISE, D.L. (1975). "Thermophilic Anaerobic Digestion of Solid Waste for fuel Gas Production, Biotechnology Bioengineering, 17, 1119-1135.
(15) AUGENSTEIN, D.G., WISE, D.L. and COONEY, C.L. (1976-1977) "Packed Bed Digestion of Solid Wastes", Resource Recovery and Conservation, 2, 257-262.
(16) WISE, D.L., KISPERT, R.G. and LANGTON, E.W. (1981). "A Review of Bioconversion Systems for Energy Recovery From Municipal Solid Waste", Part I: Liquid Fuel Production, Resources and Conservation, 6, 101-115.
(17) WISE, D.L. KISPERT, R.G. and LANGTON, E.W. (1981). "A Review of Bioconversion Systems for Energy Recovery From Municipal Solid Waste", Part II: Fuel Gas Production, Ibidem, 117-136.
(18) WISE, D.L. and KISPERT, R.G. (1981). "A Review of Bioconversion Systems for Energy Recovery From Municipal Solid Waste", Part III: Economic Evaluation, Ibidiem, 137-142.
(19) STENSTROM, M.K., ASCE, M., NG, A.S., BHUNIA, P.K. and ABRAMSON, S.D. (1983). "Anaerobic Digestion of Municipal Solid Waste", J. Env. Eng. Div. Proc. ASCE, 109, 5, 1148-1158.
(20) DE POLI F., CECCHI, F., TRAVERSO, P.G., AVEZZU',F. and CESCON, P. (1986). "Anaerobic digestion of organic fraction of garbage - Pilot Plant", in "Biogas Technology, Transfer and Diffusion", Ed. by M.M. El-Halwagi, Elsevier App. Sci. Pub., London, 446-453.
(21) CESCON, P., CECCHI, F., AVEZZU', F. and TRAVERSO, P.G. (1985). "Anaerobic digestion of organic fraction of municipal solid waste - Preliminary investigation", in "Energy from Biomass 3rd E.C.

Conference", Ed. by W. Palz, J. Coombs and D.O. Hall, Elsevier Appl. Sci. Pub., London, 572-576.

(22) CECCHI, F. and TRAVERSO, P.G. (1985). "Biogas dalla frazione organica di rifiuti solidi urbani e fanghi di supero - Studio preliminare", La Chimica e l'industria, 67, 11, 609-616.

(23) CECCHI, F., TRAVERSO, P.G. and CESCON, P. (1986). "Anaerobic digestion of organic fraction of municipal solid waste - digester performance", Total Env. Sci, (1986), in press.

(24) CECCHI, F., TRAVERSO, P.G., DE POLI, F., AVEZZU', F. and CESCON, P. (1985). "Anaerobic Digestion of Source Separated Organic fraction of Municipal Solid Waste and primary sludge", Proceeding Fourth International Symposium on Anaerobic Digestion 11-15 Nov. Guangzhou, China, 438.

(25) CECCHI, F. and TRAVERSO, P.G. (1986). "Biogas from Organic fraction of the municipal solid waste and primary sludge - Part II", La Chimica e l'Industria - Quaderni dell'Ing. Chim. Ital.", 7-9, 9-13.

(26) CESCON, P., CECCHI, F., AVEZZU', F. and TRAVERSO, P.G. (1986). "Ottenimento di biogas dalla frazione organica di rifiuti solidi urbani", in A. Frigerio "Rifiuti Urbani e Industriali - Trattamento e Smaltimento, nuovi aspetti tecnologici e normativi" Bi & Gi Ed., Verona-Italy, 57-71.

(27) TRAVERSO, P.G. and CECCHI, F. (1986). "Analysis of the particle size distributions in an anaerobic digester fed with the organic fraction of M.S.W.", Anaerobic Treatment A Grown-up technology", Proceeding of E.W.P.C.A. Water Treatment Conference, 15-19 September, Amsterdam, 600-603.

(28) MATA, J.A. and MARTINEZ, A.V. (1985). "Laboratory simulation of Municipal Solid Waste Fermentation with Leachat e Recycle", J. of Chem. Technol. & Biotechnol., in press.

(29) DE BAERE, L. (1984). "High rate dry anaerobic composting process for the organic fraction of solid waste", 7th Symposium on Biotechnology for Fuels and Chemicals", Gatlinburg, TN, May 14-17.

(30) DE BAERE L. and VESTRAETE, W. (1984). "High-Rate Anaerobic Composting with Biogas Ricovery", Biocycle, 25, March, 30-31.

(31) MARTY, B., RAGOT, M., BALLESTER, J.M., BALLESTER, M. and GIALLO, J. (1986). "Semi-solids state thermophilic digestion of urban wastes", E.E.C. Contractor Meeting Anaerobic Digestion, Villeneuve D'Ascq (Lille), 4-6 March.

(32) PAUSS, A., NYNS, E.J. and NAVEAU, H. (1984). "Production of Methane by Anaerobic Digestion of Domestic Refuse", E.E.C. Conference on Anaerobic Digestion and Carbohydrate Hydrolysis of Waste, Luxembourg, May 8-10.

(33) VALORGA, (1985). "Waste Recovery as a source of methane and fertilizer. The Valorga Process", 2nd Annual Int. Symp. on Industrial Resource Management, Philadelphia, USA, 17-20 February.

(34) CECCHI, F., TRAVERSO, P.G., CESCON, P. and BONTEMPELLI, R. (1986). "Recupero energetico dalla digestione anaerobica di rifiuti solidi urbani separati alla fonte", Convegno Gruppo Scientifico Italiano Studi e Ricerche su "Trattamento e Smaltimento di Rifiuti Urbani e Industriali, Milano, 24-25 Feb..

(35) DE BAERE, L. and VESTRAETE, W. (1984). "Anaerobic Fermentation of Semi-solid and solid Substrate", E.E.C. Conference on Anaerobic Digestion and Carbohydrate Hydrolysis of Waste, Luxembourg, May 8-10.

(36) Chiesa, G. and Bonaiuti, R. (1983). "I rifiuti domestici prodotti in Italia", Acqua Aria, 8, 807-812.

(37) C.N.R. (1980). Progetto Finalizzato Energetica I "Atti II Seminario informativo. Utilizzazione energetica dei rifiuti solidi urbani". Padova 21 Aprile.

(38) MARIMON, S.R. (1982). "Los residuos solidos urbanos, Analisis de un servicio municipal", Servicio de Estudios En Barcelona del Banco Urguijo.

(39) CECCHI, F. (1986). "Codigestione anaerobica di rifiuti solidi e fanghi di depurazione", Impublished data.

(40) MATA, J.A. (1986). "Characteristics of the MSW of Barcelona", Impublished data.

(41) DEBOOSERE, S., DE BAERE, L., SMIS, J., SIX, W. and VESTRAETE, W. (1986). "Dry Anaerobic Fermentation of concentrated substrates", Anaerobic Treatment a Grown-up Technology, Water Treatment Conference, Amsterdam, 15-19 September, 477-488.

(42) GHOSH, S., HENRY, M.P. and CHRISTOPHER, R.W. (1985). "Hemicellulose conversion by anaerobic digestion", Biomass, 6, 257-269.

(43) ALAA EL-DIN, M.N., GOMAA, H.A., EL-SHIMI, S.A. and ALI, B.E. (1986). "Biogas production from kitchen refuse of army camps of Egypt using a two stage biogas digester", in M.M. El-Halwagi, "Biogas Technology, Transfer and diffusion", Elsevier App. Sci. Pub. LTD, London, 589-599.

(44) GAMAL-EL-DIN, H., EL-BASSEL, A. and EL-BADRY, M. (1986). "Biogas production from some organic wastes", Ibidem, 463-473.

(45) METCALF & EDDY, (1979). "Wastewater Engineering Treatment Disposal", Mc Graw-Hill Co., New York, 2th Ed., 419.

(46) MEYER, F., VIENNEY, M. and TONG, S.S. (1986). "Use of synthetic fiber as packing material for methanization". "Anaerobic treatment. A Grown-up Technology", Proceeding of E.W.P.C.A. Water Treatment Conference, 15-19 September, Amsterdam, 588-591.

SESSION IV - CROP BY-PRODUCTS, PRUNING

Chairman: *J. LUCAS, Cemagref*
Rapporteur: *J.C. GOUDEAU, Min. Recherche*

BIOMASS ENERGY IN EUROPE
GENERAL REPORT ON BY-PRODUCTS OF PRUNING

J. LUCAS
Head of agricultural machinery department
CEMAGREF, France

This brings us back to wood as when we were speaking of short rotation forests or coppices.

Wood, of course, has a higher density than straw. In trying to reduce the cost delivered user or processer we need not concern ourselves with compacting, which reduces transport costs, but rather with harvesting and packing techniques.

But even for energy purposes there are two types of wood, viz. :

- thinnings (which foresters call "small wood", but which are fairly large lumps for our purposes) ;
- small pieces of wood (twigs from young coppices, undergrowth, and branches lopped off during pruning).

These two types of wood call for different methods for reducing harvesting and packing costs.

FOR THINNINGS, the best way of reducing costs seems to be to use "intelligent" equipment, i.e. a series of specialized machines, as described yesterday by Mr Morvan of ARMEF, or robots fitted with an automated arm which gathers, cuts and loads each tree. CEMAGREF is currently working on such a system.

IN THE CASE OF SMALL PIECES OF WOOD, the wastes to be loaded are too small or the species too mixed to be worthwhile sorting. Thus the machines will handle the wastes indiscriminately in batches. Any gains must come from adapting the harvesting machine to the wastes.

The summary of the paper by Mr Ortiz-Canavate and Mr J. Gil provides food for thought on this point. In trying to reduce harvesting costs the authors first of all rejected large undergrowth clearers, and subsequently smaller machines as well, and developed a machine which was just powerful enough to cope with the task.

There is a French saying to the effect that if you can do the big things you can do the small things too.

However, in view of the difficulties of making biomass competitive we should not take too much notice of this saying when it comes to the harvesting process. The smallest possible equipment should be used for each type of wood, provided that it is big enough not to break. I would ask you to bear this in mind as you reconsider some of yesterday's contributions and listen to today's.

The cheapest machine,
minimum energy consumption,
minimum labour requirement,
are the criteria for the economic harvesting of small pieces of wood.

Several of the papers presented yesterday, and several questions and discussions, showed that it was very difficult to make biomass pay as a source of energy.

Various studies, particularly those carried out by Mr J.J. Becker at CEMAGREF, have determined the threshold prices and energy yield for a number of processes, the most efficient being combustion and possibly carbonization with production of carbon/water. The conclusion was that the treshold prices were between 240 and 340 French francs per tonne dry weight, which is difficult to achieve but possible for certain residues or by-products.

We also know that the lower the density of material to be recovered the higher the harvesting cost, given the same machine.

It follows therefore that if it is difficult to make ordinary wood competitive there is little chance of making " small wood" competitive.

However, a table presented by Mr Kocsis should give us pause. In this table certain types of fuel, which are not in a very attractive form, in fact turn out to be of interest because they cost nothing to begin with and little investment is required.

Small pieces of wood have the advantage of zero or even negative cost to begin with, and since they are small they are easier to shred coarsely.

Vine branches lopped off in pruning are very small pieces of wood, fruit tree branches are slightly bigger, and short rotation coppice branches are quite large. Some branches encountered when clearing undergrowth are still bigger. Each size requires a different machine.

Yesterday we saw the "Scropion", a machine of more than 200 HP, which will shred undergrowth containing sizable branches.

Other plant, such as that illustrated below, and which has only 120 HP, are designed for less thick undergrowth.

Yesterday we also saw some machines which look more like farm machines, designed for harvesting rapid-growth coppices.

We are now going to discuss machines designed to harvest still smaller pieces of wood, i.e. the wastes from pruning fruit trees and vines.

These machines tend to resemble farm machinery rather than forestry equipment.

Mr Muratori and Mr Grilli will tell us about a "baler" for pruning wood.

Mr Amirante will describe a prototype machine designed for this same product.

Mr Ortiz Canavate and Mr Gil have developed a similar machine which they will present, and Mr Bartolelli will describe a "mechanized chain", also for recovering small pieces of wood.

However, before that, to centrate our minds still more and to point up the importance of recovering very small pieces of wood, Mr Gabrilides and Mr Martzopoulos will show that the areas involved are quite extensive in Greece alone.

ANNEX 1 : Forestry wastes tractor 1982 (CEMAGREF)

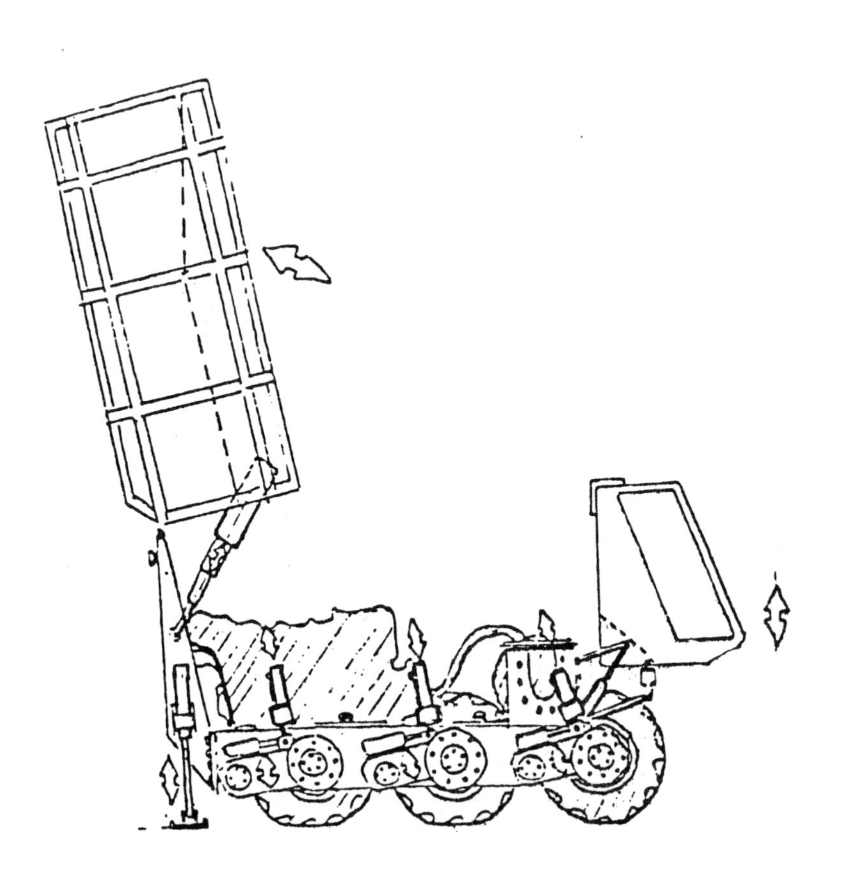

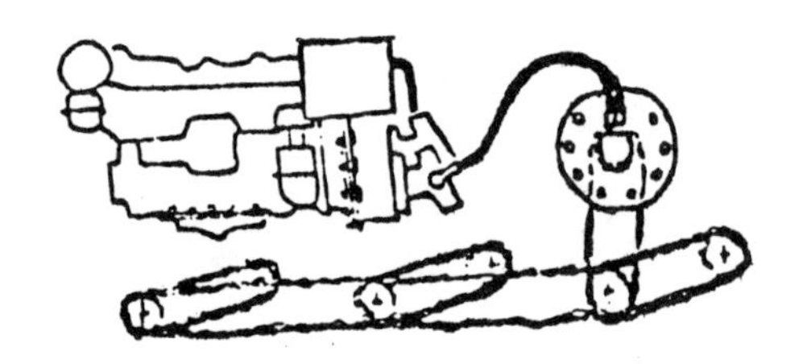

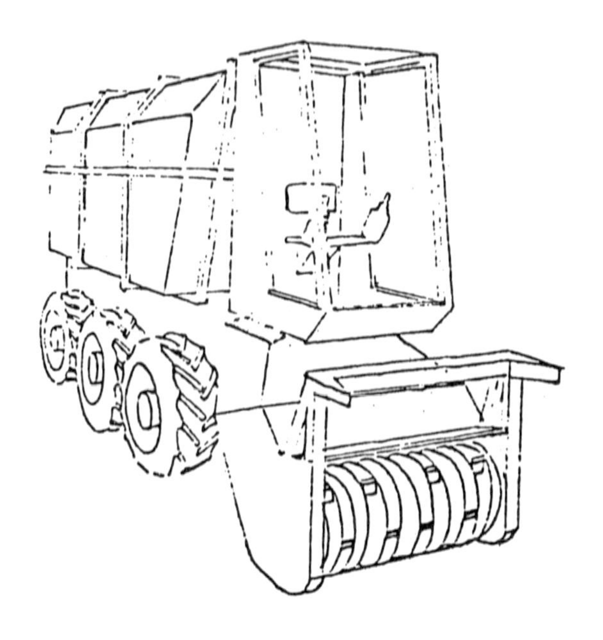

ANNEX 2 : DIAGRAM No 2

Technical data

Dimensions :	track width	: 1.3 m	- capacity : 100 kW - 2300 rmp
	wheel base	: 1.2 m between each set of wheels	- 6 non-steering drive wheels
	length	: 4 m without attachment	- hydropneumatic suspension with levelling facility
	width	: 1.6 m	
	height	: 2.6 m	- load capacity 6 m^3
			- suspended cabin
			- forestry tyres 12.0/18
Weight :	unladen	: 4 t	
	laden	: 6 t	
	useful load	: 2 t	

DEVELOPMENT OF HARVESTING AND HANDLING MECHANIZATION OF PRUNING RESIDUES IN GREECE

G.G. MARTZOPOULOS and S. TH. GABRIILIDES
Dept. of Agric. Engineering
University of Thessaloniki

Summary

The total area of Greece is 13,195,700 ha and 27 % of this represents areas under crops. The area under vineyards and trees in compact plantation is 1,022,820 ha or 29 % of the total agricultural land.

Vineyards cover 179 500 ha of which 1900 ha are covered with plants less than one year old. The main by-product of vines is the cuttings collected in vineyards, while a substantial mass of residues is collected after the grape process in wine making. The total production of vine cuttings is estimated to be about 0,18 million tonnes (d.m. 85 %) per year. The most common methods of handling or use of cuttings to date is incineration, while baling and storage has been recently introduced in this country.

Trees in compact plantation cover about 24 % of the total agricultural land. Olive trees, fruit trees, citrus trees and nut trees are the main crops in Greece. Pruning residues are the main source of by-product collected on site. The method of collection is traditionally by hand assisted by tractors and trailers. The storage of this sort of by-product is usually made in open yards or in sheds forming large low density piles. The uses of prunings are mainly incineration, composting for agricultural use and mulching. Pyrolysis is a promising process but the cost is still high. Other by-products from trees are shells and seeds collected and stored at the main product process centre. No extensive research work has been introduced, as yet, with reference to the improved methods of handling and storing pruning residues in Greece.

1. INTRODUCTION

The agricultural residues such as pruning residues are natural by-products, which, mainly for economic reasons, are not fully utilised either within the confines of the production unit or in large scale operations. Therefore, agricultural residues apply a financial surcharge to many products due to the cost of energy and labour for harvesting, handling and disposing of them, and they are often allowed to disturb the natural environment.

The agricultural residues are those coming out from arable crops, vegetables, vineyards and tree plantations. Vineyards and trees in compact plantation in Greece cover 1,022,820 ha, which is a substantial proportion (29 %) of the total agricultural area. According to the above mentioned, residues from the examined plantations have an important role in terms of good management for the benefit of Greek agriculture. The uses of vine and tree residues are mainly incineration for energy production (heat) composting for agricultural use and mulching. Pronounced difficulties, in harvesting and handling due to the irregular shape and size of these residues prohibits the efficient use of them. Therefore, the aim of this study is to expose these difficulties and to propose methods which need to be investigated, in order to permit the full utilisation of pruning residues under the Greek conditions.

2. VINEYARDS CUTTINGS

In Greece (1), 177 600 ha are covered by vines of more than one year old (table 1), and another 1900 ha are with new plantations. The distribution of vineyards within the geographic regions of Greece are shown in fig. 1.

The main source of vine by-products are the cuttings, but the residues from the vine process are considered with more interest, since they are easily collected in certain places (place of processing the product).

An optimal estimation of the vine-cuttings mass productions (2) is about 0.18 million tonnes of 85 % d.m., which in terms of energy derived from burning represents about 2500 TJ every year.

The harvesting of cuttings does not seem to be a major problem in Greece, because this is work which has to be done anyway and the small size of the average Greek vineyards facilitate this job. The cuttings have a rather uniform shape and the pieces are of a handy size. The recently introduced cutting balers are considered to be useful for large vineyards or in group farming, because they assist in mechanizing the post-pruning procedure, especially if cuttings are going to be used efficiently.

The chopping of the material before storage is believed to facilitate storage and handling while it permits mechanisation for further processing or utilization and this is under consideration for research work in Greece.

3. PRUNINGS

The total area covered by trees in compact plantation in Greece is 843 000 ha which is about 24 % of the total agricultural land.

Olive trees are the main crop covering 638 000 ha. The main fruit trees such as peach, apple, pear, apricot, cherry and citrus cover 117 600 ha and the crops of nut and dried fruit trees cover an area of 46 700 ha (1). The number of the above mentioned trees is shown in table 2. On top of this figure there are an estimated 40 000 scattered trees in Greece. A considerable proportion (25 %) of the trees in compact plantations are cultivated in mountainous areas and the rest in semi-mountainous and level communes. The distribution of these cultivations around Greece is shown in fig. 2.

The main by-products collected in the tree plantations are the prunings and fallen fruits. The latter are left anyway for manuring the plants and prunings are the only source of residues which have to be handled. Although no dendrometric data ara available in Greece the total mass of collectable prunings is estimated about 0.4 million tonnes. The collection of prunings has not been mechanized in Greece, but the short-distance of removal, assisted by tractors and trailers, and their storage in open yards or sheds is a common practice. Prunings have a total energy equivalent of 6000 TJ/year, but only about 10 % of it is estimated to be used for heating, while the rest is lost. The main reason for wasting this material is the difficulty in handling due to the irregularity of the shape and size of prunings. The cutting of prunings to small handy pieces seems to assist handling and transport of them, but this process costs labor and energy which must be judged against the profit coming out of it. Carbonization is broadly used with forest wood in Greece, but this method must be considered with some reservation if it is to be applied with pruning residues in each of the average size Greek farms. On the other hand pyrolytic or particleboards making installations could be a solution if they can serve a number of plantations within a commune and this needs to be investigated under Greek conditions.

Other by-products from trees are shells and seeds collected and stored in places where the main products of trees (nuts, fruits) are processed. These by-products are easy to handle and they have recently been used in Greece for heating greenhouses and piggeries, replacing a considerable amount of their energy needs (3).

4. CONCLUSIONS-PROPOSALS

The situation concerning the management of the available pruning residues in Greece, as described above, is in a rather unsatisfactory condition. It is believed, however, that future investigations must focus on methods or systems which will permit a) the transportation of the raw material to certain places, which will serve a broad area around them, with a low cost, b) the collection of such quantities in each place so as to assist in reducing the cost of treatment per mass unit of the material. Investigation is also needed on the selection of proper machinery to treat the material (choppers, pyrolytic plant etc.) so as to make it conveyable in order to facilitate storage, distribution over long distances and efficient use.

REFERENCES

1. National Statistical Service of Greece, Statistical Yearbook of Greece, 1974.
2. Lab. of Viticulture, Univ. of Thessaloniki, Unpublished data, 1986.
3. MARTZOPOULOS G.G., NIKITA-MARTZOPOULOU C., "Pig housing and installations in Greece", Proc. of the 36th an. meeting of E.A.A.P. Vol. 2, Sept 30 - Oct 3, 1985.

Table 1 : Areas under vineyards in Greece in thousand hectares (more than one year)

CROP (VARIETY)	TOTAL AREA	LEVEL COMMUNES	SEMI-MOUNTAINOUS COMMUNES	MOUNTAINOUS COMMUNES
Grapes for wine	93.7	42.2	29.3	22.4
Raisins (corinthians, sultanas, other)	64.1	35.4	15.7	13.0
Table Grapes	19.8	12.4	4.8	2.3

Table 2 : Areas under the main compact plantations (in thousand hectares) and number of trees in Greece.

	TOTAL AREA	LEVEL COMMUNES	SEMI-MOUNTAINOUS COMMUNES	MOUNTAINOUS COMMUNES	NUMBER OF TREES
Olive trees	638.0	251.9	219.8	166.3	100,319
Fruit trees	68.2	44.8	9.0	14.5	22,650
Citrus trees	49.4	35.4	9.0	4.7	21,196
Nut trees	46.7	19.4	11.9	15.4	10,027

FIGURE 1: Distribution of areas under vineyards in compact plantations within Greece.
% : Percentage of the agricultural land within the Region.
( ): Thousand hectares.

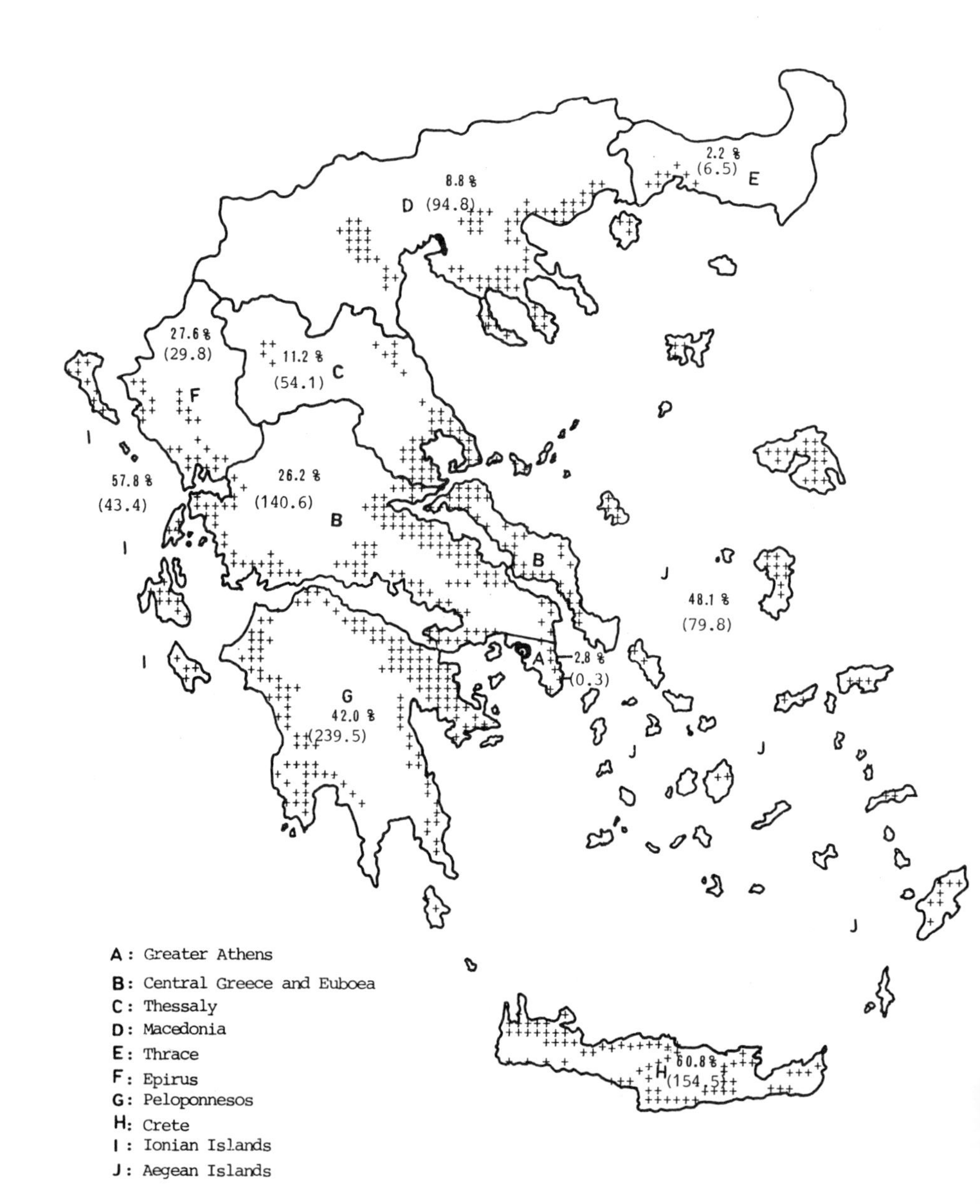

FIGURE 2: Distribution of areas under trees in compact plantation within Greece.

% : Percentage of the agricultural land within the Region.
( ): Thousand hectares.

VINE, OLIVE AND ORANGE TREES RESIDUES MECHANIZATION CHAINS IN SPAIN

J. ORTIZ-CANAVATE and J. GIL
Departamento de Mecanizacion Agraria
Universidad Politécnica de Madrid

Summary

From the middle of the 70's, some work has been done in Spain on the development of machines to manage the prune residues of vine, olive and orange trees ; the aim is to transport these products to the places of utilization in the most efficient way.

Large powered crushing machines used in forestry were tested during the first years, and they did not give good results. Later on, some smaller crushing machines specially adapted to manage pruning residues of vine, olive trees and other fruit trees have been built by Spanish manufacturers specially adapted to our conditions.

At present there are machines available in Spain and other countries that perform the following operations : piling, picking, crushing and baling. Some machines perform only one operation and others more than one. Other machines in use include front-loaders, conveyor belts and trailers to load, carry and unload the residues. Since 1980, the Spanish Department of Agriculture has tested these machines, collecting data about the performance and cost of each operation.

Most of the studies refer to olive residues. Small olive branches are given to sheep, goats and cattle which eat the leaves. Crushed branches usually are used as domestic fuel and to heat the ovens in bakery and pottery industries at a local level.

1. INTRODUCTION

In every orchard of vine, olive, citrus and any other kind of fruit trees, it is necessary periodically to separate part of the excess wood by means of pruning. Residues of pruning can be main branches of bigger or smaller size and secondary branches with leaves.

Traditionally, after pruning, large branches were separated manually with axes from the smaller ones and leaves. These large branches were used as domestic fuel while the small ones and leaves remained on the orchard until they were burned in order to get rid of them.

With the Energy Crisis in the middle of the 70's some experimental work has been carried out in Spain in order to get machines to manage the pruning residues of vine, olive and citrus orchards, and to use them as a renewable source of energy. This work has revealed that there are machines available in Spain for handling, transporting and applying this pruning wood economically.

At the beginning machines were tested for use in forestry work, but after a few years they were discarded because of their high cost of operation. The progressive introduction of national and imported machines specially adapted to work mainly with vine and olive trees pruning residues, encouraged those responsible at the Spanish Ministry of

Agriculture to organize demonstrations of operation, and practical tests of the available machines on the market. From 1983 there have been several demonstrations of machines for handling pruning residues of vine, olive and citrus trees in different regions of Spain. Some of these machines are already being used at farm level and some of them are still at an experimental stage.

In this paper the different machines utilized for these tests are described, including the practical application of some of them at the farm and industrial level.

2. BRANCH-SWATHERS

These machines are designed normally to be mounted at the rear three-point linkage of the tractor, although there is at least one that is mounted on the front.

Their function is to rake the lines where the branches are scattered. Once the volume of branches is big enough, the swath is lifted by opposite fingers, acting as a claw, and deposited in a windrow for their later transport, crushing on site or burning.

At present there are two manufacturers in Spain of branch swathers :

a) Branch-swather "Aguilar". There are two models of this machine :

- 1.50 m width, for vine and citrus trees
- 2.30 m width, for olive trees :

The lower ends of the fingers scratch the ground controling their height by means of two supporting wheels. The swath is lifted hydraulically. The 1.50 m machine needs a 30 kW tractor and the 2.30 m machine at least 40 kW.

During the tests with vine shoots, working at a speed of 5 km/h, losses were only around 2 %.

b) Branch-swather "Arenes". This machine has the catching fingers horizontal and operates backwards. The tractor with more than 15 kW of power, advances rearwards and when there are enough branches, two grip arms on the top close and sustain the swath and the tractor transports it to the chosen location.

3. CRUSHING-MACHINES

This type of machine offers a great diversity of models. There exist: stationary hand fed crushing-machines, pick-up crushing-machines and a pick-up crushing-baler machine produced by a French manufacturer. All the machines are mounted on the tractor and driven by the p.t.o.

They originated in forestry operations but they have been adapted so that the hammer or flail rotor is suited to the pruning residues of vine, olive and other fruit trees.

- Stationary crushing machines are normally hand-fed. The operation is as follows : the tractor with the crushing machine and in some cases a trailer, stops by a heap of branches ; one or two operators feed the machine which blows the crushed material into a container in the machine or a trailer. When it is finished, the train advances until the next heap. For safety the machine has a chute or a feed conveyor in order that the operators leave the branches on it and don't need to introduce their hands near the crushing cylinder. Two feeding cylinders can work in both directions in order that they can reserve the motion when there is an obstruction in the crushing machine. This mechanism is operated manually.
- Pick-up crushing-machines are mounted normally on the front of the tractor and pick branches which are windrowed between two lines of vines or trees. Similar to the previous case, the crushed material is blown to a trailer that is attached at the rear of the machine.

Existing crushing-machines manufactured in Spain are as follows :
- Crushing-machine "Cuenca", mounted on a tractor of at least 20 kW.
- Crushing-machine "Jafipes", needs a tractor of 45 kW or more. Crushing cylinder with 6 knives. Production of crushed wood : 2-2.3 t/h.
- Crushing-machine "Armentia" mounted on a tractor of more than 50 kW. The cutting disc has 3 robust knives and its diameter is 1 m. The discharge tube deflector is positioned manually. Production of crushed wood is 15-25 m^3/h.
- Crushing-machine "Dorsch", manufactured in Spain under license of a German Company. It is mounted on a 60 kW tractor. The p.t.o. drives two feeding cylinders which, rotating in opposite directions, drive the branches to the knife crushing cylinder. Crushed material is blown into a trailer attached on the rear of the machine.
- Crushing-machine "Agrator", this machine is capable of gathering crushed vine shoots on the ground. It is similar to a rotary stripper with 18 hammer-knives. This machine is not suitable for stony grounds.
- Pick-up crushing-machine "Jovimar". After picking and crushing the branches, the crushed material is scattered and left on the ground. For vine shoots, only 18-20 % remain whole, the rest is very finely divided and easily incorporated into the soil. There is a model with a container mounted on the machine to collect the crushed wood. Efficiency is 4-6 ha/d and production : 800-1.200 kg of vine-shoots/h.
- Pick-up crushing-machine "Biomasa". It needs a tractor of more than 85 kW. The pick-up and the crushing cylinder are located at the front of the tractor and the end product is transported to a container located on the rear. The pick-up is 1.20 m wide and has 2 side wheels to regulate its height. The crushing cylinder has 4 knives and rotates at 2800 r/min. Production of crushed wood is 4 t/h.

4. BALERS

Although there have been several balers for pruning residues tested in Spain, there is only one Spanish manufacturer of such machines at present :
- Baler "Jovimar", it consists of a prismatic press located at the rear of the pick-up crushing-machine of the same manufacturer. The density of the bale is about 160 kg/m^3.

Other foreign manufacturers who market their machines in Spain are :
- "Gregoire", from France ; makes prismatic bales from the material prepared by a crushing-machine of the same manufacturer.
- "Scorbati" and "Lerda" from Italy, very similar to normal balers but more robust (smaller width, pick-up fingers longer and stronger and a bigger fly-wheel).
- Roto-baler "Muratori" from Italy, also very robust, specially designed for vine-shoots. Dimensions of the bale are Ø 50 cm x 50 cm and its weight 15-25 kg that represents a density of 150-250 kg/m^3.

5. APPLICATION TO VINE PRUNING RESIDUES

Spain has a total vineyard surface of 1,650,000 ha.

When pruning the vines every year, approximately 0.8-1 kg of vine-shoots are obtained from each grapevine. Traditional vineyards in Spain have around 1,600 vines/ha, that produce 1,300-1,600 kg of vine-shoots per hectare.

Until recently, vine-shoots were collected manually, sampling in swaths the production of 8 to 12 vines. These swaths were later transported to the village to be burnt for domestic uses. Their application as fertilizer was not widespread, because of the difficulties in crushing and spreading this material.

Nowadays branch-swathers are used very extensively in Spanish vineyards. Vine-shoots are collected and deposited at the border of the field. A man with a tractor and with a branch-swather can collect the vine-shoots from 1 ha per day. Most of the farmers burn this wood at the field, but there are some who collect this biomass to be burned in local industries.

Heating power of fresh vine-shoots is 16,500-17,000 kJ/kg and 1 kg of fuel oil has a heating power of 42,000 kJ/kg. Because the furnaces with vine shoots have a lower heating efficiency than with fuel oil, we can establish an equivalence of 1 kg fuel-oil $\sim$ 3 kg of vine shoots.

When all vine-shoots in Spain were used for heating purposes, they could substitute 800,000 t of fuel-oil per year : 1,650,000 (ha) x 1,450 (kg v.s/ha) x 1/3 $\sim$ 800,000 t fuel/a.

Transport is a serious problem because of the low density of the bulk vine-shoots. A truck with a load capacity of 12,000 kg, can transport only 4,000 kg of vine-shoots this represents a 1/3 of its load capacity.

Commercial crushing-machines operate poorly with vine-shoots, because of their low density and because they entangle in the machine very easily. The solution is to prepare bales from vine-shoots, and to take these bales to a stationary chopper-blower located at the factory.

As it was explained in the previous section, there are commercial balers specially adapted for vine-shoots, but in order to get good productivity, it is more suitable that they operate stationary at the piles of vine-shoots, -going from one pile to the other-, than working on the rows picking the scattered vine-shoots. In the first case productivity can be : 4000-5000 kg/d $\sim$ 3-4 ha/d.

The cost per kg of vine-shoots in rectangular bales located at the factory resulted : 4 pta/kg.

Tests were carried out in 1982-85 at a factory of Pedro Munoz (Ciudad Real) with a stationary chopper-blower for bales from Construcciones Mecanicas Roga, El Arahal (Sevilla) with the following characteristics :

- Feed table made of wood with a slope of 45°
- Two chaff cutter knives
- Three hammers and a sieve with 40 mm Ø holes to complete the process of crushing
- Electric motor of 30 kW power.

This machine crushed around 4000 kg/h and worked 16 h/d. The crushed wood had its maximal dimension of 20-30 mm.

The economics of this process were as follows :

Cost of the chopper-blower : 5,500,000 pta

Useful life : 10 yrs

Rate of interest : 12 %

Labor : 3,400 pta/d

The cost of 1 kg of crushed wood being : 1.25 pta.

The ovens, adapted to this crushed wood, can work for bakeries, pottery industries and alcohol distilleries, the estimated cost of its use being : 1 pta/kg of chopped wood burnt.

Total cost to burn 1 kg of crushed vine shoots by this method is :

4 + 1.25 + 1 = 6.25 pta/kg

This cost is very competitive, because 3 kg of vine shoots, which cost 18.75 pta, substitute for 1 kg of fuel that costs : $\frac{42\ \text{pta/1}}{0.84\ \text{kg/1}} = 50$ pta.

This represents 2.7 times the cost of vine shoots of equivalent energy value.

6. APPLICATION TO OLIVE TREES RESIDUES

Olives are normally pruned in Spain every two years. Two parts are separated : large branches and small branches with leaves. Mean production for Jaen Region (the most important in Spain) is 9.8 kg of large branches and 24 kg of small branches and leaves per olive every 2 years.

Until the beginning of the 70's large branches were used as fuel for domestic applications and small branches with leaves were given to sheep and goats. Afterwards large branches were still burned in labour houses and villages close to the olive orchards, for winter heating, while the cost of managing and transporting small branches with leaves prevents any economical use as fuel or fodder. In recent years small branches have been generally burned in the field.

Nowadays in advanced farms the application is as follows :

After pruning large branches and small branches are separated, the large ones are taken to the rural centers to be used as fuel and the small branches with the leaves of 4-6 olive trees are piled together. A crushing-machine mounted on a tractor operates at each pile, being fed manually. The crushed material has a length of 8-15 mm and it is blown into a container or a trailer. The capacity of work of this machine is around 8,000-10,000 kg/d.

A pneumatic installation operates on the field with a cyclone to separate leaves and fine material from the coarsed chopped wood. Each component makes up 50 % of the total.

Leaves and fine material are used as fodder (61.7 % eq. starch and 9 % protein), especially for sheep or goats.

Chopped wood has a mean heating value of 13,000 kJ/kg and it is used in local ceramic and cement industries near the olive orchards in order to reduce transport costs. This chopped wood presents more advantages than using big branches or coarsed wood directly for two reasons :

- Chopped wood can be used more easily on automatic fed ovens.
- Coarse wood is used by "barrenillo" (Phloeotribus Scaraboides) for laying eggs and its storage can encourage this undesirable pest. This danger does not exist with chopped wood.

Cost of preparing the chopped wood can be calculated as follows :

Cost of crushing-machine + pneumatic installation = 3,000,000 pta.
Cost of tractor + tractor driver = 1,800 pta/h
Useful life = 10 years
Rate of interest = 12 %
Labor = 3,400 pta/d
Capacity of work = 8,000-10,000 kg/d
Utilization : 420 h/a (60 d x 7 h/d).

The cost of 1 kg of crushed wood was calculated as 3,10 pta.

To this price it is necessary to add the cost of transport from the orchard to the industry and the cost of using the furnace.

A more advanced process was introduced in 1983 by 3R Andaluza S.A. in a plant in Bailén (Jaén) for treating 25,000 t/a of biomass coming from olive and forestry residues, mixed with olive-oil wastes ("alpechin") and urban wastes.

Figure 1 shows the diagram of the process :

1) Olive and forestry crushed biomass
2) Solid urban wastes
3) Olive-oil wastes
4) Selection band
5) Magnetic separator
6) Aerobic fermentation
7) Feeder & mixer

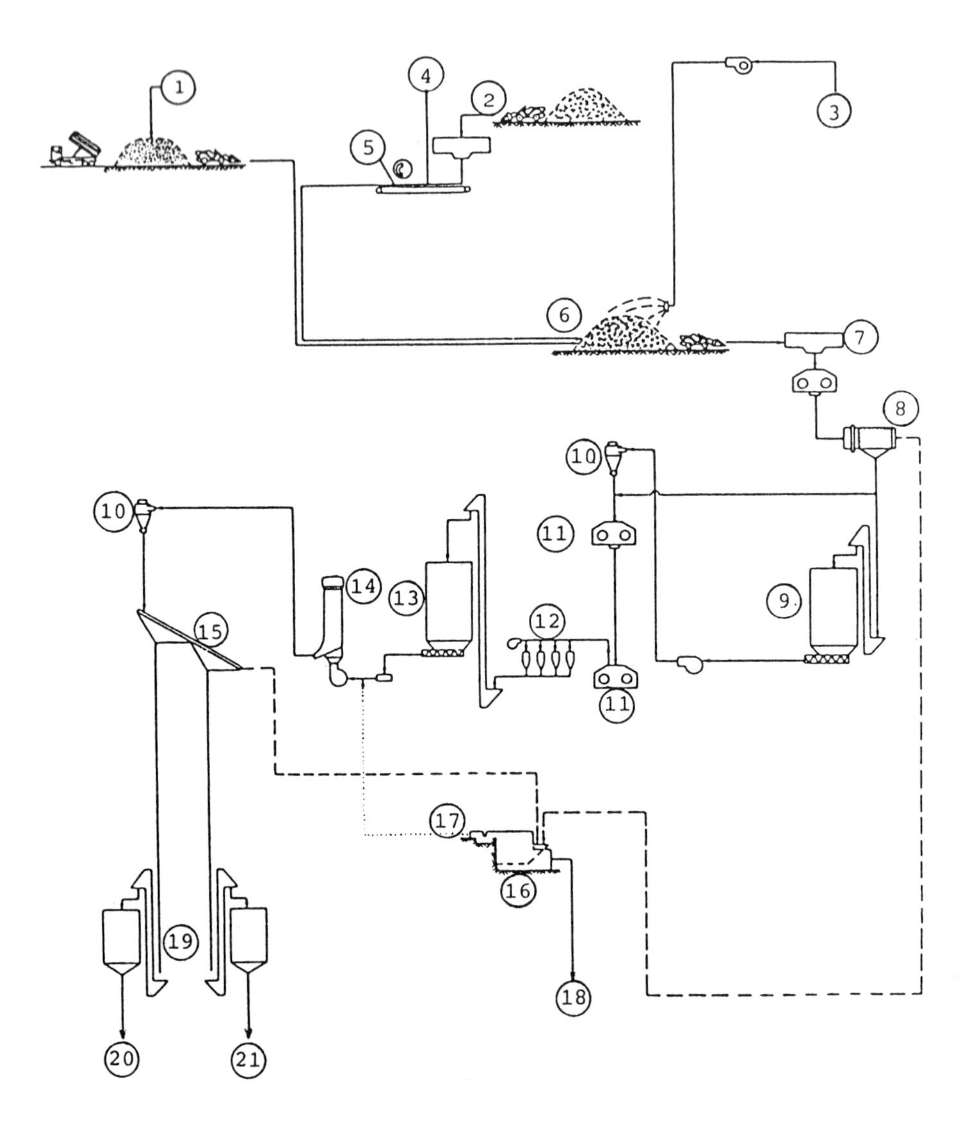

Fig. 1.- Process plant for treating olive and forestry residues mixed with olive-oil and urban wastes in Bailén (Jaén).

8) Trommel separator
9) Anaerobic digester
10) Cyclon
11) Mills
12) Cleaner
13) Hopper
14) Dryer
15) Sieve
16) Oven
17) Hot air
18) Ash
19) Storage silos
20) Fine chips
21) Grob chips

The final products, obtained after a treatment of 3 weeks are bio-chips (fine), with a size of 1-3 mm, and chips (grob) with more than 3 mm. These chips are suitable for pottery and bakery industries in the area, having a heating power of 19,000 kJ/kg and their price is 7 pta/kg which represents about 50 % of the equivalent fuel-oil cost.

7. APPLICATION TO CITRUS TREE RESIDUES

Previously machines were not generally used for the utilization of citrus residues for energy. Large branches from pruning are used restrictively at a local level and small branches and leaves are normally burnt at the border of the orchards.

In 1984 the Spanish Department of Agriculture carried out a Demonstration of Machines for the utilization of orange trees residues. There were 2 branch-swathers (both from Spain), 7 stationary crushing machines (4 from Spain, 1 from Italy, 1 from France and 1 from Switzerland) and 2 pick-up crushing machines (1 from Spain and 1 from France). All these machines are the same as the ones described previously for vine and olive residues.

8. CONCLUSIONS

- The use of vine, olive and citrus residues in Spain for energy applications and also for fodder (leaves and small branches of olives) for sheep and goats, is now at an introductory level, but we believe it is going to increase in the near future.
- With this assumption, the number of machines for picking, preparing (chop, crush, bale, etc) and transporting (and eventually for burning or distributing to animals) these residues, is going to increase.
- Existing machines, most of them coming originally from forestry operation, will improve their efficiency in agricultural applications and other machines will be developed and distributed in the coming years throughout the Spanish Fruit Industry.

ACKNOWLEDGEMENTS

The authors want to thank the help received by the Spanish Department of Agriculture, through Mr. A. Arenillas, Chief of the Section of Technology of Mechanization, who has supplied us information about the official tests with this type of machines and also the Company PROSER (Proyectos y Servicios S.A.) for its information about the plant of 3R Andaluza in Bailén (Jaén).

REFERENCES

(1) CONEJO, P.J., (1974). Eliminacion mecanica de los restos de poda. El Agricultor Practico, Oct. 74. pp. 18-21.
(2) MINISTERIO DE AGRICULTURA, (1983). Aprovechamiento de los sarmientos de vid. Dir. Gal. de la Prod. Agraria (Informe restringido).
(3) MINISTERIO DE AGRICULTURA, (1980). Aprovechamiento de los subproductos de poda del olivar. Dir. Gal. de la Prod. Agraria (Informe restringido).
(4) MORENZO, J. (1981). Aprovechamiento energético de los sarmientos. 13. Conf. Internacional de Mecanizacion Agraria. Zaragoza. April 81 pp. 57-70.
(5) PUIG, R. (1985). La biomasa como energia renovable. ERSA (Energias Renovables, S.A.). Madrid.

APPENDIX

Directions of Spanish manufacturers and representatives of machines to manage vine, olive and citrus trees residues

- Agrator S.A.
  Carretera de Murguia, km. 5
  Aptdo. de Correos, 316
  E-01080 Victoria
  Crushing-machine AGRATOR

- Agric S.A.
  Carretera Nacional 152, km. 80
  E-08080 Masias de Voltrega (Barcelona)
  Crushing-machine AGRIC

- Botas-Armentia S.A.
  Adriano VI, 20-5°
  E-01008 Vitoria
  Crushing-machine ARMENTIA

- Comercial Agricola Anfer S.A.
  Avda. de Aragon, 2
  E-31000 Carcastillo (NAVARRA)
  Pick-up crushing-machine BRUPER

- Desbrozadoras Jonnes
  Hostal de la Bordeta, 50
  E-25001 Lérida
  Crushing-machine ATILA

- Dorsch S.A.
  Travesia Sres. de Luzon, 2
  E-28013 Madrid
  Crushing-machine DORSCH

- Foresta Ibérica S.A.
  Calle Valle de Egües, 6
  E-31004 Pamplona (NAVARRA)
  Pick-up crushing-machine BIOMASA

- Indagrimec Espanola S.A.
  Rafael Salgado, 7
  E-28036 Madrid
  Vine-shoots balers (LERDA & MURATORI)

- JAFIPES
  Ctra. Artesa, s/n
  Aptdo. 26
  TREMP (Lérida)
  Crushing-machine JAFIPES

- Pedro Ortega Torrent
  Travesia Exterior, s/n
  E-43000 Santa Coloma de Queralt (TARRAGONA)
  Crushing-machine ORGO

- Promasa
  Herreros, 8
  E-16000 Las Mesas (CUENCA)
  Pick-up crushing-machine JOVIMAR

- Talleres Aguilar
  Joaquin Costa, 16
  E-22000 Graus (HUESCA)
  Branch swather AGUILAR
  Crushing-machine BARTOLUCCI

- Talleres Belloso, S.L.
  Carretera Valencia, s/n
  E-50000 Calatayud (ZARAGOZA)
  Crushing-machine BELLOSO

PLAN FOR OBTAINING ENERGY FROM WASTE (PAER)

Jesus GUERRERO
Head of the Combustible Waste Programme of the Institute for Energy Diversification and Saving (IDAE)

Summary

It is the aim of the Renewable Energies Plan (PER) to substitute primary energy for 560,000 tep/yr of biomass energy by 1988.

Of this amount 350,000 tep/yr will be obtained by using fuels produced from wood waste from forestry and agriculture. The set of actions which will allow this substitution to be made within the deadline laid down make up the IDAE's PLAN FOR OBTAINING ENERGY FROM WASTE (PAER).

1. INTRODUCTION

In order to diversify primary energy sources by substituting 350,000 tep/yr with fuels produced from wood biomass waste, the industrial sectors involved would have to manufacture and use approximately 1,000,000 tonnes of this type of fuel annually.

There follows a general outline of the measures necessary to achieve this.

a) Identification of a sufficient quantity of resources.
b) Evaluation of the cost of supplies (collection of the above resources as raw materials for fuel production).
c) Identification of the fuel products to be manufactured and specification of their basic characteristics (size, moisture content, net calorific value). Determination of the processes required and their cost. Determination of the investment needed to put these processes into industrial production.
d) Identification of the potential market.
e) Boosting for the companies who make the investment.
f) Boosting of demand for the end product.

2. BASIC STUDIES

The PAER is built on basic studies carried out on a regional level which cover the aspects referred to in points a) and d) of the preceding paragraph. These studies have been carried out by IDAE in the following Autonomous Communities :

- Galicia
- Asturias
- Cantabria
- Rioja
- Navarre
- Aragon
- Castile-Leon
- Castile-La Mancha
- Extremadura

- Andalusia
- Valencia
- Murcia
- Balearic Islands

3. COMPLEMENTARY PROGRAMMES

Measures b) and c) of paragraph 1 have been implemented as part of the Plan for Research into Wood Energy Products (PIEPMA) which consists of three specific programmes :

3.1. Resources Programme

The aim of this programme is to define a method of extraction which, when applied to the existing resources, will identify those which can be used for the production of "field chips" on the basis of the premise of the final cost of the product. Field chips are defined as chips obtained from self-propelled machines actually working in the fields. The maximum size of these chips is below 10 cm and their maximum moisture content between 35 and 40 %.

The resources are :

3.1.1. Forestry waste :

- pruning waste
- thinning waste
- scrub clearing waste
- felling waste from timber production
- sawmill waste.

3.1.2. Agricultural waste :

- pruning waste from olive groves
- pruning waste from vineyards
- pruning waste from fruit orchards.

Those resources from which field chips can be obtained for 2,000 pesetas/tonne or less will be defined as "Resources" by the programme and evaluated as such in the PAER.

The programme is to specify, test and certify the production method and the equipment to be used.

3.2. Product Programme

The task of this programme is to specify the necessary processes and equipment and the costs and investment involved in producing the products in Table 1 from field chips. These products can be summed up as follows :

- Chips : three size/moisture content classes
- Agglomerates : two size classes.

3.3. Combustion Programme

This programme is to identify the ideal combustion conditions for each of the products specified in 3.2., to which end it must test and certify the behaviour of each energy product when subjected to the different combustion methods available (fixed furnace, extension grate, injection, etc.) in conjuction with different heat recovery methods (hot air, hot water, thermal oil, different types of steam, etc.).

4. AREAS OF APPLICATION

Studies have been made of the viability of substituting PAER products for fuel oil and gas oil in the following sectors :

* Building ceramics
* Food industry
  . canned goods (meat, vegetables, fish),
  . bread and bakery products,
  . dairy products,
  . oils and alcohols,
  . industrial abbatoirs.
* Textiles
* Paper
* Rubber
* Timber
* Iron and steel (lamination furnaces)
* Domestic.

5. CURRENT SITUATION OF THE PLAN

Table 2 lists the PAER plant which have been found to be viable according to the criteria described above.

Figure 1 gives a picture of the geographical situation and Table 2 shows how far the aims for 1988 had been fulfilled by 20 September 1986.

- TABLE 1 -

P.I.E.P.M.A. - Product Programme

PRODUCT CHARACTERISTICS

A) CHIPS: DENSITY 300 Kg/m3

(*) BASIC DATUM:

GROSS CALORIFIC VALUE = 4,500 Kcal/Kg.

SIZE \ MOISTURE CONTENT	25 - 30%	10 - 20%	10%	PROCESS
2 - 5 CM	3.100 Kcal/KG 5.500 Rs/T			Naturally dried field chips
1 - 2 CM		3.700 Kcal/Kg 6.700 Rs/T		Naturally/forced dried factory chips
3 - 7 MM			4.300 Kcal/Kg 7.600 Rs/T	Forced dried factory chips

B) AGGLOMERATES: DENSITY 900 Kg/m3

* PELLETS ϕ < 10 mm
* BRIQUETTES ϕ > 5 cm.

PRICE

COMMON CHARACTERISTIC: < 1,50 Rs/therm

COMBUSTION EFFICIENCY

75–80%

TABLE 2

AUTONOMOUS COMMUNITY	NUMBER OF PLANT	CAPACITY
Andalusia	5	199,229
Castile-Leon	4	80,000
Extremadura	1	60,000
Aragon	3	90,000
Valencia	3	90,000
Cantabria	1	25,000

FIGURE 1

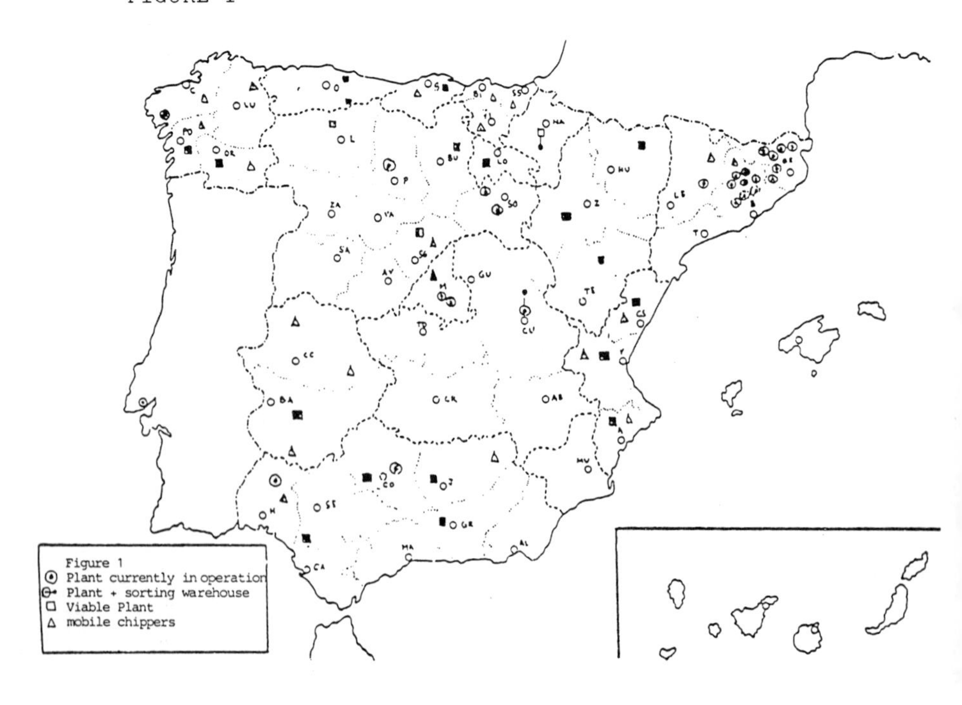

EXPERIMENTAL TESTS ON A PROTOTYPE MACHINE FOR HARVESTING FORESTRY TRIMMING WASTES

P. AMIRANTE - C.C. DI RENZO
Istituto di Meccanica Agraria
Universita' Degli Studi
BARI

1. INTRODUCTION

Over the past few years the idea has grown of using forestry trimming wastes to produce energy.

In fact, ligno-cellulose prototype fuels can be burned directly using quite simple technology, but the technical and economic situation is not always conducive to the large-scale utilization of these by-products.

Firstly, it must be borne in mind that one kg of dry biomass is in general the energy equivalent of approximately 0.4 kg of petroleum. Secondly, it should be emphasized that the efficiency of existing boilers which use alternative fuels is fairly low, varying between 45 and 75 % depending on the fuel feed system of the boiler. The real value of one kg of biomass is therefore reduced to between approximately a third and a fifth of one kg of petroleum, which means that it is important that the costs of harvesting and storing the biomass should be kept as low as possible by mechanizing these processes fully.

In strictly economic terms it should be noted that the economic value of trimming wastes is quite low and may be put at somewhere between 150 000 and 260 000 Lire/ha.

The problem must therefore be tackled on the basis that, while the economic benefit to the individual farmer is quite low, there is a significant energy benefit in view of the total quantity of biomass available.

It must also be borne in mind that for an individual holding, particularly in southern Italy, fixed energy requirements are fairly low, so that in general there is no balance between the available energy source and the energy needs of the holding concerned.

However, with regard to the technical and economic aspects direct use on a holding may simplify the problem, in that it is necessary to provide only modest facilities for harvesting the biomass, which may be stored directly on the holding so that it can dry naturally. In this way transport and handling costs are in general kept quite low.

However, even if the biomass is compacted so that it can be used at a distance from the holding, the main criterion for the use of the by-product on a medium or large scale is whether it is possible to have efficient harvesting machinery and means of transporting and handling the biomass.

This paper is concerned with the machines currently available. It defines their limitations and proposes the use of a new prototype for the mechanical harvesting of trimming wastes.

2. LIMITATIONS OF MACHINES FOR HARVESTING TRIMMING WASTES

Waste wood from production processes needs to be cut with hand-operated, pneumatic or hydraulic shears or, if the cross section exceeds 3-4 cm, hatchets or power saws, using a mobile platform.

For some crops, such as vines and citrus trees, some mechanical systems for harvesting the wastes have come into use, but these are not yet widespread.

The wastes obtained generally consist of twigs, i.e. small branches and leaves, larger branches, i.e. up to 3-4 cm in diameter, and logs, i.e. branches more than 4 cm in diameter.

Because such a diversity of material has to be harvested, it is difficult to find a single process for all categories, and yet, on the grounds of economic management of by-products, preliminary manual separation of branches and twigs from logs is out of the question.

An initial evaluation of the problem can be carried out on the basis of the maximum cutting forces as a function of branch diameter.

For diameters of up to 2-2.5 cm relatively little force is needed, but the force required increases quadratically. Above 3-4 cm diameter it is difficult to cut wood with the conventional baler cutters, unless these are strengthened.

The same conclusion may be drawn in respect of the hardness of the branches which, as is well-know, is a function of the product of the module of elasticity of the wood material and the moment of inertia of the sections of the branch. Simple calculations show that the moment of inertia of 6 cm diameter branches is approximately 80 times higher than that of branches 2 cm in diameter. Consequently, when the diameter of a branch is more than 3 cm, its increased hardness makes it impossible for the rotary baler to cut it.

Accordingly, vine shoots or trimmings from fruit trees which are less than 3 cm in diameter can be harvested using a pick-up baler of the conventional type or a rotary baler, suitably modified, whereas for less pliable branches over 3 cm in diameter new machines are required which eliminate the cutting and bending operations. Development work is therefore being carried jointly with a company in Apulia on a new prototype baler developed from a low-pressure press which eliminates the cutting and bending operations.

Before discussing the prototype under development we will look at the limitations of machines which are currently on the market, viz. :

--) pick-up balers with chambers with a rectangular cross-section ;
--) rotary balers ;
--) harvesting/shredding/loading machines.

In their present form all the above machines have a limited range of uses and require further development if their use is to become widespread. They may therefore be considered to be still at the prototype stage.

In particular we noted that :

1) The rotary balers and baler presses with rectangular section chambers can be used for the stalks of vines and for soft and flexible branches. Further efforts should be made to make the machines more reliable, since at present 20-30 % of the working time is spent on clearing the conveyor and the cutting and binding mechanisms.
2) The third type of machine at present needs substantial modifications, in particular to the conveyor assembly.
3) In addition, the use of mobile chipping machines which are primarily used in forestry, activated by the skidder operator, does not seem to be the answer to the problem of harvesting and cutting in the field. These machines in fact require manual feeding into the chipping aperture which, because of the differing sizes and shapes of the wastes, causes a great deal of difficulty and therefore results in unacceptably poor performance.

3. TECHNICAL STUDY OF A NEW PROTOTYPE LOW-PRESSURE BALER PRESS

In the light of previous experience and in view of the limitations of existing machines, a new baler press machine is being developed jointly with the CICORIA company of Palazzo S. Gervaso (PZ). It is based on a prototype of a low-pressure machine already used for straw.

At the front of the machine is mounted a new harvester with retractable teeth of the closed type, which is reinforced so that it can lift branches and logs of wood.

The wastes are fed in without cutting machinery. There is a panel which propels the wood continuously in random fashion towards the compression chamber.

The pressure achieved is quite low, with the result that the volumetric mass of the wood varies between 50 and 80 kg/m^3.

Various devices are being used to try to increase the density of the bales, which come out in irregular shapes and are similar to bundles of sticks.

The binding device was strengthened, although it was found that because of the way in which the machine operates branches which are too long and twisted may form part of two bales joined together.

Tables 1 and 2 give operating data on the first experimental prototype, comparing its technical performance and the characteristics of the bales with the results obtained from baler presses and rotary balers which are already on the market.

The first tests showed that the machine was safer to operate than existing prototypes, but most importantly it was more efficient at harvesting olive tree wastes, in that it was possible to pick up branches more than 3 cm in diameter so that no prior sorting was required in harvesting the wastes.

Following the initial results, which were encouraging, it is proposed to go on from the experimental prototype to the final machine to be put on the market.

4. CONCLUSIONS

Apulia is one of the regions of Italy which is very interested in the wood growing sector, both in percentage terms and in absolute figures, in that the specialized cultivation of this crop accounts for more than 650 000 hectares.

The main trees concerned are the olive, vines and almond trees, and it is estimated that from these three there is a total potential energy obtainable from the trimming wastes equivalent to approximately 320 000 toe.

Nevertheless, the use of these by-products for energy is somewhat restricted at present.

The factors limiting their use as energy are :

--) the land is broken up into small holdings ; in fact, more than 39 % of the holdings in Apulia are not larger than one hectare ;
--) there is little requirement for fixed heat-energy on holdings ;
--) the biomass available to a single holding is of little economic value ;
--) the cost of transporting the biomass from the holding is high.

However, farmers are very interested in collecting this waste, since at present a fair amount of equipment and labour is required to dispose of it, as it has to be cut and buried or harvested by hand and burnt at the boundaries of the holdings.

The agricultural cooperatives, which are already operating in the wine-growing and olive-growing sectors, are therefore very interested in resolving this problem, and are setting up centres for gathering and compacting the biomass so that it can also be used for energy outside the individual production areas, which saves the individual producers their present expenditure on disposing of the wastes.

Plans have therefore already been made for compacting centres in areas where olive- and wine-growing are predominant, so that the biomass can be harvested and compacted. Investment of the order of 600 to 900 million Lire is to be made for constructing the plants, with annual turnover expected to be in the region of 400 to 600 million Lire.

A market survey carried out in various centres in Apulia has also shown that there are definite possibilities for disposing of the compacted product.

The main limiting factor in the use of this waste is therefore the unreliability of the machines, which is discouraging the farmers from moving in this direction.

It is therefore to be hoped that research will be intensified on developing efficient machines for harvesting trimming wastes and that finance for the research work will be obtained from authorities concerned with increasing the use of all forms of energy.

Table 1 - Mean values for the performances of the experimental machine and machines for harvesting, baling, loading and transporting olive and vine trimming wastes on trailers to holdings

Crop	Harvesting and baling			Loading and transport			Total labour required	
	Baler press			Loaded by bale loader and transported by trailer to holding				
Olive	No of bales / h	ha / h	t / h	No of bales / h	ha / h	t / h	hrs - labour / ha	hrs - labour / t
	50 - 60	0.6-0.6	1.7-2.1	80 - 90	0.8-0.9	2.7-3.0	6.7-7.8	2.0-2.3
Vine	Rotary baler			Manual loading and transport by trailer to holding				
	35 - 50	0.3-0.5	0.8-1.2	55 - 65	0.5-0.6	1.3-1.6	9.0-12.6	3.5-4.8
Olive	Low-pressure baler press (experimental prototype)							
	105	1	2.3	80 - 90	0.8-0.9	2.7-3.0	2.0-3.0	0.9-1.3

Table 2 - General features of the trimming wastes used in the harvesting and baling trials carried out with the experimental machines and machines already on the market

	Vine		Olive	Random bundles (experimental machines)
Bales	cylindrical	rectangular	rectangular	average sizes
- measurements (m)	d 0.5, l 0.5	0.75x0.37x0.45	0.75 x 0.37 x 0.42	1.1 x 0.30
- average weight of bale (kg)	24	25	34	18.5
- humidity (%)	38	38.5	42.0	40.0
- volumetric mass (kg/m^3)	244	182	295	57

TREE CROPS PRUNING MECHANIZED CHAINS IN PUGLIA REGION

V. BARTOLELLI and D. METTA
RENAGRI - Centro per la promozione dell'uso delle energie rinnovabili in agricoltura

Summary

Mechanization plays a basic and priority role on the possibility and type of utilization of agricultural by-products. Profitable by-products recovery without using suitable machines is unthinkable (1). At present, the machines for harvesting residues that are available on the market have been developed. In practice, prunings are collected and subjected to a first treatment (in the shape or manufacture), that is determined by their future use. The choice of a pruning pick-up machine, is based on nature and characteristics of the pruning and utilization of the residues. A further important element is the site where the collected by-products are to be utilized : if transportation involves high costs, the pick-up machine is required to reduce the volume of the collected residues. Once again it is pointed out how picking up and conditioning operations are in close connection. This paper refers in particular to residues from tree cultivations.

1. PICKING UP OF RESIDUES FROM HERBACEOUS CULTIVATIONS

The herbaceous sector, especially as regards straws, is traditionally linked to the utilization of residues in those farms with a mixed production system. The use of cereal straws as stable bedding and of some residues (from beets or vegetables) as fodder integrators is rather common.

The picking up techniques for these herbaceous residues are very similar to the ones employed for fodder crops ; for this reason, after suitable modifications, the same machinery can be utilized.

No data and experiences about residue collection methods are available ; but in this case, the picking up operation can be carried out by means of conventional machinery.

The adaptability of self-loading vehicles and of cutter-loaders or mower-cutter-loaders allows the operation to be performed on by-products of different type.

An interesting study, moreover, could fit the conventional combine-harvesters with a device for mowing (when the harvesting machine of the principal product does not do it) and for the direct loading of stalks and cobs into the tank. This allows a reduction in number of the operations in the field and a cleaner product. This is useful when the product is stored in silo.

2. PICKING UP OF BY-PRODUCTS FROM TREE-CULTIVATIONS

The residues from multi-year cultivations are the prunings, generally of wooden nature as shoots and branches of different size. The non-deciduous cultivations, or those subjected to summer lopping, produce a considerable volume of leafy branches, which in some cases (olive-tree) may be utilized as fodder or for combustion (often at the field side) or chopped and plowed under the soil.

Independently of the species, the principal factor influencing the choice of the machine and of the site for the recovery of these by-products, is the diameter of the wooden residues.

For the treatment of shoots and branches not exceeding 3-4 cm in diameter, the employment of balers or cutters by modifying the conveyance system, the compression chamber or the cutting elements, is possible. The available technology does not allow the use of such machines with larger residues. For these, a conditioning system, based on the use of chopping machines, similar to the ones used in the forestry and wood manufacturing sectors, can be employed. The by-product recovery involves a considerable energy saving, since the outlay for conditioning and transportation to the farm are calculated to be a few percentage units of the energy content of the by-products.

In optimum conditions (slightly undulating land, regular planting space, few hundreds meters from the farm center) the incidence of the energy cost on income is equivalent to 1-2 %, increasing up to 3,5-4 % in more disadvantageous conditions, but never exceeding 10 % even where high power is required, as for cutting operations by chopping machines.

Among the tree cultivations grown in Italy, the most suitable one for the total mechanization for the by-product picking up operation is the vine, the lopping of which consists of cutting out the twigs from the 1 or 2 years old branches.

An immediate consequence of the facility of mechanization is the availability on the market of many machines fit for this use and developed on the basis research and tests.

As regards the other cultivations, the pruning matter is very heterogeneous and the ratio between leafy branches (diameter 3-4 cm) and wood (diameter 3-4 cm) depends on species, type of breeding and type of pruning (fruit formation pruning, regeneration pruning, etc.). For this reason, the separation between the two categories and the machinery feeding are made by man.

Further studies and experiments will be carried out for the development of machines for the collection of residues from different cultivations.

The machines and working systems for the picking up of the following by-products will be analyzed :

- vine
- olive-tree
- almond-tree
- other fruit-trees
- citrus-trees

2.1. Vine

2.1.1. Season and type of operation for the vine-branches collection

The dry lopping of vine is carried out during a period starting from the end of autumn till the end of winter. Generally, the branches are cut out from the plant and left in the inter-row. It is necessary that the distance between two rows is more than 3 m in order to allow the pick-up machines to pass through. By contrast, the branches are usually piled up at the edge of the vineyard by means of tractor-drawn sweep rakes. The best way is to windrow the branches in the center of the inter-row, according to the smallest or equal width of the rake, in order to avoid the use of side-delivery rakes with the consequent economic and energy outlay.

The rationalization of such operation involves a considerable energy saving. The use of sweep rakes for piling up branches at the edge of the vineyard consumes 70 % of the total energy need of mechanical means employed for the complete picking up, conditioning and baling operations, and 50 % of the human labour requirement (6 hours/worker/ha).

In the case of mechanized collection by means of balers and considering a prunings production of around 3-4 tons/ha on flat grounds, 3-4 hours per worker/ha (with a productivity rate equal to 1 ton per hour/worker) are necessary to perform the picking up, transportation within 100 meters and piling up operations.

2.1.2. Machinery for vine-branches collection and conditioning operations

The machines available on the market for picking up and conditioning of such residues comprise four principal categories :

- pick-up-balers
- pick-up-cutter-balers
- pick-up-cutter-loaders
- pick-up-press-balers

2.2. Olive tree

2.2.1. Season and modalities for the picking-up of by-products

The lopping of olive-trees covers a long period, from January until the first week of June, but it is not carried out every year so that the cutting operation gives a greater quantity of wood rather than foliage. There are two typologies of picking up and conditioning, which different preliminary operations correspond to.

2.2.2. Machinery for by-products collection and conditioning operations

In case where the machines are able to treat both wood and foliage, the material must be windrowed or piled up in order to allow the chopping machine to operate. This type of machine always requires manual feeding ; it can work with every type of by-product having a size smaller than 90-150 x 100-400 mm (the size is relative to the inlet opening of the machine), chopping them into fragments of 5-30 mm by means of disks or cylinders fitted with blades. The pick-up balers, necessary when the by-product is utilized for zootechnical purposes in order to facilitate its transfers, requires a manual separation of foliage from wood, since they cannot work with residues larger than 3-4 cm and often need a man to feed them.

As an alternative to pruning, picking up and conditioning in the field, collection can be carried out together with lopping by employing a trailer on which the prunings are loaded ; the subsequent conditioning (fascine binding, baling, brush wood cutting up, etc.) will take place within the farm itself.

Considering this latter case and referring to flat oliveyards with regular planting space and with the use of cutting machines with a working capacity of 0.6-0.8 tons/h, the energy outlay will be 35-40 kWh per ton of dry matter.

2.3. Almond-tree

We are considering this species because of its importance in the Region Puglia. As regards both the picking up modalities and the other considerations, they do not differ from the ones concerning olive-tree cultivation.

2.4. Other fruit-tree cultivations

An intermediate situation between vine and olive/almond-tree cultivations is that regarding apple-trees and other stone fruits. In this case pruning takes place every year during January and February since a great quantity of by-products with a diameter smaller than 3-4 cm is obtained ; low power is required for wood conditioning and the smallest residues can be subjected to treatments very similar to the ones employed for vine-branch wood.

The recovery of such by-products is the only alternative to combustion at the field side, since the grass covering practice, frequent in fruit-orchards, does not allow ploughing into the soil.

In this case, both residue collection and processing or just collection together with mechanical lopping can take place on the field. If collection is done together with mechanical lopping it is necessary to use a trailer for foliage recovery.

The expected energy outlay will be about 25-30 kWh per ton of dry matter.

2.5. Citrus-tree cultivations

The lopping of this species generally occurs late in the winter or early in the spring, before blooming. The residues are placed in the inter-row and subsequently piled up and burnt at the field side or chopped and ploughed under the soil.

Because the residue are soft, they can be easily subjected to recovery after they have been chopped or just picked up and baled. Such a technology must be tested in real conditions which make the machine completely suitable to the purpose.

The expected energy outlay for the picking up and lopping operations is 35-40 kWh per ton of dry matter.

3. MECHANICAL METHODS FOR THE COLLECTION OF PRUNINGS FROM OLIVE-TREE AND VINE IN PUGLIA : AN AREA/RENAGRI TEST

The extremely variable weather conditions and soil in Italy are such that the tests carried out in different areas of the country present typologies of their own. These experiences only give an indication value of the by-product picking up and processing operations, but it is necessary to concentrate the attention on local situations in order to obtain more reliable data.

The case under consideration relates to a site reflecting in part the typical agricultural and climatic conditions of the Region Puglia, where the machines have been tested on impervious soils with surface rocks, irregular plant distance and mixed cultivation system.

In view of a future agreement between ENEA (National Committee for Alternative Energies) and RENAGRI (Center for the Promotion of Renewable Energies in Agriculture - a non profit Association seated in Roma) a test comparing the different methods and machines for pruning collection took place in the spring 1984. The species considered were vine and olive-tree because of their great incidence on the agriculture of the Region.

At the same time, studies about the possibility of burning these by-products as alternative sources to traditional fuels, have been developed.

3.1. Vine prunings picking up

3.1.1. Experimentation site

The research took place in a farm near Brindisi, on a 8-ha medium textured and good fertility soil, in an area where the farm units are considerably split up : 4 lots at a distance ranging between 500 m and 9 km from the farm center.

The farm is typical of the area, with a representative crop system, being covered with vines and olive-trees.

3.1.2. Working system and means employed

For collection, a 26 kW tractor and a "Autorac 100 Muratori" pick-up-press-baler have been utilized. The latter is a medium size machine fitted with a conveying device, composed of two toothed rollers collecting the material which, through rotating rolls, is conveyed into a suitable chamber. Inside the chamber, once the optimum pressure is reached, the material is bound and expelled in the form of cylindric bales having an average weight of 18 kg each, a diameter of about 0.5 m and a width of approximately 0.6 m. During the binding and ejection phase, the collection operations do not take place.

For the transportation of the bales to the farm, a further tractor with a capacity of 51 kW and 3 trailers was required. The picking up and processing operations carried out by the pick-up-press-baler (Tab. 1) were preceded by the vine-branch windrowing, manually carried out on alternate inter-rows during lopping operations.

Bale transportation has been subdivided into two phases :

- transportation on field - from the inter-row to the edge of the vineyard using barrows trailed by the 26 kW tractor ;
- transportation to the farm by means of trailers drawn by the 51 kW tractor.

3.1.3. Testing results and considerations

The operations recorded low values for the working system efficiency (0.13 t/h), the forward speed of the machine (1.2 km/h) and, for the productivity of labour (0.12 t/h for the collecting and baling operations - 0.08 t/h for the whole working procedures).

The clear inferiority of this value is due to the still wide utilization of manpower (about 13 hours per worker/ha) in the vine-branches windrowing and during collection, where the continuous presence of a worker is required to place the residues at the inlet opening of the conveying device.

Further reasons can explain why this value is low : the first one refers to the time during which the machine is not used. The actual collection time, covered about 30 % of the total time including transfers, damages, flooding, turns, cleaning, maintenance, sinking in the mud, repair work and pauses in general. Moreover, the experimental results of the unitary production shows a value corresponding to 50 % of the average production values registered by national literature. This reason also contributes to reduce the values.

A further problem is the cost of transportation from the field to the farm center, sometimes reaching distances of about 9 km.

Even in the presence of these problems which need to be solved in order to establish the economic convenience of the whole process, the energy balance confirms the validity of the by-products recovery, the cost of which (195 kWh per ton of dry matter), including transportation to the farm, is 3.6 % of the energy content of the collected and conditioned residues.

The first suggestions to be carried out concern the efficiency of the pick-up-baler : the possibility of flooding must be reduced by modifying those parts stopping the machine working (conveying system, binding and cutting devices) and by making the machine suitable to the materials employed. This can be achieved by enlarging the head width (1.20 m) which is too small for the usual minimum planting space (2 x 2 m) and thus involves frequent material losses with the consequent repercussion, as already mentioned, on human labour requirements.

3.2. Olive-tree prunings picking up

In the Region Puglia the two different methods of olive-tree pruning collection and processing have been tested :

- chopping of material by means of chipping machine and collection, transportation and storage into the farm of the chopped material ;
- pick-up-baling of the leafy branches, subjected to separation, with transportation to the farm and consequent storage.

3.2.1. Picking up, conditioning, transportation and storage by means of chipping machine "CHIPPER 100 GANDINI"

The test took place in a farm near Brindisi (Castellana Grotte) in a lot 2 km from the farm center, in a hilly area covered with ripe olive-trees and with cherry and almond trees as association crops. The low productivity soil presented surface rocks which interfered in the regular planting space.

For collection, a 26 kW tractor and chipper 100 Gandini pick-up-cutter have been utilized. The latter is a chipper used in the forestry sector, fitted with an inlet opening inside which toothed rollers chip the residues. These are subsequently conveyed, through a vertical duct, into a trailer placed in the back.

For transportation to the farm, a tractor and trailer have been used. The chipping machine worked on residues previously piled up, stopping every 4-5 trees.

None of the operations suffered from problems linked to the working of the machine, but all the evaluation parameters presented low values (Tab. 2) because of the scarce unitary production of by-products (due to the large space among the olive-trees thus involving dead time for transfer) and the wide employment of manpower, which is necessary for piling up material and feeding the machine.

The branches, in fact, must be suitably prepared for their introduction into the machine loading mouth by cutting the leafy branches and eliminating foliage.

The item "working-capacity" has increased in the plots with a higher unitary production. This was due to a higher efficiency of the chipping machine which, therefore, was not exploited to its maximum power on the other occasions.

The energy outlay of the whole operation system (258 kWh/ton of dry matter) represented 4.7 % of the energy content of the recovered by-products. It must be emphasized that direct costs weigh only 2.15 %, while the remaining percentage depends on indirect costs relevant to the energy employed to construct the means utilized.

3.2.2. Pick-up-baling by means of "LERDA 900 L" machine, transportation to the farm and storage

The disadvantageous conditions of the plot under examination and the positive results reached (Tab. 3) make the use of this machine especially suitable for the olive leafy-branches pick-up-baling.

The testing plot is located at 7 km from the farm center, near Monopoli (BA), covered with olive-tree cultivations, irregular planting space, surface rocks and with almond-trees as companion crop.

The collection has been carried out by a 29 kW tractor and a pick-up-press-baler LERDA 900 L. This is a forage baler modified for the picking-up of olive-tree prunings. It produces parallelepiped bales of about 35 kg each. The pick-up and compression systems have been considerably reinforced because of the very hard operating conditions the machine is designed for.

For transportation, the 29 kW tractor, together with a trailer on which the bales were loaded, have been employed. As in the other cases, the bales have been stored manually.

The promising results regarding the working capacity of the machine and of the whole working system has encouraged further experimentation in order to increase the efficiency of the operations. In particular, the only modification to be carried out on the machine regards the string cutting system, which is often clogged up by residues in the slot of the needles.

Another factor negatively influencing the productivity depends on the excessive distance between windrows, due to the irregular planting space and the presence of almond-trees, and on the distance from the farm center which adds to the transportation costs. This is anyway a general problem regarding the need of picking up and processing systems suitable for any production situation.

The picking-up-baling by LERDA 900 L machine is undoubtedly the best solution among those experimented. These promising results are also due to the small incidence (1.21 %) energy outlay (66 kWh/ton of dry matter) in comparison to the energy content of the recovered by-products.

REFERENCES

(1) E. NATALICCHIO - C. SEMENZA "Possibilità produttive, tecniche di raccolta, utilizzazione attuale e alternativa dei sottoprodotti agricoli" CNR, Progetto finalizzato Meccanizzazione agricola, Quaderna 23, Roma - 1981.

(2) A. ARRIVO - E. DI CANDIA "Sottoprodotti e residui delle colture in Puglia : possibilità di meccanizzare le operazioni di recupero e trasformazione" Informatore Agrario 31/83.

(3) P. GUARELLA "Raccolta e condizionamento in balle di residui di potatura di vite e olivo" Informatore Agrario 39/84.

(4) M. AMIRANTE - G.C. DE RENZO "Sottoprodotti vegetali : una fonte di energia da recuperare" Terra e Vita 40/85.

(5) M. BARTOLELLI "Studio socio-economico sulla utilizzazione dei residui agricoli per la conversione in biocombustibili nella Puglia" Commissione delle Comunità Europee, Direzione generale della Scienza, Ricerca e Sviluppo, Contratto di Studio ESE-R-080-I (S) - Bari, 1984.

(6) RENAGRI "Sperimentazioni di metodi di raccolta meccanica dei residui legnosi della potatura della vite e dell'olivo in Puglia" contratto ENEA-RENAGRI 33979/10-12-1985 - Roma 1986.

(7) C.R.B. Centro Ricerche Bonomo "Predisposizione del piano operativo per una campagna da effettuare in Puglia concernente la raccolta, il condizionamento, la distribuzione e l'utilizzo dei sottoprodotti di lavorazioni agricole - potatura della vite, dell'olivo e del mandorlo - e della industria del legno" Contratto ENEA-CRB 23/84 del 22-12-83 - Bari, 1984.

Tab. I - Summarizing data relative to picking up, conditioning, transportation and conditioning operations of vine-branches.

OPERATIONS	MACHINERY	WORKERS n.	ADVANCING SPEED OF THE OPERATING MACHINE (km/h)	BALE WEIGHT (kg)	PRODUCTION (t/ha)	OPERATING MACHINE OR PICK-UP SITE WORKING CAPACITY (t/h)	(ha/h)	HUMAN LABOUR OUTPUT (t/h x man)	TOTAL TIME REQUIRED
Pick-up-baling + loading	Tractor 26 KW + Pick-up-press baler AUTORAC 100	2-4	1.2 (1)		1.57	0.46(1)	0.29	0.12 (1)	74h 40'
Transportation	Tractor 26 kW + Tractor 51 kW	3	4.5	18	-	0.92(1)	-	0.3	12h 23'55"
Storage	Manual	3				6.1 (1)	-	2.21	2h 02'20"
TOTAL						0.13(2)	-	0.08 (2)	94h 06'15"

(1) Calculated on actual picking up time .

(2) Calculated on total time, wastes of time included .

Tab. II - Summarizing data relative to picking up, chopping transportation and storage operations of the olive-tree prunings.

OPERATIONS	MACHINERY	WORKERS n.	ADVANCING SPEED OF THE OPERATING MACHINE (km/h)	CHOP PRODUCTION (t/ha)	OPERATING MACHINE OR PICK-UP SITE WORKING CAPACITY (t/ha)	HUMAN LABOUR OUTPUT (t/h)	TOTAL TIME REQUIRED
Chopping + loading	Tractor 26 kW + Chipping machine CHIPPER 100	3-4	-	6	0.18 (1)	-	45h 19'
Transportation + Storage	Tractor 26 kW + Trailer	1	12	-	3.43	-	2h 20'
TOTAL	-	3-4	-	-	0.168	0.075	47h 39'

(1) Calculated on the actual chopping time.

Tab. III - Summarizing data relative to picking up, transportation and storage operations of the olive-tree prunings Farm: "Angiulli".

OPERATIONS	MACHINERY	WORKERS n.	ADVANCING SPEED OF THE OPERATING MACHINE (km/h)	PRODUCTION (kg/plant)	OPERATING MACHINE OR PICK-UP SITE WORKING CAPACITY (t/h)	HUMAN LABOUR OUTPUT(1) (t/h)	TOTAL TIME REQUIRED
Pick-up-baling + loading	Tractor 26 kW + Pick-up-baler LERDA 900 L	2	0.31	26	1.61 (2)	-	2h 43'
Transportation + Storage	Tractor 29 kW + Trailer	1	12.3	-	3.86 (3)	-	1h 08'
TOTAL	-	2	-	-	0.95 (4)	0.62	4h 36' (5)

(1) Production/total working time.
(2) Bale production/picking up and waste time.
(3) Bale production/transfers.
(4) Bale production/picking up time.
(5) The total time required for the picking up, transportation and unloading operations includes also the waste of time, due to inconveniences, which has not been calculated in the evaluation of the single operation parameters.

BALER FOR THE HARVESTING WITHOUT SHREDDING OF PRUNING-WOOD FROM VINEYARDS AND FRUIT TREES IN GENERAL

Walter GRILLI
MURATORI SpA - Castelnuevo Rg. - MODENA

Summary

Muratori S.p.A., one of the leading Italian manufacturers of agricultural machines, including steerage hoes, rotary harrows and cutter/choppers, which are used throughout the E.E.C., in the United States and in several African and Asiatic countries, explain why they decided to develop a machine for harvesting pruning-wood. A brief description of the machine follows, including the various construction and testing phases. Several concise charts present technical and economical data obtained from trials in different countries and on different types of pruning wood.

The conclusion gives future prospects for utilizing the machine, which depend mainly on promoting full use of all viable secondary energy sources. The report then considers the intrinsic qualities of the machine, which are clearly quite exceptional, as demonstrated by the unqualified enthusiasm of its purchasers. Furthermore, this concerns an important political, economic, and educational problem, on which the E.E.C. can make a considerable contribution.

1. INTRODUCTION

Recovery of an agricultural by-product such as pruning-wood is an important factor on the energy front since it can reduce energy costs and involve a great number of farmers who stand to benefit directly from this activity.

Manual harvesting of pruning wood has been abandoned for some time now, since it is not economical by traditional methods. The pruning wood is typically shredded and left in the field, and very little energy is regenerated.

As an alternative, existing machines used in the recovery of pruning wood cut it into :

a) small pieces which are compressed in plastic net bags ; the fragility of the bags makes them impractical to transport and store ;
b) pieces of a preset maximum length which are compressed into square bales of a maximum weight of 12 kg. These are not very practical to transport and store, since the cuttings can easily slip out of bale-ends ;
c) minute pieces (chips) which are fed directly from the machine into trailers or containers ; since the amount of biomass is reduced, transport is easier and the number of consumers increased.
This also applies to storage. For practical use of the chips, however, it is necessary to compact them (so as to obtain a product which is more suitable for conversion into heating fuel), which involves a rise in costs.

These machines are expensive to buy and operate. They are too large or shaped in such a way that they cannot easily pass between the rows of plants which are kept as narrow as possible in intensive cultivation for obvious reasons.

2. AIM OF PROJECT

The machine under study was designed to harvest pruning wood from vineyards and fruit trees in general without shredding it, and to gather the cuttings into round bales which are easy to transport and store, and are practical for use as fuel for central heating on a farm or in the home, or for the production of hot water or steam in food-processing industries, preferably near the harvesting area.

The made-up bales keep the product intact, and reduce to a minimum the loss of cuttings from bale-ends. The compact baler can be used for rows of varying widths, on different kinds of terrain, and for many different types of fruit trees as well. Maintenance is straightforward, and both towed and tractor-mounted models have been produced, suitable for use with small or medium-sized tractors. Power (N = 35-50 hp) is taken from the tractor PTO at 540 rpms.

3. MACHINE DESCRIPTION

Looking at Fig. 1, starting from the rear, the baler is constructed as follows :

1) pruning-wood collection assembly, composed of two toothed rollers, placed one above the other. One of the rollers, at ground level, picks up the cuttings, and the two rollers feed them into the pick-up funnel;
2) pick-up funnel, which feeds the harvested cuttings into the baling-chamber ;
3) baling-chamber, containing rollers which roll and press the harvested cuttings together to form the bales ;
4) pressure-gauge located inside the baling-chamber, which indicates when the bale has been compacted sufficiently ;
5) a system for tying up the bales, triggered when the bale has been compacted to the described level ;
6) a system for ejecting the made-up bales from the rear of the baler ;
7) a robust frame, upon which all the above parts are mounted ;
8) two free-running wheels ;
9) tractor towing attachment ;
10) a system of safety devices incorporated into the drive-shaft, the pick-up roller and the main transmission.

The machine for making-up bales 50 cm. in diameter, 50 cm. long, and weighing 20 kg., has a rating of only 30 hp, and is extremely compact. Its total width is 120 cm. and total length 190 cm. Another model is now available, which makes bales 50 cm. in diameter, 100 cm. long and weighing 40 kg.

The baler operates by stopping for each bale. When the bale is formed, the tractor stops, the bale is tied up and then ejected from the rear. Once the bale is ejected, the tractor continues and the baler begins the process of making up the next bale.

4. MACHINE CONSTRUCTION

The planning and design of the machine were carried out directly by Muratori, with the help of several external consultants. The various parts were produced by a number of small engineering firms, used frequently in the past by Muratori, and in close collaboration with Muratori. The machine was assembled in Muratori's own factory. Construction of the machine did not present any particular problems.

The first prototype was developed and tested with financial support from the EEC.

Five machines were constructed and put through a series of trials, which were organized in the vicinity of the Muratori S.p.A. works (88 tests), as well as in various regions of Italy (27 tests), various E.E.C. countries (43 tests), and Egypt (5 tests). The trials were aimed at :

- preparing the baler thoroughly for operation with different types of pruning-wood and under different working conditions ;
- determining clearly the baler's productivity under varying conditions ;
- defining occurately the properties of the end-product : its heating value, its effectiveness for use in boilers, ect.

5. MACHINE START-UP

As a result of numerous tests, the following modifications were made to the machine so that it could handle a greater variety of pruning-woods :

a) two rollers were added, and their relative positions adjusted to improve take-up of the cuttings and formation of the bales ;
b) the rollers near the pick-up area revolve at a different speed from that of the bale-chamber rollers, thus leaving the pick-up zone free ;
c) modifications to feed platform and chute ;
d) the hydraulic pressure-sensor system in the baling chamber, and consequently the hydraulic controls which operate the opening of the bale-ejection flap were modified ;
e) the safety shock-absorber was modified.

Finally, after a general reappraisal, it was decided to make the machine wider, in order to facilitate the feeding of cuttings into the baling-chamber. Consequently, the length of the bale was increased from 50 cm. to 100 cm., with the diameter remaining unchanged.

6. MEASUREMENTS

Test results are given in tables I, II, III, IV and V, which chart the various types of pruning-wood and the different bale-formation and harvesting operations, in relation to machine productivity, number of personnel used, fuel consumption and amount of time necessary for operations.

7. RESULTS

a) Technical considerations

Comparing the data given in tables I, II and III with the forecasts made during the planning stage, it can be seen that the margin of error is generally 10 %.

The principal difficulties in harvesting the various types of pruning-wood (apart from vine or olive wastes) lie in the size of the cuttings, their flexibility, and above all, in the fact that during the pruning operations, the end of the branch, which is considerably thicker than the cutting, is often cut off along with the part to be pruned. This piece of branch attached to the cutting creates difficulties when it enters the baler, and interferes with bale formation.

To deal with this problem, during the pruning operations the cutting should be thrown towards the center to form a "windrow", in preparation for the subsequent mechanized harvesting. The piece of branch must be cut separately from the pruning-twig itself and, if possible, put directly into a small basket or bag carried by the person doing the pruning.

In others words, little by little, the worker must be brought to accept the idea that pruning-wood is a product which has to be harvested mechanically and cheaply, rather than just rubbish to be left on the ground among the trees.

This is a question of creating awareness, and will require a great deal of effort at all levels.

b) Economic considerations

The data in tables IV and V show that the economics of making-up and collecting the bales produced are interesting, and that they are influenced by the type of pruning-wood and by the operating conditions for the machine.

The costs of the fuel used for running the tractor which tows the baler and harvesting trailer are quite reasonable.

The Kcal and ept costs are competitive compared with fuel oil, and therefore the product is of interest from the point of view of energy consumption.

For these reasons, the harvesting and utilization of pruning wood proves to be definitely advantageous.

The possibilities the machine offers for use on small farms or in farming consortia are considerable in the light of the data evaluated and the estimated amount of energy available from the pruning-wood produced annually in Italy from several fruit tree plantations (table VI) and the data relating to the area under vines and the corresponding production in several other countries (table VIII).

Although this data is incomplete, it demonstrates the considerable economic importance of harvesting pruning-wood.

8. CONCLUSIONS AND PROSPECTS FOR PROJECT UTILIZATION

The machine, which is fairly cheap both to buy and to operate, does not present any particular technical problems, apart from those relating to continuous improvement of its effectiveness on various types of fruit-tree pruning-wood and on different types of terrain, etc.

The real problem lies in the large-scale utilization of the bales produced, and this can be overcome by :

- compacting the bales in appropriate balers to make them suitable for use even in common domestic stores ;
- using them for the production of biogas and fuel gas for internal-combustion engines ;
- granting concessions to users of the bales.

Solutions of this type require time, finance, etc.

It is furthermore necessary to launch a promotional campaign for the machine using all possible means, including any assistance which may be forthcoming form the E.E.C., to aid the marketing of the baler in all the E.E.C. countries and non-member countries too.

9. SYMBOLS AND UNITS

y	:	year
cm	:	centimetres
dm	:	decimetres
rpms	:	revolutions per minute
h	:	hours
ha	:	hectare
HP	:	horse power
Kcal	:	kilocalories
Kg	:	kilograms

L	:	total cost of forming + collecting bales
lt	:	litres
N	:	horse power
No	:	number
PTO	:	tractor power take-off
Qli	:	quintals (one hundred kilos)
t	:	tons
ept	:	equivalent petrol tons

10. BIBLIOGRAPHY

(1) Baldini E. - Analisi preliminare delle destinazioni energetiche alternative dei prodotti e sottoprodotti agricoli : il possibile contributo dell'arboricoltura da frutto italiana. Publication No 187 of the CONSIGLIO NAZIONALE DELLE RICERCHE - Centro di Studio per la tecnica frutticola - Bologna - March 1982.

(2) QUID 1982 p. 1504.

(3) Aranda-Heredia E. - Récolte mécanique de la biomasse à destination énergétique : le bois de taille de l'olivier. Atti Simp. Intern. Meccanizz. Agricola, Bologna, 15-16 November 1981.

(4) Baldini E. - Residui di potatura e risparmio energetico. L'Informatore Agrario, 4, 1982.

(5) Baldini E., Alberghina O., Bargioni G, Cobianchi D., Iannini B., Tribulato E., Zocca, A. - Analisi energetiche di alcune colture arboree da frutto. Rivista di ingegn. agr., 2, 1982.

(6) Pellizzi G., Bodria L., Sangiorgi F., Castelli G. - Agricoltura e crisi energetica. Progetto finalizzato meccanizzazione agricola del C.N.R., Quaderno di sintesi No 18, Renagri, Roma, 1981.

(7) Piccarolo P. - Possibilità di recupero energetico dei sottoprodotti dell'oleificio. L'Informatore Agrario, 48, 1980.

(8) Pimentel D. - Handbook of energy utilization in agriculture. CRC Press. Boca Raton, Fla., 1980.

TABLE I : Average operative conditions of bale-formation

Type of cutting	N° tests 1	total hours worked 2	bales made 3	average N° bales per hour 4=3:2	average weight kg. 5	size of bales dm. 6	qts. of cuttings harvested 7=3x5	power used (HP) 8	fuel consumed lt. 9	personnel used for baling 10	personnel harvestng &trnsport 11
VINE	69	698	23,574	34	20	0.5x0.5	471,480	50	3,480	1.30	2
APPLE	26	196	6,032	31	19.8	0.5x0.5	119,164	50	975	1.76	2
PEAR	20	139	4,329	31	20.7	0.5x0.5	89,610	50	680	2	2
PEACH	13	108	3,536	33	19.9	0.5x0.5	70,537	50	545	1.76	2
PLUM	10	66	1,704	26	19.5	0.5x0.5	33,228	50	330	2	2
OLIVE	12	105	3,311	32	21.3	0.5x0.5	70,623	50	525	1	2
COTTON	5	47	1,771	38	19	0.5x0.5	33,649	50	235	1	2
AVERAGE	155	1,359	44,257	33	20	0.5x0.5	888,291	50	6,770	1	2

TABLE II : Summary of bale-formation results

type of cutting	speed of baling			bale size	average weight of bale	quantity harvested		personnel (man h / ha)		fuel consumption		
	ha/h	bales/h	bales/ha	dm	kg.	t/h	t/ha	baling	harvesting	lt/h	lt/ha	lt/t
	1	2	3=2:1	4	5	6=2x5/1000	7=3x5/1000	8	9	10	11=10:1	12=11:7
VINE	0.35	34	87	0.5x0.5	20	0.68	1.94	1.30	2	4.97	14.2	7.3
APPLE	0.41	31	76	0.5x0.5	19.8	0.61	1.5	1.76	2	4.97	12.1	8
PEAR	0.39	31	80	0.5x0.5	20.7	0.64	1.65	2	2	4.89	12.5	7.5
PEACH	0.38	33	87	0.5x0.5	19.9	0.66	1.73	1.76	2	5	13.1	7.6
PLUM	0.39	26	67	0.5x0.5	19.5	0.50	1.3	2	2	5	12.8	9.85
OLIVE	0.375	32	85	0.5x0.5	21.3	0.68	1.81	1	2	5	13.3	7.35
COTTON	0.36	38	106	0.5x0.5	19	0.72	2	1	2	5	13.9	6.95
AVERAGE	0.37	33	89	0.5x0.5	20	0.66	1.78	1	2	4.97	13.45	7.52

TABLE III : Summary of operative conditions of bale harvesting

type of cutting	bales made	hours of harvesting	average N° of bales harvested	N° bales / ha	ha/h	personnel used	fuel consumed		
							lt/h	lt/ha	lt/t
1	2	3	4=2:3	5	6=4:5	7	8	9	10
VINE	23,577	174	135	97	1.4	2	4.97	14.2	7.3
APPLE	6,032	51	118	76	1.5	2	4.97	12.1	8
PEAR	4,329	33	131	80	1.6	2	4.89	12.5	7.5
PEACH	3,536	26	136	87	1.6	2	5	13.1	7.6
PLUM	1,704	18	95	67	1.4	2	5	12.8	9.85
OLIVE	3,311	26	127	85	1.5	2	5	13.3	7.35
COTTON	1,771	11	161	106	1.5	2	5	13.9	6.95
AVERAGE	44,257	339	130	89	1.46	2	4.97	13.45	7.52

<u>TABLE IV</u> : Summary of costs of baling and harvesting

type of cutting	N° tests	total hrs worked for baling and harvesting	N° bales made and harvested	tons made & harvested	cost of baling	cost of harvesting	total cost (L)	cost L/t	cost/ha
1	2	3	4	5	6	7	8	9	10
VINE	69	872	23,574	471.480	22,339,400	2,610,000	24,949,000	52,917	27,277
APPLE	26	247	6,032	119.164	7,659,600	782,500	8,442,100	70,844	47,229
PEAR	20	72	4,492	89.62	5,049,700	485,000	5,534,700	61,757	37,428
PEACH	13	134	3,536	70.53	4,075,300	390,000	4,465,300	63,310	96,595
PLUM	10	84	1,704	33.228	2,399,600	250,000	2,649,600	79,740	61,338
OLIVE	12	131	3,311	70.623	3,276,000	390,000	3,666,000	51,909	28,679
COTTON	5	58	1,771	33.649	1,466,400	165,000	1,631,400	48,483	24,421
AVERAGE	155	1698	44,420	888.294	46,266,000	5,072,500	51,338,500	57,794	32,469

TABLE V : Results in Kcal/Kg. produced by combustion (in heavy boiler Mod. MO45) of product having 10 - 15% haumidity, after 2-3 months of harvest and natural drying under cover

TYPE OF CUTTING	KG. OF FUEL IN BALES	EPT	TOTAL Kcal. PRODÙCED	Kcal/t	L/t	L/EPT
1	2	3	4	5=4:2	6	7
VINE	15,000	3	66,000,000	4,400,000	52,917	264,585
APPLE	10,000	2	41,500,000	4,150,000	70,844	394,220
PEAR	5,000	1	20,875,000	4,175,000	61,757	308,785
PEACH	5,000	1	21,150,000	4,230,000	63,310	316,550
PLUM	5,000	1	20,750,000	4,150,000	79,740	398,700
OLIVE	10,000	2	42,500,000	4,250,000	51,909	259,545
TOTAL	50,000	10	212,775,000	4,255,500	60,906	304,530

NOTES : COLUMN N° 6 - SEE COLUMN N° 9, TABLE V
COLUMN N° 7 - VALUES COL. 6 x 5 (EX. 52,917 = 164,585 L/EPT)
THE HEAT GENERATED WAS USED TO HEAT WATER FROM AN EXTERNAL TEMPERATURE OF 20° TO 70° C.
TESTS WERE CARRIED OUT NEAR THE MURATORI WORKS.

TABLE VI : Estimation and evaluation of energy production from pruning wood annually produced by several types of fruit-tree cultivation

CULTIVATION	Usable surface (000 ha)	BIOMASS PRODUCED		AVAILABLE BIOMASS (000 t)	ENERGY EQUIVLENT (EPT)
		per hectare (t)	total (000 t)		
VINE	1,260	2,5	3,150	2,205	450,000
OLIVE	1,035	2,5	2,590	1,810	355,000
OTHER FRUIT TREES	487	2,6	1,270	894	172,000
TOTAL	2,782	2,52	7,010	4,909	977,000

TABLE VII : Estimation and evaluation of the surface cultivated as vineyard in several countries, corresponding annual production and relative biomass.

COUNTRY	SURFACE CULTIVATED (000 ha)	ANNUAL PRODUCTION (000 t/a)	BIOMASS eq. (000 t/a)
SPAIN	1,719	3,155	1,550
ITALY	1,379	7,169	3,600
FRANCE	1,225	6,766	2,380
WEST GERMANY	100	808	400
PORTUGAL	390	862	430
GREECE	194	423	210
U.S.A.	310	1,496	750
ARGENTINA	366	2,424	1,200
TURKEY	800	37	18
ALGERIA	206	254	127
U.S.S.R.	1,254	2,903	1,450

SESSION REPORT

The three main aspects developed by the different speakers are related to :

1. The kind and characteristics of the by-products before harvesting.
2. The techniques for harvesting.
3. Utilization of the by-products.

Several by-products are considered :

- wastes from vineyard, olive trees, fruit trees.

The main utilization mentioned is direct combustion principally to produce heat. It was noted that the basic price should not exceed 35 to 50 ECU by ton of dry matter (price established relatively to a price of 25 dollars/barrel for oil).

All prices proposed by the different speakers seem different : GREECE, ITALY, SPAIN.

MAIN PROPOSITIONS

- Priority for optimization of harvesting techniques with low cost including such parameters as reduced energy consumption, handling costs, and increased utilization time during the year.
- Necessity to produce machines appropriate to the further utilization of the by-product.
- Determination of the appropriate conditions for utilization : balls, pellets, chips.
- A better diffusion of experiments and demonstration effected in several countries and cooperation between experimenters.

In conclusion, a very important potential use of these by-products in Mediteranean Countries (ITALY, GREECE, SPAIN and FRANCE) was highlighted.

DISCUSSION

Questions by Dr ORTIZ

- Proposed prices for the by-products used as combustibles seems very different from one country to one another, the machines are not built on a large scale.

Reply from Dr LUCAS

- The essential problem is the basic price (35 to 50 ECU) which is the cost of biomass valorization in comparison with the equivalent oil price of 25 dollars for one barrel ; conditions which seem likely to obtain for the next few years.

It is necessary to take into account the possibility to have negative prices for some by-products (like vine shoots for example).

SESSION V - ECONOMIC EVALUATIONS, PUBLIC INCENTIVES

Chairman: *V. BARTOLELLI, Renagri*
Rapporteur: *H.E. WILLIAMS, Ecotec Research and Consulting Limited*

The economic analysis of different mechanisation steps from harvesting to storage of crops and crop by-products

The use of wood chips from forest harvest residue for energy production potential - Potential yield , quality, harvesting techniques and profitability

An approach to the economic evaluation of biomass energy production projects

THE ECONOMIC ANALYSIS OF DIFFERENT MECHANISATION STEPS FROM HARVESTING TO STORAGE OF CROPS AND CROP BY-PRODUCTS

E. AUDSLEY
Operational Research Group, AFRC Engineering
Wrest Park, Silsoe, Beds MK45 4HS, United Kingdom

Summary

Because of the effect of harvesting on succeeding operations such as delaying drilling, it is unrealistic to calculate an absolute cost. Costs are therefore relative and farm specific.

The first step in analysing the effect of changing systems of mechanisation is a model of a farm incorporating the relevant aspects of the system such as workrate, labour and power requirement which affect other farm operations. Other logistics models can be used to calculate the effect of travel distance carting on overall system performance. The general model of an arable farm selects the cropping, timeliness penalties, labour and machinery which maximise farm profit. The change in farm profit measures the cost of one mechanisation system relative to another. Each system is evaluated on three farm types (cereal, arable, arable with roots) with three soil types (light, medium, heavy). The second step is to add the effects which do not affect other farm operations, notably differences in storage cost and losses, to produce an overall relative cost. Other farms can be rapidly modelled by selecting appropriate data from a database of crop definitions, operation definitions and machinery costs.

As an example this paper describes the analysis of three existing straw harvesting systems, and one proposal, compared with field burning and soil incorporation of straw. The cost, relative to field burning, of producing straw bales by the existing systems is £11-22/t depending on the soil type and farm type. The system of incorporation costs £26-33/ha. If briquettes are valued at £50/t in bulk, then an in-field briquetting machine working at 1 h/ha would need to cost less than £60-70 000 to be more profitable and at 1.5h/ha it will need to cost less than £30-50 000. An additional effect is a change in cropping, in the former case maximising straw yield by changing to winter wheat and in the latter changing to winter barley.

By considering the distribution of values from different uses for the straw versus the cost of harvesting and transport, the effect on the demand for straw can be estimated.

1. INTRODUCTION

Straw is burnt in the field because it is of no use. A farmer has to consider the economics on the farm of the alternative methods of straw disposal available. The decision whether it is more profitable to burn or incorporate depends solely on the farm but the decision to bale depends on the value of the straw when it has been harvested. If it is used on the farm, then its value depends on the cost of the alternatives, for example its value as a fuel depends on the cost of oil or coal. If it is sold for use off the farm, then its value depends on (i) the cost of the alternatives to its proposed use, (ii) the cost of transport which is a function of bale type and distance, (iii) the balance of supply and demand which will influence the price the user is prepared to pay.

The introduction of new uses for straw, new methods of baling and new rules for burning will affect the balance of supply and demand, the cost of transport and relative cost of alternatives. Consequently the value of baled straw will alter and with it farmers' decisions whether to bale, burn or incorporate. This paper describes a study(1,2) which estimated the effect of various possible changes on the amount of straw baled and incorporated, on farm profitability, on the price of straw and on the uses for straw.

The above ground straw yield depends on the weather, crop rotation and soil fertility as well as the variety. The amount of loose straw behind the combine depends on the height of cut which depends on the condition of the crop. This study assumes a loose straw yield of 4.12 t/ha with winter wheat and 2.93 t/ha with barley.

2. THE ARABLE FARM MODEL

In order to evaluate the effect of changes to operations on a farm, we have developed, over a number of years, our arable farm model. Fuller descriptions of the model can be found in references (3, 4, 5) and I will here describe a very simplified version, which illustrates its main characteristics.

Consider just 2 crops requiring 3 operations. The time is divided into six periods and the workrates and timeliness penalties are shown in Fig. 1.

The objective of the farmer is to maximise his profit by selecting the number of men, machinery and cropping. His constraintss can be constructed as follows:

1) Labour: the period 2 constant is constructed from the second row of the figure:

$$10m \geq 1.5\ xh_2 + 0.5\ xp_2 + 2.0\ yh_2 \qquad \ldots (1)$$

and similarly for all six periods. Machinery constraints are formulated in the same way.

2) Sequence constraints: ploughing following crop Y can only be carried out after harvesting crop Y, thus there are two constraints:

$$yp_3 \leq yh_2 + yh_3 \qquad \ldots (2A)$$

$$yp_3 + yp_4 + yp_5 \leq yh_2 + yh_3 + yh \qquad \ldots (3A)$$

and similarly for other sequences.

Fig. 1 A simplified arable farm

Period	Time available per man	Time needed for unit area														
		Crop X							Crop Y						Crop X	
		Harvest		Plough				Drill	Harvest			Plough			Drill	
1	10	1.5														
2	10		1.5	0.5					2.0							
3	8				0.5					2.0		0.5				
4	7					0.5					2.0		0.5		0.3	
5	5						0.5							0.5		0.3
6	10							0.6								
Profit	-15	1.0								2.2						
Timeliness cost		0.0	0.1					0.0	0.4	0.2	0.0				0.0	0.1
Variables	m	xh_1	xh_2	xp_2	xp_3	xp_4	xp_5	yd_6	yh_2	yh_3	yh_4	yp_3	yp_4	yp_5	xd_4	xd_5

3) Rotational constraints: crop X can follow crop X or crop Y but will yield less if it follows X. Similarly for crop Y, leading to a rotational penalty matrix, such as:

	X	Y
X	0.1	0
Y	0	1.0

If r_{abk} is the area of crop a going into crop b after harvest in period k, then the sequence constraints are modified:

$$yp_3 + r_{yy2} + r_{yy3} \leqq yh_2 + yh_3 \qquad \ldots (2B)$$

and

$$xp_2 \leqq xh_1 + xh_2 + r_{yy2} \qquad \ldots (4)$$

$$xp_2 + xp_3 \leqq xh_1 + xh_2 + r_{yy2} + r_{yy3} \qquad \ldots (5)$$

4) Total area

The total area constraint is:

$xh_1 + xh_2 + yh_2 + yh_3 + yh_4 \leqq T$ where T is the farm area.

5) Objective

The profit is the crop gross margins less the cost of men and machinery, the timeliness penalties and the rotational penalties;

$$P = 1.0\ xh_1 + 0.9\ xh_2 + 1.8\ yh_2 + 2.0\ yh_3 + 2.2\ yh_4 - 15m - 0.1 \sum_k r_{xxk} - 1.0 \sum_k r_{yyk}$$

The full model uses many more crops and operations; the time available is divided into different classes for jobs ranging from ploughing to spraying; the area constraint is formulated to allow several crops in one year; and there are numerous other features to accurately describe the farmer's constraints. A large database contains details of crops, operations and machines which allows farms to be rapidly modelled by selecting the appropriate data.

To calculate the effect on labour, machinery and cropping of changing the straw disposal methods using the model, three types of arable farm are defined:

(i) Arable farm with roots - cereals, oilseed rape, sugar beet and potatoes.
(ii) Arable farm without roots - cereals and oilseed rape.
(iii) Cereal farm - winter wheat, winter and spring barley only.

Three types of land are considered, termed light, medium and heavy, which are a reflection of the workable hours available during autumn and winter.

A typical sequence of operations have been selected to construct nine standard farms. To determine the effect of a change, the relevant operation is replaced by the new operation along with any other consequent changes, and the new optimum farm plan is calculated. The difference in profit is thus the value or cost of the new operation.

3. THE FATE OF STRAW

3.1. STRAW BURNING

Field burning of straw has become very unpopular off the farm for a wide variety of reasons, although numerous agricultural advantages have been demonstrated by experiments. Straw burning is normally carried out 9-13 days after combining. A man (with tractor) is needed to clear the straw and prepare a barrier then supervise the burn.

Let d h/ha be the rate at which the straw is cleared from the side of the field, p h/ha the rate for ploughing or cultivating the fire barrier and b the man hours needed to burn the field (which is largely independent of the size of the field). If the field is x ha, assumed square, and a fire barrier w metres wide is constructed then the time needed is:

$$\frac{0.04\ w\ (d + p)}{\sqrt{x}} + \frac{b}{x} \quad \text{man h/ha}$$

Suppose clearing the straw takes 3 ha/h, cultivating takes 2 ha/h and 2 men take 2 hours to burn the field. If 10 ha is the maximum that can be burned at any one time, with a 25 m fire break this corresponds to a maximum field size of 13.4 ha, which needs 0.53 man h/ha.

3.2. USES FOR STRAW

Once burning has been banned, there are only two alternatives for a farmer to handle the straw after combining - baling and incorporating. A third alternative is not to combine but adopt another harvesting technique - whole crop harvesting. The straw can then be used to dry the grain but even an inefficient drier would only use about 30kg/t grain so this does not dispose of the straw. If the straw is to be used, then it must be harvested which basically means baling. An alternative is to produce briquettes in the field which can be handled like grain and sold in competition with coal. AFRC Engineering are working to develop such a machine.

3.2.1. BALING

Although there are a large number of variations and numerous proposals for new machines with new sizes particularly at the higher densities, four main sizes of bale can be considered:

	Wt, kg	Dimensions, cm	Density, kg/m^3
Conventional	20	41 x 46 x 91	118
Round	345	170 dia. x 155	98
Hesston	600	122 x 122 x 244	160

Baling is generally carried out soon after combining (2-4 days) in order to produce good quality straw for feeding and bedding. Baling takes 12 man min/t for conventional and round bales and 4.0 man min/t for Hesston bales. A logistics model for farm transport can be used to calculate the time needed to cart the straw to the farmstead over different distances by different means. For the analysis average values are used. Assuming a flat 8 system, carting and stacking takes 31 man min/t for conventional (53 man min/t for a non-mechanical system), 18 man min/t for round, and 8.5 man min/t for Hesston bales. Therefore to clear a field of wheat takes 2.95 h/ha (conv), 2.06 h/ha (round), and 0.86 h/ha (Hesston). Twine can be an important element of the cost of baling. Typical costs are 1.8p/conv. bale, 7.1p/round bale, 50.6p/Hesston bale giving £0.90/t, £0.21/t and £0.84/t respectively.

Once the straw has been baled and carted to the farmstead there are a number of uses for the straw. For all uses, however, the straw must be stored. This can be uncovered, covered - for example using PVC, or in a building but including ruined straw the costs are similar. Typical annual costs are £5.1/t for conventional bales, £6.0/t for round bales and £3.0/t for Hesston type bales.

3.3. INCORPORATING

Incorporating all the straw into the soil has only become of interest in the UK relatively recently and a large number of possible systems have been proposed, but very few have yet stood the test of time. There is a general concensus that the straw should be firstly chopped and spread, then mixed with the soil, after which normal cultivations can be carried out. Although these can be carried out as separate operations it is possible to combine them. Two typical sequences are considered here. The first operation is to chop the straw on the combine harvester, reducing workrate by 10%, followed by

1) 2 passes with a tined implement, then plough, rotavator, drill.
2) 2 passes with a tillage train, then spray, drill.

3.4 SOIL NITROGEN EFFECT

One advantage claimed for incorporation is that it increases the soil organic matter content, but experiments have shown a relatively small increase in soil organic matter due to straw incorporation. There is also ample experimental evidence that soil nitrogen is locked up by microbes. A model of soil organic carbon turnover(6) demonstrates this effect. The model identifies six types of organic carbon. Two are material added to the soil, two types of biomass and two types are stable organic matter. A monthly transition matrix describes the decomposition of material from one state into material in other states and carbon dioxide. The rates of transition are affected by the temperature, moisture, whether a crop is present and the soil type, measured by its cation exchange capacity (CEC).

3.5. CHEMICAL COSTS FOR WEED CONTROL

A blackgrass weed population model (R. Cousens, AFRC Arable Crops, personal communication) is used to estimate the levels of control needed with and without burning and with and without soil inversion. Assume that the best control that could be achieved by an expensive spray is 95% and that doubling the cost of spray has a multiplicative effect on control, i.e. if a spray costing £y/ha achieves control x then a spray costing £ny/ha will give control c such that $(0.95 - x)^n = (0.95 - c)$. Then to achieve a control level c will cost:

$$y \frac{\ln (0.95 - c)}{\ln (0.95 - x)} \text{ £/ha}$$

The additional weed control cost of not burning is thus £22/ha for non-inversion and £12/ha for inversion assuming a typical value of 33% kill from burning. It is assumed that the 3% loss of yield for non-inversion shown by experiments could be cancelled out by these extra spray costs.

TABLE I

Reduction in farm profit from baling instead of burning

	Arable with roots		Arable without roots		Cereals	
	£/ha	£/t straw	£/ha	£/t straw	£/ha	£/t straw
Heavy land						
Conventional - non-mechanised	55	21	71	25	65	19
Conventional - mechanised	39	14	47	17	44	14
Round	30	11	36	12	35	11
Hesston (Note 1)	49	18	52	16	51	15
Hesston (Note 2)	13	5	15	5	15	4
Medium land						
Conventional - non-mechanised	40	16	69	24	73	21
Conventional - mechanised	26	9	45	15	46	13
Round	21	7	33	11	36	11
Hesston (Note 1)	48	17	51	17	52	14
Light land						
Conventional - non-mechanised	24	10	69	22	77	21
Conventional - mechanised	16	6	45	15	50	14
Round	15	5	33	11	37	10
Hesston (Note 1)	46	13	52	17	53	14

Notes
1. The 200 ha farm is assumed to run one Hesston baler
2. The farm is assumed to fully use one Hesston baler (approx. 1000 ha)

The model can be used to estimate the amount of nitrogen taken from the crop due to incorporating straw. Fig. 2 shows the yearly totals released to the crop. On the Woburn soil, there is always more N released to the crop. On the Rothamsted soil, the first year shows a reduction of 10 kg N/ha but by the fifth year, this loss becomes a net gain. The reduction in nitrogen released to the crop ranges from 0 - 30 kg/ha in the first year, falling to ±10 kg/ha after seven years. Fig. 2 also shows the effect of adding 1t C/ha which is typical of baling followed by cultivating. The effect is approximately a third of the full incorporation effect. This was rarely considered before incorporation and is probably too small to measure in field experiments. In addition baling removes the straw which contains phosphate and potash worth about £11/ha in wheat. One important point which the model demonstrates vividly, is that the change in total soil organic matter is very small.

The effect of the reduction in nitrogen on yield depends on the amount of nitrogen added to the crop. At 130 kgN/ha, a reduction of 15 kg/ha will cause a drop in yield of 2%.(5) At 70 kgN/ha, the same reduction causes a drop in yield of 7%. Older experiments on incorporation at lower nitrogen levels are therefore likely to show a bigger yield penalty due to incorporation.

Fig. 2 Effect of straw incorporation on N release from soil

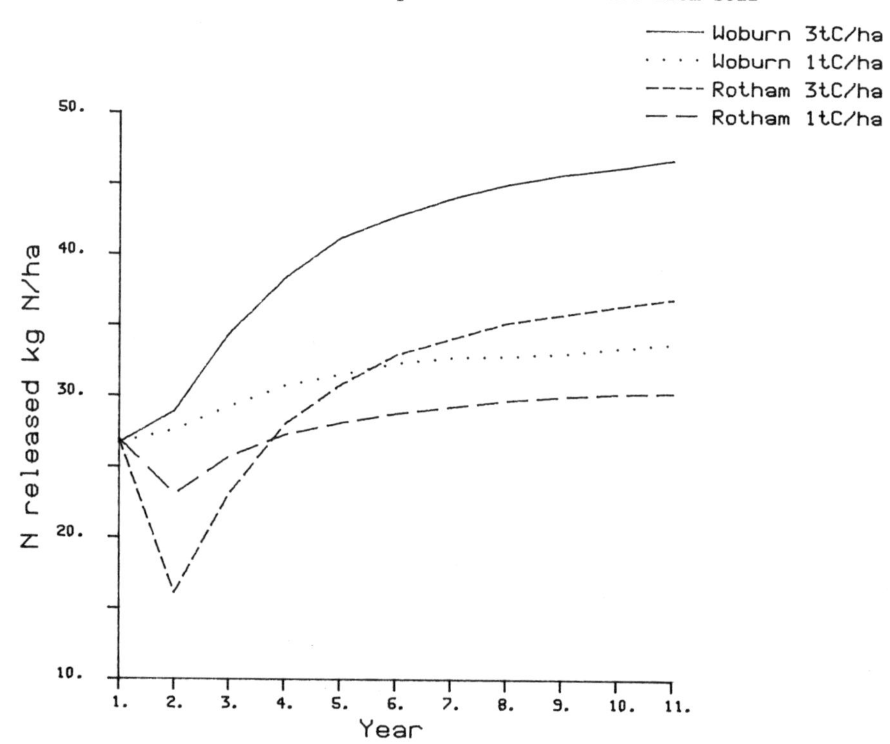

4. ECONOMIC COMPARISON OF STRAW DISPOSAL METHODS ON THE FARM

4.1. BALING

Table I lists the reduction in the farm profit from baling relative to burning due to labour, machinery and crop timeliness effects to which must be added the cost of storing the bales, and extra sprays and fertiliser. Most costs are in the £15-20/t range.

The effect of baling instead of burning on a cereal farm is to increase the area of winter barley grown, with only a marginal effect on the amount of labour employed. The effect on a root farm is to increase the labour by approximately one man and increase the area of root crops.

4.2. INCORPORATION

Table II lists the change in profit from incorporation relative to burning. To this must be added the cost of extra sprays and in the case of the non-inversion techniques an additional cost due to weed seeds not killed by inversion. In the first year one should also add the reduction in fertiliser, taken as an average value of 15 kgN/ha. After 5 years incorporation, the profitability relative to burning and baling will change. The cost relative to burning will be reduced by about £20/ha. Note that the price of straw is thus also likely to rise over the years by £5/t to remain competitive.

With System 1 on both heavy and light land the effect of incorporation is to increase the labour needed by about 10%. The area of root crops is increased by about 20%, more on light land. This is due to the increase in their profitability relative to cereals, partly because the increased labour needed for roots can now be used on the cereals.

TABLE II

Extra cost of incorporation relative to straw burning

System	Change in farm profit due to labour & mach., £/ha		Extra spray cost, £/ha	Extra fert. cost at 15 kgN/ha, £/ha	Total extra cost £/ha	
	Heavy	Light			Heavy	Light
Arable with roots						
1	-14	-8	12	6	32	26
2	+32	+14	71	6	45	63
All cereals						
1	-15	-13	12	6	33	31
2	+52	+24	71	6	25	53

4.3 IN FIELD BRIQUETTING

An alternative approach to incorporating or baling the straw is to produce briquettes from the straw in the field. This is under development and therefore no workrates for actual machines are available. An alternative is to plot breakeven capital cost versus workrate for different land types.

Assume that the briquettes will be stored on the farm, with an enclosed storage cost double that for conventional bales, hence £10/t and will be sold in bulk at £50/t. Also assume the machine uses a 100 kW tractor plus two extra men for carting. Fig. 3 shows the breakeven capital cost for the briquettes relative to burning. If the time needed is 1 h/ha, the machine needs to cost less than £60 - 70,000. At 1.5 h/ha, it must cost less than £30 - £50,000. On all farm types at the slower rates, there is a strong tendency to switch from winter wheat to winter barley. At the fastest rate, the tendency is to go for the most straw and grow winter wheat.

Fig. 3 Breakeven capital cost of in-field briquetting machine

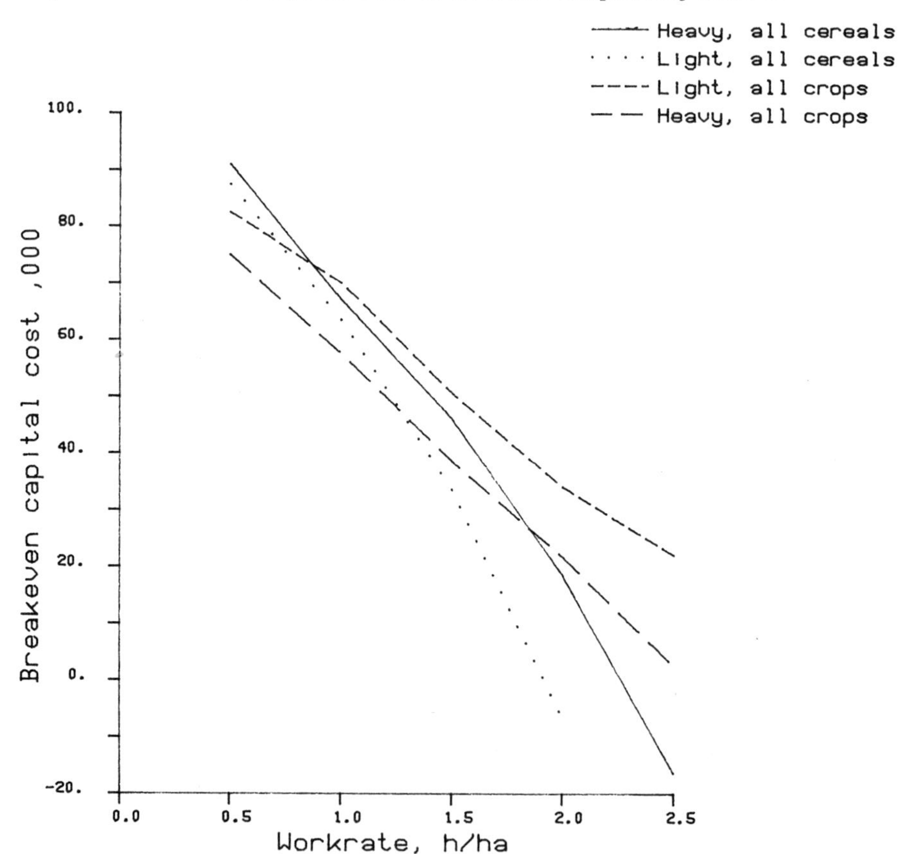

5. DETERMINING THE PROPORTION OF FARMERS CHOOSING TO BURN, BALE AND INCORPORATE

The farmer has the choice of burning the straw in the field, incorporating or baling it. If he bales the straw, it may be used on the farm or sold. It is assumed that each farmer has a utility from a distribution and makes the choice which, in his view, maximises his profit. Collectively therefore, for a given set of conditions, one can predict the proportion of farmers who will burn, incorporate or bale.

In the case where farmers sell the straw they have baled, the price is determined by external demand. In the same way that the farmer has a distribution of costs, the customer has a distribution of values. The straw price at the farm, plus tranport cost is the straw cost arriving at the user, which from the user distribution of value determines the proportion of users purchasing straw or, collectively, the demand for straw at this price. So given a particular set of circumstances there is an equilibrium price for straw at which supply equals demand. By relating the effect of a change in circumstances to the equilibrium price for straw, the long-term effect on the proportion of straw burned, baled or incorporated on the farm and amount of straw purchased for other uses is determined.

Different regions of the country have varying amounts of straw available and demands for it. The regional price of straw therefore reflects the price in other regions, taking into account transport costs. The problem can be modelled mathematically and the equilibrium price for each region calculated.

5.1 DETERMINING THE VALUE DISTRIBUTIONS FOR STRAW

The distributions of value or utility for straw are estimated for each possible use and is assumed to be a normal distribution, characterised by its mean and standard deviation. Details of the full analysis are given in the report (2).

5.1.1 BEDDING

One of the traditional uses for straw is as bedding for livestock. The difference between the cost of the straw based system and the slurry system is the value of straw for bedding. Up to 100,000t could be used for poultry litter in competition with other materials such as wood shavings, but this is a minor use relative to cattle and certainly within the expected error on the estimate of straw use and is therefore neglected.

	Value, £/t	Quantity
cubicles	36.0 ± 8.9	0.25t/unit per winter
dairy cows	14.4 ± 25.9	0.63t/unit per winter
beef cattle	24.8 ± 23.6	0.63t/unit per winter
pigs	36.0 ± 8.9	0.30t/breeding pig/yr

5.1.2 STRAW AS FEED

The price that farmers would be prepared to pay for straw of a known feed value can be determined from the price of the alternative feeds by a minimum cost ration formulation with and without the straw. The value of the straw can be expressed as linear function of ME, m and crude protein, p depending on the type of ration in which it is included:

300 kg + 1.0 kg/day: Value, £/tDM = 12.7 m + 0.254 p - 7.02
300 kg + 0.5 kg/day: = 11.0 m + 0.248 p - 53.3
100 kg + 0.75 kg/day: = 14.3 m + 0.078 p - 66.7

Using these equations with the typical feed values found for barley straw, a range of monetary values for straw can be found.

Ammonia treatment can improve the digestibility (DOMD) of straws by 10-15 percentage points. Comparing the value of treated straw with untreated straw the profit from treating is £-4.5/t ± 18.9 . One would therefore expect about 40% of the users of barley straw for feeding to opt for ammonia treatment. This increases the overall average value of barley straw for feeding by £2.3/t. Improving the reliability of the method of treatment would increase the amount of straw treated and hence increase the amount used. For example, increasing the value of the worst treatments by 1 MJ/kg DM, raises the profit to £-0.8/t ± 14.6 which means that 48% of straw users will use ammonia treatment. However this only increases the value of straw by another £0.3/t.

	Value, £/t	Quantity
barley straw	30.2 ± 10.9	0.632t/yr/unit
wheat straw	19.0 ± 7.5	0.534t/yr/unit
ammonia & barley	25.7 ± 23.3	0.767t/yr/unit
best barley option	32.5 ± 10.9	0.687t/yr/unit
		0.180t/yr/sheep (10%)
caustic soda	24.3 ± 6.5	0.37 Mt

5.1.3 STRAW AS A FUEL

Weight for weight straw contains about half as much energy as coal and a third as much as fuel oil. As straw is similar in convenience to coal, it seems reasonable to assess its fuel value in comparison with coal, not oil. Straw bales need large amounts of storage and special boilers to burn them, whereas if straw can be converted to coal-like briquettes it can directly compete with coal. A number of machines exist with production costs ranging from £22-66/t.

	Value, £/t	Quantity
bales	21.7 ± 4.8	20t/yr/holding
		1.57 Mt/yr glasshouses
briquettes	- 0.8 ± 11.0	16.6 Mt/yr

5.1.4 FIBRE USES

Straw has been used for paper making in the UK and although now all paper is made from wood, there are numerous plants in Europe and elsewhere successfully using straw. Technology cannot therefore be considered a major obstacle. The straw must be clean and dry and so must be baled very soon after harvest, placing an extra cost on baling the straw. Combining the operation of harvesting and baling increases the cost by up to £2/t on all cereal farms.

	Value, £/t	Quantity
paper	-20 ± 16	1.0 Mt/yr
board	- 9 ± 13	1.4 Mt/yr

5.1.5 CHEMICALS

In all cases it is difficult and very expensive to extract chemicals from straw and the product yield is often low. The value of straw has to be compared with the original main product - cereals. Whereas straw has about 40% carbohydrate which is difficult to extract, cereals have about 70% carbohydrate which is easy to extract. Consider a plant costing £L/t processed, with a yield s for straw and c for cereals, with cost prices a and b £/t respectively.

The cost per tonne of output from the straw is $(L + a)/0.4s$
whereas the cost per tonne from cereals is $(L + b)/0.7c$
If the yields are the same, the equivalent price of straw

$$a = (0.4b - 0.3L)/0.7$$

If cereals are £100/t and the process costs a typical £100/t, straw is worth £14/t. However for ethanol, the yield from straw is 70% of the yield from cereals (s = 0.7c), thus the equivalent price is a = (0.28b - 0.42L)/0.7 ie. straw is worth £-20/t.

5.2 INCORPORATION AND BALING COST DISTRIBUTION

Combining the cost of lost yield from incorporation with the reduction in profit from labour, machinery and timeliness, the cost of incorporation relative to burning is 14.8 ± 7.5 £/t. By itself this indicates that 2.5% of farmers would choose incorporation rather than burning. Similarly the cost of baling and storage £/t, is 22.8 ± 8.5 (barley), 22.0 ± 7.0 (wheat)

5.3 COST OF TRANSPORT

Surveys of the transport costs for grain give a good estimate of the cost of using lorries to transport seasonal agricultural produce. Converting to straw loads, the cost of transporting a one-way distance, x miles, £/t is:

0.09x + 7.80 - conventional bales, 7t load
0.13x + 11.38 - round bales, 4.8t load
0.06x + 4.96 - Hesston bales, 11t load

Suppose a bale of density ρ t/m³ is formed and assume it is at least as solid as a Hesston bale so that a lorry can be well packed (unlike round bales). Then a lorry load weighs 76ρ t. Note that 25t is typically the heaviest load a lorry can legally carry so that the bale density cannot usefully be more than 330 kg/m³. The generalised cost of transporting straw is thus $\frac{0.72}{\rho} + \frac{0.008x}{\rho}$, £/t.

If the straw is produced as small briquettes and handled loose in bulk, the lorry capacity will be similar to that of grain lorries, and thus the transport cost is $\frac{1.56}{\rho} + \frac{0.018x}{\rho}$ £/t

where in this case ρ is the bulk density of briquettes. The maximum useful bulk density is 714 kg/m³.

6. OVERALL STRAW USE

6.1 PROCEDURE TO ESTIMATE STRAW USE

The price of straw and the use of straw on and off the farm are determined for a number of scenarios to indicate the effect of possible changes. The basic data are taken to be demand for straw for cubicle, dairy cow and cattle bedding, for feed and for farmstead heating. Alternative scenarios consist of different straw values, transport costs and/or additional demands for straw.

The first step is to determine the amount of straw that will be baled for on-farm use. The alternatives are burning and incorporating. Taking a large sample of farmers gives the proportion of all farmers that one would expect to bale, burn and incorporate. Thus for example, when the value of dairy cow bedding less baling costs is -7.6 ± 26.8, 59% of farmers will burn, 39% will bale and 2% will incorporate. Table IV lists the calculated values for the basic data and the scenarios considered later. Deducting the straw used from the straw produced on each farm gives the total potential supply of straw for sale in each county. The unsatisfied demand is the corresponding potential demand.

Table III

Scenario	Cubicle bedding	Dairy cow bedding	Other cattle bedding	Feed	Fuel
	Straw used on farm of production, 000t				
Basic data	464	118	182	350	323
L) Burn + £5/t	491	135	207	437	472
N) Burn + £25/t	503	181	255	468	614
P) Incorp + £25/ha	454	115	182	385	309

A supply/demand model then calculates the equilibrium prices and the corresponding transported supplies and demands. Table IV lists the proportion of the total demand for straw which is met with straw at these prices. Also listed is the amount of straw for which no off-farm use is economical. These farmers choose incorporation or burning, for example 6.47 Mt burned in the basic scenario.

6.2 MODEL VALIDATION

Models such as this cannot be verified by direct experimentation. The results have been compared with a survey of straw use in each county, the market prices of straw in each region and with straw surveys by the Ministry of Agriculture in 1983 and 1984. Overall the model's predictions are within the errors one would expect from the data and suggest that predictions based on the model will be a useful guide to what would happen in practice.

Table IV

Proportion of total demand for straw met for given scenarios and amount of field burning and incorporation

Scenario	Straw field burned Mt	Straw incorp Mt	% of potential demand using straw: Bedding Cubicle	Bedding Dairy cow	Bedding cattle	Feed	Fuel	Other
Basic data	6.47	0.16	83	38	55	70	31	-
A) Hesston transport	6.02	0.15	89	40	58	76	36	-
B) Densest transport	5.53	0.14	94	43	61	82	43	-
H) Briquetting	6.29	0.16	82	37	54	69	28	2[3]
I) Briquetting + £10/t	5.55	0.11	80	36	52	65	22	9[3]
L) Basic, burn + £5/t	5.24	0.56	88	41	59	76	43	-
N) Basic, burn + £25/t	0.38	3.92	95	51	68	87	67	-
J) All demands + briq	5.60	0.14	80	36	51[1]	64[2]	23	30[4]/1[3]
P) Incorp + £25/ha	4.73	1.14	80	36	51[1]	65[2]	23	32[4]
Q) P + dense transport	3.91	0.94	92	41	55[1]	78[2]	33	35[4]

Notes.
1. Includes use and demand by pigs for bedding
2. Includes use and demand by sheep for feed
3. Demand for briquetting
4. Demand for glasshouse heating

6.3 EFFECT OF ALTERNATIVE SCENARIOS

6.3.1 TRANSPORT COST

The first two scenarios consider reducing the transport cost by producing denser packages of straw. At a density equivalent to Hesston bales (Scenario A) there is a small reduction in the amount of straw burned which is reflected in a small increase in all the uses for straw. Producing the densest package which reduces transport costs (B), approximately doubles this saving but only reduces the amount burned by 15%.

6.3.2 NON-AGRICULTURAL USES

There are a large number of non-agricultural uses and the majority are clearly uneconomical.

a) Briquetting

A demand proportional to the amount of surplus straw in each county was considered with a total potential demand of 16.6 Mt (more than the total surplus straw) and with the value of briquettes for use as fuel(Scenario H). At this value there is very little demand (2% of the total) although there are notable areas such as Norfolk where the price justifies over 8% of the supply going to briquettes.

Increasing the value of straw as briquettes by £10/t increases the demand to 9% (I). This not only reduces the amount of field burning by 12% but also reduces the amount used for other uses because the price increases . 31% of the Norfolk surplus will now go to briquettes, 19% in Cambridgeshire but less than 1% in Cornwall.

This illustrates the problem which any large potential use for straw will face. Compared with prices of £14/t it appears an attractive proposition but once the demand exists the price increases to about £20/t. For example, a single demand for 0.5 Mt was sited in Essex, valuing straw at £20 ± 10/t. 39% of the demand used straw but the straw price increased by £4. Sited in Somerset only 9% of the demand used straw and there was no effect on the price.

b) Paper and board

Straw use for paper has even less effect than briquetting. Only 0.4% of the demand is used and 3% more straw is field burned. Increasing the value by £10/t, increases use to 2% of potential but has no other significant effect. Straw use for board is very similar, 1% of potential demand being used at the estimated value, rising to 4% with a £10/t increase.

6.3.3 RESTRICTIONS ON BURNING

Additional restrictions on burning increase the labour and machinery needed and hence the cost. They therefore reduce the relative cost of baling and incorporation. Increasing the cost of burning by £5/t(Scenario L) (equivalent to about £18/ha) reduces the amount of field burning by 19% and increases the amount of straw incorporated by 0.40 Mt (equivalent to 0.1 Mha). The amount of straw used on the farm of production increases (Table III) by 21%. The increased availability of straw causes prices to fall and consequently the use of straw to increase, for example cubicle bedding from 83% to 88% of potential demand. The largest increase is for farm heating from 31% to 43%.

Increasing the cost by £25/t(N) (£88/ha) which is almost equivalent to a complete ban on burning, the analysis suggests that a few diehards (0.38 Mt) will still burn straw but 3.92 Mt will be incorporated. An extra 2.33 Mt of straw is used within agriculture - almost half of it for farm heating. There is a 41% increase in the use of straw on the farm of production (Table III). However 50% of dairy cows will still not be kept on straw.

6.3.4 IMPROVED INCORPORATION

Although a number of sound scientific and experimental reasons are put forward for incorporation reducing yields compared with burning, it may be that providing incorporation is properly understood and carried out, the loss may not be as bad as anticipated. Suppose incorporation costs £25/ha less than anticipated (Scenario P). The amount of field burning is reduced by 16% and the majority of this straw is incorporated. There is only a small increase in other uses. The effect of dense straw packages for transport(Q) is very little altered although the combined reduction in field burning is 30%.

7. CONCLUSIONS

1. The reduction in profit from baling straw with a selling price of £15/t relative to straw burning is £6-28/ha for arable farms with roots and £20-31/ha for cereal farms.
2. The reduction in profit from straw incorporation relative to straw burning initially is £26-33/ha. After 5 years this should have reduced by about £20/ha.
3. If burning is banned the price of straw will initially stabilise around £10-15/t and as incorporation becomes established will increase by £5/t. This suggests that few straw uses not presently economical will become economical due to a ban on straw burning.
4. Increasing straw (bulk) density for transport to 330 kg/m³ with present agricultural uses of straw will increase the amount used (and hence reduce the amount of field burning) by 1 Mt.
5. Increasing the value of straw converted to briquettes by £10 reduces straw burning in the field by 12%. However it has the effect of increasing straw prices and reducing the amount for present agricultural uses.
6. An in-field briquetting machine costing £60000 would have to work faster than 1 h/ha. At £40000, it must work faster than 1.5 h/ha.

8. REFERENCES

(1) AUDSLEY, E. The economic consequences on the farm of a ban on straw burning: Farm profit and practices. Div. Note, DN. 1234, natl. Inst. agric. Engng, Silsoe, June 1984

(2) AUDSLEY, E. The effects of field burning restrictions, straw incorporation and alternative uses as fuel, fibre or feed on the amount of field burning and the supply and demand for straw. Div.Note, DN. 1307, natl Inst. agric. Engng, Silsoe, December 1985

(3) AUDSLEY, E. An arable farm model to evaluate the commercial viability of new machines or techniques. J. agric. Engng Res., 1981 26

(4) AUDSLEY, E. Planning an arable farm's machinery needs - a linear programming application. The Agricultural Engineer, Spring 1979

(5) ENGLAND, R.A. Reducing the nitrogen input on arable farms. J. Agric. Econ. Vol. XXXVII, No. 1,,Jan 1986
(6) HART, P.B.S. Effect of soil type and past cropping on the nitrogen supplying ability of arable soils. Ph.D. Thesis, Soil Science Dept, Univ. of Reading, June 1984
(7) MAFF The storage of farm manure and slurries : farm waste management. Booklet 2273, 1984

THE USE OF WOOD CHIPS FROM FOREST HARVEST RESIDUE FOR ENERGY PRODUCTION POTENTIAL YIELD, QUALITY, HARVESTING TECHNIQUES AND PROFITIBILITY

G. BECKER
Institut für Forstbenutzung und Forstliche Arbeitswissenschaft
der Universität Freiburg i. Br.
Fed. Rep. of Germany

Summary

Facing a worldwide increase in utilization of fossil fuels raises questions about the role of renewable resources in the middle and long term energy supply. Under which conditions, with what techniques and what economic situations can renewable resources help cover the growing future energy need ? In the area of forestry, the over-riding question is about the quantity of biomass available from woody plants, both from traditional forestry practices (small diameter and waste wood) and that which can be produced in special short-rotation energy wood plantations. To date, handling and utilization of these renewable resources has been both difficult and expensive. One possible technique to effectively and economically supply energy wood is through homogenization of the product coming from the forest through the process of chipping.

A study partly financed by the European Economic Community (EEC) has recently been completed. Both extensive field and laboratory studies were undertaken to provide information about wood chipping for energy use in the following areas :

1. Amount and structure of the yield and surplus yield of biomass available through prevailing forest harvesting techniques
2. Wood quality characteristics of this biomass
3. Production techniques
4. Production costs
5. Market potential
6. Overall evaluation

The following report should provide useful information in these main areas.

1. AMOUNT AND STRUCTURE OF ADDITIONAL BIOMASS YIELD AVAILABLE FROM CONVENTIONAL FOREST HARVESTING TECHNIQUES

Firstly, it is most important to have information about the additional yield of wood biomass available from chipping, before studying the problem further. It has been necessary, because of the difficulty in recording wood volume and differing wood-chip moisture content, to record all yield data in the form of absolute dry weight. This unit of measurement is most useful for wood utilizing industries and energy producers, giving information on overall yield and surplus yield, and on an individual tree basis giving highly specific yield data. In our research the two ways were studied concurrently. Table I and Figures 1, 2, 3 and 4 show, for typical cases, the amount of currently available yield and the potential increased yield.

Results of the single-tree biomass research can also be used to provide diameter specific information on additionally utilizable biomass.

Our research results also allow overall or stand level estimation of the expected additional realizable yield from generally available tree and stand data (D.B.H., volume stand density or weight of normally utilized tree volume).

The research has shown the highest amount of chips produced (per hectare) is 46 metric tons dry weight. This is achieved through chipping previously unused parts in clearcutting hardwood stands. The highest increase in utilization of harvest material is achieved through complete chipping of small diameter thinning material in pine or hardwood stands with 73 % of the total biomass or 28 metric tons dry weight.

2. QUALITY OF THE WOOD CHIP MATERIAL

The moisture content of the wood chips is a very important quality characteristic when used for energy production but also important for industrial utilization such as chipboard production. The moisture content may vary from 85 % to 130 % during the vegetation period in the so called "hot logging system" where each phase of the harvesting operation follows the previous phase with minimum lost time.

This range of moisture content holds regardless of the parts of the trees being tested. Both conventionally utilized pulp wood as well as crown and branches have similar moisture contents. Therefore, increased utilization of the entire tree should not significantly increase the moisture content. By separating the phases of the harvesting system, especially letting the felled trees lie on the site for a few weeks, will help reduce the undesirable high moisture content and make the material more suitable for industrial use.

Another quality characteristic that is important in industrial utilization (not in energy production) is the amount of mineral impurities in the material. The distance between the site of felling and the chipper location greatly influences the amount of mineral impurities. Especially bad, but unavoidable, is the amount of soil picked up by the tree during skidding in wet (muddy) conditions. Our studies show that the amount of mineral impurities falls within the range of 0,01 and 0,1 percent of absolute dry weight. This is in every case a tolerable range. (These figures were arrived at through ashing the chips and will therefore also contain the amount of naturally present mineral content of the wood material).

Shape and size of the chips is also an important chip characteristic, both for industrial use and for energy production (especially burning systems using automatic feeding). Through the process of screening the chips, the amount of chips in different size classes was found. When studying individual tree parts, we found that the amount of fine and large chips was significantly higher from crown, branches and twigs. However, the amount of material in the desired middle size class remained at over 80 percent of the total.

The normally produced chip has a prismatic form (30 - 40 mm long, 10 - 15 mm wide and 4 - 6 mm thick), which is also the form desired by most users.

The amount of wood and non-wood parts in the material is also an important characteristic. The chip material was separated into wood, bark, twigs and leaves or needles (a time consuming process). Our results show that for whole tree chipping, wood made up 80 to 93 percent of the total chip material. Wood content is more important in industrial use than in energy production. The remaining material was composed of 5 to 15 percent (absolute dry weight) bark, and a negligible 1,5 to 2,5 percent twigs (< 5 mm, and not separated into wood and bark) and leaves or needles.

A separate analysis of only the previously unutilized tree parts showed a decrease in the wood content of the chip material (compared to whole tree) due to the small diameter of the tree parts. Correspondingly, there was an increase in bark content. Overall, there wasn't a significant increase in the leaf (or needle) percent. Only when looking at the small size class (< 4 mm), leaf content made up a significant part (14 to 35 percent) of the material. While on sight the amount of leaf material contained in chips from whole-trees appears high, analysis shows that it actually is a small component of the total chip mass. This difference may be attributed to the low density of leaf material.

3. OUTPUT OF DIFFERENT CHIPPING SYSTEMS

The differences in output were studied for differing harvesting systems using different chippers with varying output capacities. Usually output is most dependent on tree size, in particular diameter. Because of the complexity of combining differing systems with differing tree size, only a range of outputs for chippers is reported. The lowest output was from chipping material from pine thinnings with, on the average, 6,5 m^3 of chips or 2,07 tons of moist chips (0,91 tons dry weight) per hour.

First outputs from spruce thinnings were not significantly higher. A moderate level of output resulted from chipping hardwood material on the site (10 - 12 m^3 chips or 4 tons fresh chips or 2 tons dry chips per hour). The highest output is from small diameter (thinning material) pines either skidded to the access road or concentrated in large bundles on the landing. This output was 16 - 17 m^3 chips or 5,2 tons fresh chips or 2,7 tons dry chips per hour.

The time study analysis shows that the primary influencing factors to output were the size of the material chipped, the volume of material per hectare and the amount of travelling required by the chipping machine.

4. COST OF THE DIFFERENT HARVESTING SYSTEMS

We found a wide range of average costs per m^3 of chips in the different trials. The differences were due to different levels of output and in particular to different operating costs of the systems.

The lowest costs (DM 27/m^3 or DM 122/ton of dry chips) occured when chipping hardwood crown material in clearcut operations. Only slightly higher costs were observed when chipping small diameter hardwood stands or spruce thinnings (using small hand fed chippers). Significantly higher costs (DM 35 - 40/m^3 or DM 240 - 260/ton dry chips) were found in harvesting systems which use thinning material from small diameter pine stands and operate the chipper in the stand or at the access road. Most expensive (DM 55/m^3 or DM 360/ton dry chips) are thinning operations, both in small diameter pine and spruce stands, in steep terrain using cable logging for extraction.

It becomes obvious, that chipping costs were not as important in determining costs of the entire harvesting system as were the other components (felling, prebunching and intermediate transport).

5. MARKET POTENTIAL AND MARKET PRICES OF GREEN CHIPS

Both new sawmill technologies, which produce a higher output of chips, as well as increased recycling of fiber material add to the already large supply of fiber from industrial roundwood. The result of this situation is stagnating prices for industrial roundwood and declining prices for wood by-products of wood utilizing industries and recycling material. The introduction of additional amounts of green chips to the market may cause great problems unless the wood manufacturing industries and energy producers enlarge their capacities to utilize the additional material. Market potential is also negatively influenced by the lack of grading rules and detailed information about the quality of green chips. Overall, the market prices for green chips (15 - 20 DM/m^3 or 105 - 150 DM/ton dry weight) are lower than production costs when comparing only direct costs with selling prices.

6. OVERALL EVALUATION

A total evaluation (macro scale) must take into account additional aspects of chip production beyond the previously discussed direct production costs and profits. Better use of locally available energy supply results in a more desirable independence from energy imports. The reported cost evaluations are based on the cost situation of public forest enterprises with high wages and personnel benefits together with machine costs solely born by the forestry enterprise. Alternative economic evaluation models are possible. Farmers may have lower production costs due to : seasonal availability of already owned equipment which can be used in chip production, seasonally available labor, ability to produce facing prices just above marginal costs and ease of entry and exit from production with fluctuating prices (lower risk).

Full use of the tree elements previously left in a logging operation results in a cleaner site, beneficial for regeneration, pest management and future logging activities. In West Germany, clean-up of the site by concentrating the branches and other parts, either by machine or manually, and then burning results in costs between 1000 and 3000 DM.

In evaluating the economic situation for energy production, the potential of fast-growing plantations (both for biomass and energy wood) on marginal farmlands should be considered. The production costs of such plantations can be much lower than conventional forestry practices.

Also to be considered is the effect of removal of nutrients from the site due to the increased utilization in whole-tree harvesting. Much of the nutrients in the tree are concentrated in the previously unused parts. This may result in additional long-run costs due to fertilization, especially on poorer sites.

Future research should apply cost-benefit analysis to the overall problem. This could provide interesting insights through the overall combination of points made in this paper.

AN APPROACH TO THE ECONOMIC EVALUATION OF BIOMASS ENERGY PRODUCTION PROJECTS

M. DI PALMA and G. BARBIERI
Ecoter - Rome

Summary

The approach to the economic evaluation of biomass energy production projects does not differ significantly from other energy production projects, though the analysis presents some peculiarities which deserve a closer examination. Generally speaking from an economic point of view (i.e., from the point of view of the domestic economy as a whole, regardless of institutional operators and of income distribution considerations) a biomass energy production project is to be considered a "productive unit", which consumes resources (project inputs) and produces different resources (project outputs). Project inputs represent the costs of the project, while project outputs are its benefits. Costs, in this approach, are measured as the opportunity costs : since economic resources are limited, the undertaking of an investment diverts resources from alternative uses. Market prices do not always reflect opportunity costs. This is relevant in our case for the raw material itself - biomass - which is seldom marketed. Another viable approach is to consider the production costs of biomass (though always in terms of opportunity costs). The economic evaluation of benefits follows the same guidelines : here, too, market prices may be misleading. Moreover, energy prices are often heavily taxed, represent a tariff or are statutorily controlled. One possible solution is to consider the opportunity cost of producing the same amount of energy in an alternative plant (e.g. the production cost of electricity- not its price ! - in a conventional thermo-electrical plant) after comparing the calorific power of biogas (or biomass) with conventional fuels. Further benefits - such as the social value of waste disposal or the use of by-products as fertilizers - may be introduced after careful discussion and net of associated costs. The economic feasibility of biomass energy production projects is conditioned by financial constraints, which have effect on the "willingness to pay" of project beneficiaries : we will consider - with reference to the case of Italy - the influence of market regulations, of purchase prices of excess electricity, of the legislation in favour of integrative energy sources.

1. IDENTIFICATION AND DEFINITION OF THE PRINCIPAL DOMAIN OF THE PROJECT

The economic evaluation of biomass energy production projects is not significantly different to that of other energy production projects.

The first step in the analysis and evaluation of a project is to identify and define its principal domain. Although this is a key phase, in particular where a highly practical approach is adopted, it generally receives insufficient attention in the now abundant literature on cost-benefit analysis, perhaps because it does not lend itself to sophisticated formal analyses.

From a theoretical standpoint, it is indeed obvious that any project forms part of a network of interrelations which, when taken as a whole, make up the economic system. It is the analytical process itself which, in focusing attention on the project in question, places it at the centre of the network of relations with the rest of the economy. What is involved here is merely a shift of perspective which, while certainly providing the analysis with new avenues of interpretation, has no substantive influence on the facts.

From the practical point of view, however, if we are to avoid having to follow ad infinitum the project's repercussions on the whole network of interrelations, these same analytical requirements also call for a certain degree of decontextualization. Although this procedure is necessary, it is nonetheless both arbitrary - because based on an a priori choice by the analyst - and somewhat problematic, due to the intrinsically complex nature of the system of interrelations in question.

In delimiting the scope of the project covered by the analysis, sectoral, territorial and temporal variables are brought into play.

Consideration of the sectoral aspect involves identifying the main direct links which the project establishes - both upstream and downstream - with the production system. From this point of view (but also from others, as will become clear in what follows), it may be useful to consider the project in question as a productive unit which consumes resources (project inputs) and produces different resources (project outputs). The project inputs are, in turn, either "primary factors" (conventionally speaking, labour and capital) or the products of other productive units, which serve to identify the sectors upstream of the project. Similarly, the project outputs either go towards meeting end demand or are used as inputs by other productive units, thus identifying the sectors downstream of the project.

As regards the territorial aspect, it is likewise necessary to distinguish the spatial domain within which the project's effects will be most keenly felt from the rest of the territory, where its effects are either non-existent or negligible. Although this is also, in theory, an arbitrary procedure, it is relatively easy in practice to delimit the project's main sphere of influence. In this regard, it must be stressed that projects often have a territorial connotation specific to them. For example, an irrigation project is implicitly linked to its own particular "territory", which corresponds to the area delimited by the irrigation infrastructure. This feature is shared by the energy production projects under discussion in this paper in as far as direct use of the energy produced (provided this has not been fully integrated into the national grid) reflects a well defined spatial distribution of the effects.

Two further aspects should be clarified here :

- this territorial delimitation, hitherto exemplified on the side of the project outputs, can also be carried out on the inputs side by taking into account the territorial origin of the resources employed, provided this origin seems important within the definition and characterization of the project. An example here would be the excavated material required for major infrastructure works or the catchment area of a sewage treatment plant or disposal unit for solid urban waste ;
- this assumption of a reference territorial domain is quite distinct from the assumption, common in cost-benefit analysis, of the "national economy standpoint", which is in fact intended to distinguish the business perspective characteristic of financial analysis in the strict sense from the broader perspective specific to economic analysis, whereby costs and benefits are assessed from the point of view of society as a whole on the basis of criteria shaped by the welfare economy. Despite the possible ambiguity which may arise with regard to use of the term "national economy", it is independent of any concrete territorial characterization.

Finally, as regards the temporal variables, it is important to consider first of all the useful life of the project in question, which in turn parallels the various useful lives of its constituent operations.

Indeed, while for some categories of works and installations it is necessary to take account of a limited useful life (with obsolescence added to physical wear and tear), the functional timespan for public works is generally of the order of decades.

A further useful consideration is the economic life of the project, i.e. the period of time beyond which increases in the net annual marginal benefit, adjusted to the current social discount rate, bear no relation to the project's current net value. Assuming a social discount rate of 8 %, in line with that suggested by the Italian authorities responsible for projects funded by the FIO (Investment and Employment Fund), this timescale may be considered equivalent to 30 years, after which time the value discounted from the benefit falls below 10 % of the corresponding non-discounted value.

2. DETERMINATION AND QUANTIFICATION OF COSTS

In order to determine and quantify costs (and, moreover, benefits), it is again useful to see the project in terms of a productive unit. In making use of certain inputs (be they intermediate inputs or primary factors), the project diverts resources from alternative uses. In this approach to the problem, which is central to cost-benefit analysis, the costs of the project are those borne by society as a whole in giving up the above-mentioned alternative uses of the resources employed in the project (opportunity costs). In other words, the costs of the project inputs correspond (net of the costs of any other inputs) to the value of the goods and services which those inputs could have produced in a different productive unit. Whether, in a purely hypothetical market economy, these costs would tally with the market prices of those goods and services, is a question of strictly academic concern, which may safely be left to scholarly debate. The fact is that in most, if not all, cases the prices of goods and services sought and offered on the market do not reflect the relative scarcity of the resources.

To sum up, the inadequacy of the system of market prices can be viewed from three different angles :

- market prices are expressed gross of indirect tax and since such taxes are levied at different rates on the various goods and services, a distorting factor is introduced into the system of relative prices. Furthermore, from the point of view of society as a whole, indirect taxation can be considered as a simple transfer ;
- the system of market prices may contain other distortions in cases where domestic prices do not reflect consumers' willingness to pay (e.g. in the case of rationed goods and politically determined prices) ;
- finally, the system of market prices may not reflect the system of priorities established by economic policy.

Another category - that of non-traded goods and services - quite simply has no market price and it should be stressed here that at least one of the inputs to a biomass energy production project, i.e. the biomass itself, usually belongs to this category. Thus, even assuming the existence of market forces (the "unseen hand") capable of aligning prices with the relative scarcity (i.e. the opportunity cost) of the resources, this would only hold true for freely traded goods and services.

Amongst the techniques of deriving the opportunity cost from the project inputs, particular attention should be paid to the methods of marginal cost and marginal consumption value, since these provide an appropriate way of evaluating the shadow price of the biomass. When a project uses a good or service as an input, one or other of the following will hold true :

- if the supply of that good or service is elastic, the demand expressed by the project will result in an increase in the supply of that good or service, which will in turn increase demand for the goods and services required for its production. If the market prices of such goods and services reflect their relative scarcity or if their shadow price value is known, the marginal cost thus obtained will represent the desired shadow price of the project input ;
- if, on the other hand, the supply of that good or service is non-elastic, the demand expressed by the project will result in decreased consumption of that good or service. In this case, the desired shadow price of the project input will be represented by the marginal consumption value, provided substitute goods or services exist whose market prices do not reflect the relative scarcity (or whose shadow price value is known).

As far as biomass energy production projects are concerned, the second hypothesis seems at first sight rather implausible. However, a brief examination of this hypothesis may be worthwhile, if only to illustrate an alternative method of estimating the shadow price of the biomass. Given, then, that we are still dealing with theoretical simplifications and for clarity's sake presenting only extreme cases, let us assume that the supply of biomass is strictly limited and cannot readily be adapted to demand stimuli. In such a case, the biomass required as input to the energy production process would be diverted from other uses, for example from use as agricultural fertilizer. Here, then, the market price of the substitutes (for example, chemical fertilizers), appropriately adjusted and corrected, can be taken to represent the shadow price of the biomass.

In the first hypothesis, which seems more realistic, the supply of biomass is assumed to be elastic, so that here it is the marginal cost of the biomass which must be evaluated. Leaving aside the need to evaluate the shadow prices of the production inputs (with some simplification, a not unduly difficult task, given that such shadow prices can usually be derived by applying appropriate conversion factors to the market prices of the inputs), the only major problem involved in this estimation is whether the biomass itself carries with it a cost for society as a whole. There is no straightforward answer to this question. Let us consider two extreme cases.

Let us first assume that biomass is produced exclusively as a raw material for energy production projects. In this case, it is clear that the production cost of the biomass itself is exactly equivalent to the shadow price of the input to the energy process and is, as such, borne in full by society.

Let us now assume, instead, that society expresses a preference regarding the disposal of particular waste products (a collective preference of the type which, in this context and in any democratic system, is generally made explicit in the form of a law or regulation). In this case, the biomass which may be produced in the disposal process is considered a by-product, whose cost may either be nil (in which case the method of disposal employed admits of no alternatives or itself represents the most economical alternative) or equivalent to the difference between the average incremental cost of disposal in the biomass-producing plant and the average incremental cost of disposal in the economically most efficient plant.

3. DETERMINATION AND QUANTIFICATION OF BENEFITS

The same line of argument will be adopted for the determination and quantification of benefits as for costs, namely that here too market prices may be misleading.

The price of energy is not in fact a market price but a "political" price, which is administered or controlled. Apart from the fact that the various energy sources are not fully interchangeable, it must also be borne in mind that electrical energy is not a primary source and that the parallel generation of steam is also possible in power stations. Considerations such as these, whose detailed analysis would be out of place on this occasion, lead one to favour the adoption of a shadow price for electrical energy produced in biomass-fuelled power stations.

Conceptually speaking, the problem posed is analogous to that already discussed with regard to the estimation of costs.

Since in all cases the plants involved in biomass energy production projects are small in size, the additional production of electrical energy will not have any appreciable effects at national level on overall supply and demand as regards energy. The problems which arise in determining the shadow price of electrical energy in the face of shifts or fluctuations in supply and demand can therefore be disregarded here and the shadow price of the electrical energy produced may be taken as the average incremental cost.

In Italy, where over 70 % of the electrical energy produced by the ENEL is generated in conventional power stations, this cost of production in a large-scale, oil-fired power station (according to ENEL estimates, the economically most efficient conventional power station is, under current conditions, an oil-fired station with four 640 MW generating units).

In line with the literature on cost-benefit analysis, average incremental cost (AIC) is taken to mean the unit cost of production expressed by the formula

$$AIC = \frac{\Sigma\,(CI_i + CE_i)\,(1 + r)-1}{\Sigma\,(QP_1)\,(1 + r)-1}$$

where

CI_i represents the investment costs incurred in year i, expressed in constant values ;

CE_i represents the operating costs incurred in year i, expressed in constant values ;

QP_i represents the physical quantities produced in year i ;

i represents the designated year in the economic lifespan considered ;

r represents the social discount rate adopted.

Three observations may be made regarding this method of estimation.

The first and most obvious is that this evaluation is made on the basis of shadow prices. This is particularly important in our case with regard to the operating costs and, above all, determination of the price of fuel oil. It is well known that in Italy a composite tax is levied on petroleum products (VAT and production tax). This tax varies according to the product concerned (either as regards the rate of VAT or the amount of production tax) and can also be adjusted in line with changes in prices on the international markets. By convention, however, the shadow price of the fuel is taken to be its frontier price, which represents the true cost to the national economy of supplying this resource.

The second observation concerns the crucial part which the social discount rate plays in determining the shadow price. Where installations with high initial investment and an extended construction period are concerned, the higher the social discount rate, the higher - all other things being equal - the AIC. This observation is far from irrelevant when one considers that the ENEL, on the basis of an international agreement brought about on account of the strategic importance of the energy infrastructure, bases its evaluations on a 5 % discount rate, whereas, as we have seen, the FIO has adopted a social discount rate of 8 %.

It is also worth mentioning, in passing, that the adoption of different social discount rates is at the root of the debate concerning the economic expedience of nuclear energy, since, compared with conventional plants, nuclear power plants - despite their lower operating costs - are characterized by just such a high initial investment and extended construction period and can therefore only appear advantageous given a social discount rate which is not too high.

The third and final observation reintroduces and concludes the discussion of the economic expedience of biomass energy production plants. Since the shadow price of the energy produced is calculated with reference to the currently economically most efficient production methods and since, anyway, the calorific power of the biomass is lower than that of natural gas or fuel oil, it should by now be clear that the economic benefits of biomass energy production must be viewed from a twin perspective :

- that of the costs of the inputs and, in particular, the possibility of assigning to the biomass a very low or zero opportunity cost if, as suggested above, the biomass is seen as a by-product (or associated product) of the service production process "disposal of solid urban waste" or "sewage treatment" ;
- that of the benefits and the possibility of assigning other benefits to such plants (e.g. the recovery of fertilizers) or positive spin-offs (e.g. as regards the environment), albeit with the caveat that the inclusion of such benefits carries with it the risk of duplications and overestimates and therefore requires accurate evaluations and proper consideration of the associated costs.

4. FEASIBILITY REQUIREMENTS

The final argument in this short paper on the economic evaluation of biomass energy production projects concerns the feasibility requirements for any action undertaken. This argument is distinct from economic analysis in the strict sense and concerns the surrounding conditions which determine the feasibility of the proposed project solutions.

A positive assessment as regards the economic expedience of a project from the point of view of the domestic economy as a whole is, indeed, a necessary but not sufficient condition of that project's actual feasibility.

First of all, it must be appreciated that the operator responsible for the construction and management of the energy production plant is also subject to the constraints of economic viability. This requirement is particularly acute in the case of the private operator aiming to show a profit but is also keenly felt by the public operator, who is publicly accountable for the allocation and use of resources. In both cases, costs and benefits are regarded by the authority responsible for construction and management from what may be termed a "business" perspective, i.e. they are defined and accounted for on the basis of current prices and tariffs. As was stressed above, where the problem was approached from the opposite perspective, "current" prices, tariffs and costs may differ very appreciably from the shadow prices used in economic analysis. This means that a project which is viable from the point of view of the domestic economy as a whole may turn out to be impractical from a business perspective.

Over and above the fact that "current" prices and shadow prices may not tally, "financial" feasibility (we are applying this term to the notion of feasibility in order to distinguish this level of analysis from that of economic analysis) is determined by elements relating to upstream factors, technology and market outlets.

As regards the upstream factors, it is obvious that supply sources and conditions play a more vital part for the operator responsible for managing the plant than that which we advanced in relation to economic analysis. It has already been stressed that as things stand, there is no market price for biomass for the simple reason that it has no market. We may simply add here that since in this situation the marketing costs would be such as to outweigh any economic advantage, the key role lies with integrative processes, in which the recovery of energy from the biomass is simply the final phase in a process of sewage treatment and/or solid waste disposal (solid sewage, the organic portion of solid urban waste, effluent from the agri-foodstuffs industry, excrement from livestock rearing). Also subsumed in this approach is the problem of supply quality, i.e. the adoption of mutually dependent methods and techniques of disposal and energy recovery, and that of the quality and concentration of the biomass.

As regards outlets, too, the correct approach would seem to be that of integration, albeit in this case more for legislative and institutional than for strictly technical and economic reasons. Despite some faint signs of improvement, the legislative and institutional framework in Italy does not appear conducive to the production and distribution of energy. What glimmer of hope there is concerns the benefits foreseen by Law No 308/1982 on the production of thermal, electrical and mechanical energy from renewable sources, which provides for non-repayable finance and interest subsidies. However, there still remains the "paradox of the rigidity-based wastage inherent in the organization of the whole energy sector", as spelled out in the CUEIM report published in Sinergie 4/84 :

> "This paradox consists in the fact that although the technology is available which would allow energy to be produced at competitive costs and hence competitive prices, there exists no matching organization of energy demand capable of economically exploiting this cost differential. Our energy system is, it seems, unable to conceive of what lies downstream of mere production technology, namely an energy service, unable to actualize the potential savings, unable to ensure that energy is used more flexibly, unable to recover the surplus in relation to the requirements of the producer of renewable energy"

As regards the technology, let it suffice that we submit to the experts the basic conditions - determined by the socio-economic and institutional context - required for the operation of biomass energy production plants : low break-even point, high productivity, flexibility, a willingness to recycle energy (not necessarily electrical) into the production process itself (more widespread acceptance of the notion of energy saving in the purifying phase) and scope for the utilization of by-products (fertilizers etc.).

CONCLUSIONS AND RECOMMENDATIONS

1. The workshop recognised the need for a regularly updated catalogue of existing wood harvesting machines, listing both machine and product specifications.

2. The workshop identified a need for international collaboration in the development of new classes of multiple use harvesting and collection machines and to assess the potential for using or adapting existing machines for new situation. For example it verified a need to develop small machines suited to Mediterranean countries and also recognised the need for machines suitable for use on small timber holdings.

3. The workshop recognised a need for new co-operative, municipal or specialist contracts services to fully utilise harvesting and collecting machines.

4. The workshop raised the question as to whether it is advisable to use subsidies on the machines, or the product, as a means of stimulating their use.

5. The workshop identified a need to develop projects which integrate harvesting/storing/transport and utilization of the final product.

6. The workshop identified a need for the cooperation of multidisciplinary feasibility studies so that the wider impacts on the environment, not only of biomass, but of alternative uses can be considered. In the case of energy users, the energy balance as well as an economic analysis is required.

7. The workshop did not answer the question about the social benefits of biomass. Therefore, there is a need to determine whether there are potential environmental benefits from the collection/harvesting/ management of forests in relation to fire prevention and to determine what data are available on such benefits.

8. It is necessary to be aware that biomass harvesting must be undertaken in a fashion that maintains the quality of the forest/land etc.

9. Attention should be paid to the fact that there are a number of potential uses for wood in addition to use as a biomass fuel.

10. Co-ordination by the Commission in the wood harvesting R & D programme would be useful in ensuring links with the International Energy Association.

LIST OF PARTICIPANTS

ABEELS, P.
U.C.L. Fac. Sciences Agronomiques
Dép. Aménagement du Terr.
Place Croix du Sud 3
B - 1348 LOUVAIN-LA-NEUVE

ALFANI, F.
Università Dell'Aquila
Facoltà di Ingegneria
Poggio Roio
I - L'AQUILA

AMIRANTE, P.
Ist. Mecc. Agraria
Università Bari
Via Amendola 165/A
I - 70126 BARI

ANTONELLI, L.
Consorzio ALTEN
Via Monte Carmelo 5
I - 00166 ROMA

ASSENZA, D.
Premeco SRL
Via A. Friggeri 43
I - 00136 ROMA

AUDSLEY, E.
National Institute of
Agriculture Engineering
Wrest Park Silsoe
UK - BEDFORD MK 45 4 HS

BALDELLI, A.
Itabia
Via Archimede 161
I - 00197 ROMA

BALDELLI, C.
Itabia
Via Archimede 161
I - 00197 ROMA

BALDINI, S.
Istituto Del Legno Del C.N.R.
Piazza Edison 11
I - FIRENZE

BARTOLELLI, V.
Renagri
Corso Vittorio Emanuele 173
I - 00186 ROMA

BECKER, G.
Inst. Für Forestbenutzung
Albert-Ludwigs-Universität
Holzmarktplatz 4
D - 7800 FREIBURG

BELLETTI, A.
A-Biotec
I - 47100 FORLI

BERTI, S.
C.N.R.- Firenze
Piazza Edison 11
I - FIRENZE

BIANCHI, G.
ENEA - FARE
I - ROMA

BODRIA, L.
Ist. Ing. Agraria
Università Milano
Via G. Celoria 2
I - 20133 MILANO

BONINO, G.
BES
Via Piani Della Madonna 127
I - 17017 MILLESIMO (SV)

BRUNETTI, L.
Ass. Prov. Olovic. Brindisi
Via A. Fogazzaro 65
I - 72015 FASANO DI PUGLIA

BRUNETTI, N.
ENEA
Str. Prov. Le Anguillarese Km.1.300
I - 00060 ROMA

CACACE
Italtekna
Piazza Buenos Aires 17
I - ROMA

CALDERARO, V.
Università La Sapienza Roma
Via Gramsci 53
I - 00195 ROMA

CANONACO, P.
Via C/Da Crosetto Pal. D
I - 87068 ROSSANO CALABRO (CS)

CANTARELLA, M.
Università Di Napoli
Dipart. di Ingegneria Chimica
Piazzale Tecchio
I - 80125 NAPOLI

CARUSI, E.
Regione Abruzzo
Via La Marmora 16
I - PESCARA

CECCHI, F.
Dip. SC. Ambientali
Università Venezia
Calle Larga S. Marta 2137
I - 30123 VENEZIA

CIAPANNA, C.
Regione Abruzzo
Ente Regionale di Sviluppo
Rocca S. Maria
I - TERAMO

CIMINO, V.
Regione Marche
Via Gentile Da Fabriano
I - ANCONA

COSENZA, G.
S.G.B. Ingegneria
Via Ardeatina 59
I - 00042 ANZIO

CRESTA, E.
Dip. Chimica Univ. di Siena
Via Mantelli 44
I - 53100 SIENA

CURRO, P.
SAF
Via Casalotti 300
I - ROMA

CUSTODERO, S.
S.E.S
Piazza L. Cerva Eur
I - ROMA

CUTOLO
SAF
Via di Casalotti 300
I - 00166 ROMA

D'AMICO, M.G.
Regione Abruzzo
Via Orientale 30
I - CIVITELLA M.R. (CH)

DANISE, B.
Regione Campania
Via Oberdan 32
I - NAPOLI

DE MARINIS, L.
ERSA
Via G. Ambrosio 1
I - ORTONA (CH)

DENIS, M.A.
CIMAF
Boulevard Carnot prolongé 7
F - 51310 ESTERNAY

DI PAOLA, F.
SV.I.M. Service
Via Tommaso Fiore 62
I - 70123 BARI

ECCHER, A.
Centro di Sperimentazione
Agricola e Forestale
della S.A.F.
Via Casalotti 300
I - 00166 ROMA

EISEN, L.
C.C.E.
Direction Générale
"Télécommunications"
Industries de l'Information &
Innovation"
L - 2029 LUXEMBOURG

ERBAGGI, C.
Via Due Ponti 200 A
I - 00191 ROMA

FARINA
SAF
Via di Casalotti 300
I - 00166 ROMA

FERRERO, G.L.
C.C.E.
Direction Générale "Energie"
Rue de la Loi 200
B - 1049 BRUXELLES

FINASSI, A.
CNR
Ist. Mecanizzazione Agricola
Ten BORASO
I - 13100 VERCELLI

FONZI, F.
ALTEN
Via monte Carmelo 5
I - 00166 ROMA

GABRIILIDES, S.
Aristotelian University
of Tessaloniki
Lab. of Agricultural Engineering
GR - 540 06 TESSALONIKI

GAILLARD, M.
Cemagref
Parc De Tourvoie
F - 92160 ANTONY CEDEX

GANDINI, C.
Gandini Meccanica
Via Della Valletta 1
I - 46040 GUIDIZZOLO (MN)

GARAVOGLIA, S.
Fiat Tratori Spa
Direzione Prodotto
Corso Ferrucci 112
I - 10138 TORINO

GHERI, F.
Consulente Ersa
Poggio Ugolino
I - 50015 GARASSINA

GIOLITTI, A.
Università Siena
Istituto di Chimica
Pian dei Mantellini
I - 53100 SIENA

GIUDIDDA, R.
Largo Forano 4
I - ROMA

GOUDEAU, J.C.
Min. Recherche
F - POITIERS

GRASSI, G.
C.C.E.
Direction Générale "Science,
Recherche & Développement"
Rue de la Loi 200
B - 1049 BRUXELLES

GRILLI, W.
Muratori SPA
Machine Agricole
I - 41100 MODENA

GUAZZIERI, L.
Regione Abruzzo
Via Livenza 9
I - MONTESILVANO (PE)

GUERRERO, J.
Instituto Para La Divers.
y Ahorro de la Energia
Min. Ind. Y En. - DG - IDAE
C/Augustin De Foxa 29
E - 28036 MADRID

GUGLIELMINO, C.
Ersa
Via G. Spataro 18
I - VASTO (CH)

HARTOG, J.
Biomass Development Europe
Avenue De Tervueren 36
B -1040 BRUXELLES

HEINRICH, R.
FAO - FOI-F817
Via Delle Terme di Caracalla
I - 00100 ROMA

HERNANDEZ, C.
I.D.A.E.
Via Castellana 95
E - MADRID

HUMMEL, F.
Forestry Investment
Management LTD
The Ridgeway 8, Guildford
UK - SURREY GUI2DG

IACQUANIELLO, G.
Alten
Via Monte Carmelo 5
I - 00166 ROMA

IANNOTTI, A.
Cafcol
Via Le Vega, 32/A
I - ROMA

INSUNZA, S.
Team
Via F.S. Nitti 15
I - ROMA

KELLER, P.
Statens Jordbrugstekniske
Forsog
BYGHOLM
DK - 8700 HORSENS

KOCSIS, K.
FAO, REUR, F.141
Via Delle Terme di Caracalla
I - 00100 ROMA

LAUFER, P.
Service Agr. Forets. Biom.
Agence Française pour la
Maîtrise de l'Energie
Rue Louis-Vicat 27
F - 75737 PARIS CEDEX

LEONE, M.
Team
Via F.S. Nitti 15
I - ROMA

LISA, L.
CNR - Ist. Mecc. Agr.
Via O. Vigliani 104
I - 10135 TORINO

LUCAS, J.
CEMAGREF
Parc de Tourvoi
F - 92160 ANTONY

LYONS, G.
Agricultural Institute
An Foras Taluntais
Oak Park Research Centre
IRL - CARLOW

MARTZOPOULOS, G.
Aristotelian University
of Tessaloniki
Lab. of Agricultural
Engineering
GR - 540 06 TESSALONIKI

MARZETTI, P.
ENEA CASACCIA
Via Anguillarese Km.1,300
I - ROMA

MAZZANTI, U.
C.R.F. SRL
Via Salaria 434 B
I - ROMA

MITCHELL, C.P.
University of Aberdeen
Dept. of Forestry
St. Machar Drive
UK - ABERDEEN AB9 2UU

MONTAGNANI, G.
Dip. Energ. - Univ. Pisa
Via Diotisalvi 2
I - 56100 PISA

MORANDINI, R.
Istituto Sperimentale
Per la Selvicoltura
V. Le Santa Margherita 80
I - 52100 AREZZO

MORVAN, J.
Armef
Rue de la Fayette 85
F - 75009 PARIS

NADE', S.
C.C.E.
Direction Générale "Energie"
Rue de la Loi 200
B - 1049 BRUXELLES

NATALICCHIO, E.
A.I.G.R.
Via Celoria 2
I - MILANO

NEBURE, W.
ENEA
St. Prov. Anguillarese Km1.300
I - 00060 ROMA

NUNZI, L.
Agrimont
Via Reno 5
I - ROMA

ORTIZ TORRES, L.
Univ. Madrid
Avda. Complutense 22
E - 28040 MADRID

ORTIZ-CANAVATE, J.
E.T.S. Ing. Agronomos
Ciudad Universitaria
E - 28040 MADRID

PAIRE, G.
Amn. Prov. di Cuneo
C.SO Nizza 21
I - CUNEO

PALZ, W.
C.C.E.
Direction Générale "Science,
Recherche & Développement"
Rue de la Loi 200
B - 1049 BRUXELLES

PASQUARELLA, A.
Via E. Arborea 31
I - ROMA

PECA, G.
AGIP Petroli
Via Laurentina 449
I - 00142 ROMA

PELLEGRINI, G.
Università Roma
P.le Aldo Moro 5
I - 00187 ROMA

PELLICCIOTTI, A.
Ersa
Via S. Sebastiano 69
I - CASALBORDINO

PELLIZZI, G.
Aigr
Via Celoria 2
I - MILANO

RAZZETTI, L.
Ersa
S.Vito Fraz. Valle Castellana
I - TERAMO

RICCI, L.
A. Biotec
Via Matteotti 115
I - 47100 FORLI

RIEDACKER, A.
Agence Française
Pour la Maîtrise de l'Energie
Rue Louis Vicol 27
F - 75015 PARIS

ROSSI, G.
Renagri
C. So Vittorio Emanuele 173
I - ROMA

ROSSI, M.
Enichem
Milano Fiori Strada 2, Palazzo F4
I - 20090 ASSAGO (MI)

SANNA, P.
Eni Ricerche
Via Ramarini 32
I - 00015 MONTEROTONDO

SCARAMUZZI, G.
SAF
Via dei Crociferi 19
I - 00187 ROMA

SCARZELLA, L.
BES
Via Piani Della Madonna 127
I - 17017 MILLESIMO (SV)

SIBONO, A.
Vismar Italia S.R.L.
Via Caffaro, 1/23
I - GENOVA

STREHLER, A.
Techn. University Munich
Vottinger strasse 35
D - 805 FREISING

TIEFENBACHER, H.
Agrarische Rundschau, Waldbau
Institut, Adolf Ciela-Haus
Peter Jordan-strasse 70
A - 1190 WIEN

TOOMEY, B.J.
FAO-ACR. SERV. DIV.
Via Delle Terme Di Caracalla
I - 00100 ROMA

TRAVERSO, P.
Univ. Venezia
Dorsoduro 2137
I - VENEZIA

TRONCI, M.
Università Roma,
Facolta Ingegneria
Istituto di Machine
Via Endossiana
I - 00185 ROMA

TROSSERO, A.
FAO
Via Terme di Caracalla
I - ROMA

ULGIATI, S.
Univ. Di Siena Dip. Chimica
Pian dei Mantellini 44
I - 53100 SIENA

VAING, G.
Ceemat
Parc De Tourvoi
F - 92160 ANTONY CEDEX

VAN LANDEGHEM, E.
Cimaf
F - PARIS

VERANI, S.
Soc. Agricola e Forestale
Via di Casaloti 300
I - ROMA

VERANI, S.
SAF
Via Casalotti, 300
I - ROMA

VITO, A.
Regione Campania
Servizio Sperimentazione
in Agrocoltura
Via Oberdan 32
I - NAPOLI

VOLPE, F.P.
Regione Abruzzo
Via IV Novembre 24
I - LANCIANO (CH)

WILLIAMS, H.E.
Ecotec Research and
Consulting Limited
Priory House
Steelhouse Lane 18
UK - BIRMINGHAM B4 6 BJ

WIRYOND
Ministry of Forestry
Manggala Wanabakti Building
Via Gatot Subroto
INDONESIA - JAKARTA

ZITO, U.
C.C.E.
Direction Générale "Energie"
Rue de la Loi 200
B - 1049 BRUXELLES

INDEX OF AUTHORS